J. MICHAEL BISH
G. W. HOOPER FO
University of California, San Francisco
San Francisco, CA 94143

Virology in Medicine

Virology in Medicine

EDITED BY

Henry Rothschild, M.D., Ph.D.
Professor of Medicine and Anatomy
Louisiana State University School of Medicine

J. Craig Cohen, Ph.D.
Associate Professor of Medicine and Biochemistry
Louisiana State University School of Medicine

COORDINATING EDITOR

Charles F. Chapman

New York   Oxford
OXFORD UNIVERSITY PRESS
1986

Oxford University Press

Oxford   New York   Toronto
Delhi   Bombay   Calcutta   Madras   Karachi
Petaling Jaya   Singapore   Hong Kong   Tokyo
Nairobi   Dar es Salaam   Cape Town
Melbourne   Auckland

and associated companies in
Beirut   Berlin   Ibadan   Nicosia

Published by Oxford University Press, Inc.
200 Madison Avenue, New York, New York 10016
Oxford is a registered trademark of Oxford University Press.

Library of Congress Cataloging in Publication Data
Main entry under title:
Virology in Medicine.
   Includes index.
   1. Virus diseases.   I. Rothschild, Henry, 1932–
II. Cohen, J. Craig.   III. Chapman, Charles Frederick,
1929–
[DNLM: 1. Virus Diseases.   2. Viruses. WC 500 M484]
RC114.5.M417 1984   616'.0194   84-20574
ISBN 0-19-503559-3
ISBN 0-19-504017-1   (pbk.)

NOTICE

The indications and dosages of all drugs in this work have been recommended in the medical literature and conform to the practices of the general medical community at Louisiana State University
School of Medicine in New Orleans. The medications described do not necessarily have specific ap-
proval by the Food and Drug Administration for use in the situations and dosages for which they are
recommended. Consult the package insert for use and dosage as approved by the FDA. Because standards for usage change, it is advisable to keep abreast of revised recommendations, particularly those
concerning new drugs.

Printing (last digit): 9 8 7 6 5 4 3 2 1
Printed in the United States of America

To Shoshana, Jamin, Amy, and Karen

Preface

The viruses, instead of being single-minded agents of disease and death, now begin to look more like mobile genes. Evolution is still an infinitely long and tedious biologic game, with only the winners staying at the table, but the rules are beginning to look more flexible. We live in a dancing matrix of viruses; they dart, rather like bees, from organism to organism, from plant to insect to mammal to me and back again, and into the sea, tugging along pieces of this genome, strings of genes from that, transplanting grafts of DNA, passing around heredity as though at a great party. They may be the mechanism for keeping new, mutant kinds of DNA in the widest circulation among us. If this is true, the odd virus disease, on which we must focus so much of our attention in medicine, may be looked on as an accident, something dropped.

LEWIS THOMAS, *The Lives of a Cell*

This "something dropped," a perhaps incidental corollary of viral activity, forms the subject of this book. Research on viral diseases has yielded new information and improved methods of diagnosis and treatment at such a swift pace that our body of knowledge is now doubling in little more than a year. The physician thus faces the peril of not knowing what he or she should—indeed *must*—know.

Great care was therefore given to the selection of topics for this volume. The objective is to present basic and current information that is both useful and pertinent for medical students and clinicians—particularly house staff, family physicians, pediatricians, and internists—information having applicability in practice. We have attempted to respond to the Association of American Medical Colleges' report "Physicians for the Twenty-First Century," prepared by the Panel on the General Professional Education of the Physician and College Preparation for Medicine. That report recommends limiting the amount of factual data the students are expected to commit to memory and replacing such data with information (and self-study experiences) from which they can learn usable concepts. We have cut down on the content of the book by eliminating historical or tangential materials and emphasizing the information needed for the effective treatment of patients with virological diseases. Some diseases, such as smallpox, have been virtually eliminated from the Western world, and so were not included in our considerations. They have been superseded by subjects of greater currency and interest or by those in which recent developments are more definitive and reportable.

To meet its purpose, the book begins with the underlying essentials of virology, and then explains how the host responds to viral infection and the disease changes that result.

Chapter 4 clarifies the factors and problems underlying community infections and considers the implications for epidemiology. The middle chapters focus on diagnoses and treatments of viral diseases by organ site, including infections of the upper respiratory tract, brain, liver, and gastrointestinal system. Chapters 8 through 12 cover subjects of special importance in medical virology: viral infections of the fetus and newborn infant, herpesvirus infections, new and emerging viral diseases, and the role of viruses in cancer. The final three chapters discuss methods of diagnosis and treatment, and immunization against viral diseases.

Because any discussion of modern virology requires an understanding of the related fields of molecular genetics, cytology, biochemistry, and immunology, some of the specialized information may be new to the clinician. We have therefore supplied a glossary as an appendix to the text. We hope that the technical and theoretical information will be helpful for understanding recent developments and the mechanisms by which viruses apparently cause disease. Our intent has been to illuminate significant findings, changes, and trends in basic and applied medical virology.

We gratefully acknowledge the administrative assistance of Rebecca Carl, the editorial help of Virginia Howard, and the preparation of final manuscripts by Norma O. Badeaux and Angela M. Jones.

To Calderon Howe, M.D., emeritus professor and former head of the Department of Microbiology, Louisiana State Unviersity School of Medicine, New Orleans, we owe a great debt—for his thorough review of the book's contents, his extensive comments, and his helpful suggestions—a debt beyond our power to repay, except with heartfelt thanks and affectionate regard.

New Orleans   H.R.
May 1985   J.C.C.
                                                                                           C.F.C.

Contents

Chapter 1 / Basic Virology   3
Purnell W. Choppin

Viruses: Definition and Importance   3

    Impact   3
    Historical Notes and Definition   4

Viruses: Structure, Composition, Classification   5

    Properties of Virions   6
    Viral Nucleic Acids   6
    Morphology   8
    Classification of Viruses—Principles and Criteria   17

Assays of Viruses   17

    Assay of Infective Units   17
    Assays Based on Physical or Chemical Properties   20

Steps in the Virus Replication Cycle   24

    Adsorption   24
    Penetration   26
    Uncoating   30
    Biosynthesis of Viral Nucleic Acids and Proteins   31
    Aspects of Biosynthesis of Selected RNA Viruses   33
    Assembly of Virions   39
    Release of Virions   43

Conclusions   44

Recommended Readings   44

Chapter 2 / Host Response to Viral Infection   46
Jerry W. Smith and Calderon Howe

Immune Mechanisms   46

Viruses as Antigens   47

Immune Responses   49

Elimination of Infectious Virus   52

Elimination of Infected Cells   55

   Antibody-Dependent Cellular Cytotoxicity   57
   Cytolytic Antibody   57
   Antibody-Independent Mechanisms: Direct Cytotoxicity   57

Immune Clearance   59

Interferon   61

Virus-Induced Immunosuppression   63

Conclusion   63

Recommended Readings   64

Chapter 3 / Pathology of Virus Infection   65
Murray B. Gardner, William Sullivan, and Steven Hinrichs

Pathogenesis of Viral Infection   65

   Enveloped Viruses   66
   Nonenveloped Viruses   68
   Interpretation of Virus Assays   70

Light Microscopic Changes of Virus-Infected Cells   71

   Cell Lysis and Inclusions   71
   Cell Fusion   73
   Cell Hyperplasia   74

Correlations Between Pathology and Virus Detection in the
Laboratory   77

Histopathology of Host Immune Response to Viral Infection   81

Immunopathology of Viral Infection   83

   Viral Antigen and Antibody   83

Inflammatory Response 85

Molecular Detection of Viral Genomes 85

Recommended Readings 86

Chapter 4 / The Epidemiology of Viral Diseases 89
Neal A. Halsey

Surveillance 89

Transmission 89

Descriptive Epidemiology 89

> *Time 91*
> *Place 97*
> *Person 99*

Analytic Epidemiology 99

> *General 99*
> *Causality 100*
> *Prospective Studies 102*
> *Retrospective Case-Control Studies 103*

Epidemiologic History 104

> *Purposes 104*
> *Times 104*
> *Place 105*
> *Person 105*
> *Animal Exposures 105*
> *Food, Water, and Drug Exposures 105*
> *Other 106*

Latent, Persistent, and Recurrent Infections 106

Recommended Readings 109

Chapter 5 / Respiratory Tract Infections 111
Thomas R. Cate

Common Cold 112

> *The Syndromes 112*

Rhinoviruses   113
Coronaviruses   114

Laryngotracheobronchitis, Bronchiolitis, and Pneumonia in
Children   115

The Syndromes   115
Parainfluenza Viruses   116
Respiratory Syncytial Virus (RSV)   118

Influenza Syndrome and Atypical Pneumonia in Adults   120

The Syndromes   120
Influenza Viruses   121
Adenoviruses   126

Diagnosis   128

Treatment   129

Prevention   130

Recommended Readings   132

Chapter 6 / Viral Diseases of the Central Nervous System   133
Robert E. Shope

Acute Clinical Syndromes   133

Aseptic Meningitis   133
Encephalitis   134
Poliomyelitis   134

Pathogenesis of Viral Infections of the CNS   135

Specific Causes of Viral CNS Disease   137

Enteroviruses   137
Childhood Disease Viruses   139
Arboviruses   140

Diagnosis   149

Recommended Readings   150

Chapter 7 / Hepatitis and Diarrhea Viruses   152
Sarah H. Cheeseman and Neil R. Blacklow

Viral Hepatitis 153

The Disease 153
Hepatitis B 155
Hepatitis A 162
Non-A, Non-B Hepatitis 164
Other Viruses Causing Hepatitis 165

Gastroenteritis Viruses 165

Clinical Syndromes 165
Norwalk and Norwalklike Viruses 166
Rotavirus 168
Viral Gastroenteritis Agents of Uncertain Medical Importance 170

Recommended Readings 170

Chapter 8 / Viral Infection of the Fetus and Newborn Infant 172
John F. Modlin

Intrauterine Viral Infections 173

Evaluation of Suspected Congenital Infection 173
Cytomegalovirus 175
Rubella Virus 178

Perinatal Viral Infections 181

Evaluation of Suspected Perinatal Viral Infections 181
Herpes Simplex Virus 182
Enteroviruses 186
Varicella-Zoster Virus 189
Hepatitis B Virus 191

Recommended Readings 194

Chapter 9 / Childhood Exanthemas 196
John F. Modlin

Measles (Rubeola) 196

Agent 196
Transmission, Pathophysiology and Clinical Manifestations 197

Diagnosis   197
Complications and Mortality   198
Measles in the Altered Host   199
Epidemiology   199

Mumps   199

Agent   199
Pathophysiology   200
Complications   201

Rubella Virus   202

Recommended Readings   203

Chapter 10 / Herpesvirus Infections   205
Joseph S. Pagano

Biologic Features   206

Virus Structure and Antigenicity   207

Pathogenesis   209

Laboratory Diagnosis   212

Immunity   214

Epidemiology   215

Conclusion: Prevention and Treatment   217

Recommended Readings   218

Chapter 11 / New and Emerging Viral Diseases   220
Neal Nathanson and Francisco Gonzalez-Scarano

Spongiform Encephalopathies   220

Kuru   221
Creutzfeldt-Jacob (C-J) Disease   222
Properties of the Spongiform Agents   223
Viroids   225

Chronic CNS Disease Due to Conventional Viruses   226

Progressive Multifocal Leukoencephalopathy (PML)   226
Subacute Sclerosing Panencephalitis (SSPE)   228

Chronic Diseases that May Be Caused by Viruses   229

    Diabetes Mellitus   230

    Multiple Sclerosis   231

    Acquired Immune Deficiency Syndrome (AIDS)   233

    Delta Hepatitis   233

Hemorrhagic Fevers   234

Future   Prospects   Using   New   Molecular   and   Immunologic
Techniques   235

    Nucleic Acid Probes   235

    Enzymatic Digestion   236

    Monoclonal Antibodies   237

    Comparison of Agents   237

Conclusion   238

Recommended Readings   238

Chapter 12 / Viruses and Cancer   240
Henry Rothschild and J. Craig Cohen

Viruses Known to Transform Cells   242

    Papovaviruses   243

    Adenoviruses   245

    Herpesviruses   246

    Hepatitis B Virus   246

    Retroviruses   247

Viruses and Human Cancer   252

    Cervical Carcinoma   252

    Burkitt's Lymphoma   252

    Nasopharyngeal Carcinoma   253

    Hepatocellular Carcinoma   254

    T-Cell Leukemia   255

Conclusion   255

Recommended Readings   256

Chapter 13 / Diagnostic Virology   257
Evelyn W. Keller and Martin S. Hirsch

Standard Techniques   259

    Virus Isolation   259

    Virus Detection   260

    Virus-Host Systems   263

    Identification   264

    Practical Considerations   266

    Conventional Serologic Methods   268

Rapid Diagnostic Techniques   272

    Antigen Detection   272

    Antibody Detection   275

Present Realities and Future Expectations   278

Recommended References   280

Chapter 14 / The Status of Antiviral Therapy   281
Fred Allison, Jr.

Current Status of Antiviral Agents   282

    Variola Virus (Smallpox)   282

    Influenza A Virus   282

    Rhinovirus   289

    Varicella-Zoster Virus   290

    Herpes Simplex Virus   291

    Hepatitis B Virus   296

    Other Viral Diseases   297

Recommended Readings   298

Chapter 15 / Immunization Against Viral Agents   229
James D. Cherry

History   299

Properties of Viral Vaccines   300

Principles of Immunization   301

    Target Groups   301

Herd Immunity 302
Risks and Benefits 302
Informed Consent 303

Prevention of Specific Viral Diseases by Immunization 304

Poliomyelitis 304
Measles 308
Rubella 312
Mumps 314
Rabies 315
Influenza 317
Hepatitis B 318
Yellow Fever 320
Adenoviral Acute Respiratory Disease 320
Smallpox 321

Experimental Vaccines and Viral Illnesses for which Vaccines Should Be Developed 321

Respiratory Viruses 321
Herpesviruses 322
Enteric Viruses 323

Conclusion 324

Recommended Readings 324

Glossary 326

Index 330

Contributors

Fred Allison, Jr., M.D.
Professor and Head
Department of Medicine
Louisiana State University School of
Medicine
New Orleans

Neil R. Blacklow, M.D.
Professor of Medicine and
Microbiology
Director, Division of Infectious
Diseases
University of Massachusetts Medical
School
Worcester

Thomas R. Cate, M.D.
Professor of Medicine, and
Microbiology-Immunology
Department of Microbiology and
Immunology
Influenza Research Center
Baylor College of Medicine
Houston

Sarah H. Cheeseman, M.D.
Associate Professor of Medicine,
Pediatrics, and Microbiology
Division of Infectious Diseases
University of Massachusetts Medical
School
Worcester

James D. Cherry, M.D.
Professor of Pediatrics and Chief,
Division of Infectious Diseases
The Center for the Health Sciences
University of California at
Los Angeles
School of Medicine
Los Angeles

Purnell W. Choppin, M.D.
Professor of Virology
Vice-President, Academic Programs
The Rockefeller University
New York

J. Craig Cohen, Ph.D.
Associate Professor of Medicine and
Biochemistry
Louisiana State University School of
Medicine
New Orleans

Murray B. Gardner, M.D.
Professor of Pathology
Director, Anatomic Pathology
Department of Pathology
University of California, Davis Medi-
cal Center
Sacramento

**Francisco Gonzalez-Scarano,
M.D.**
Assistant Professor
Departments of Neurology and
Microbiology
University of Pennsylvania School of
Medicine
Philadelphia

Neal Halsey, M.D.
Professor
Departments of Tropical Medicine
and Pediatrics
Tulane University Medical Center
New Orleans

Steven H. Hinricks, M.D.
Assistant Professor
Department of Pathology
University of California, Davis
Medical Center
Sacramento

Martin S. Hirsch, M.D.
Associate Professor of Medicine
Infectious Disease Unit
Massachusetts General Hospital
Harvard Medical School
Boston

Calderon Howe, M.D.
Professor Emeritus
Louisiana State University School of
Medicine
New Orleans

Evelyn W. Keller, B.A.
Infectious Disease Unit
Massachusetts General Hospital
Boston

John F. Modlin, M.D.
Assistant Professor of Pediatrics
Division of Infectious Disease
Department of Pediatrics
Johns Hopkins University School of
Medicine
Baltimore

Neal Nathanson, M.D.
Professor and Chairman
Department of Microbiology
University of Pennsylvania School of
Medicine
Philadelphia

Joseph S. Pagano, M.D.
Professor and Director, Cancer Re-
search Center
University of North Carolina
School of Medicine
Chapel Hill

Henry Rothschild, M.D., Ph.D.
Professor of Medicine and Anatomy
Chief, Section of Genetics
Louisiana State University School of
Medicine
New Orleans

Robert E. Shope, M.D.
Director, Arbovirus Research Unit
Yale University
New Haven

Jerry W. Smith, Ph.D.
Director, Research and Development
DIFCO Laboratories, Inc.
Research and Development Center
Ann Arbor, Michigan

**William F. Sullivan, M.S., S.M.
(A.A.M.)**
Virologist
Sutter Memorial Hospital
Sacramento, California

ABBREVIATIONS

AraA	adenine arabinoside	**IDU**	idoxuridine; 5-iodo-2′ deoxyuridine
AraC	cytosine arabinoside		
ARD	acute respiratory disease	**IPV**	inactivated poliomyelitis vaccine
BKV	a human papovavirus		
cDNA	complementary DNA	**ISG**	Immune serum globulin
CF	complement fixation	**JCV**	a human papovavirus
CJD	Creutzfeldt-Jakob disease	**LCM**	lymphocytic choriomeningitis
CMV	cytomegalovirus		
CNS	central nervous system	**mRNA**	messenger ribonucleic acid
C-onc	cellular or proto-oncogene		
CPE	cytopathic effect	**OPV**	oral poliomyelitis vaccine
CSF	cerebrospinal fluid	**PML**	progressive multifocal leukoencephalopathy
DNA	deoxyribonucleic acid		
EBV	Epstein-Barr virus	**RNA**	ribonucleic acid
HAV	hepatitis A virus	**RSV**	Rous sarcoma virus; respiratory syncytial virus
HBcAg	hepatitis B core antigen		
HBeAg	hepatitis B e antigen	**SGOT**	serum glutamic oxaloacetic transaminase
HBIG	high-titer hepatitis B immune serum globulin		
		SSPE	subacute sclerosing panencephalitis
HBsAg	hepatitis B surface antigen		
HBV	hepatitis B virus	**SV40**	simian virus 40
HI	hemagglutination inhibition	**T antigen**	tumor antigen
		TME	transmissible mink encephalopathy
HPV	human papilloma virus		
HSV	herpes simplex virus	**TMV**	tobacco mosaic virus
HSV-1	herpes simplex virus type 1	**tRNA**	transfer ribonucleic acid
HSV-2	herpes simplex virus type 2	**V-onc**	viral oncogene
HUIFN	human interferon	**VZV**	varicella-zoster virus

1

Basic Virology

PURNELL W. CHOPPIN

VIRUSES: DEFINITION AND IMPORTANCE

Impact

Viruses are enormously important as causative agents of human diseases. They are the most common cause of acute disease, and evidence of their involvement in a variety of chronic diseases is increasing. About a quarter of patient visits to physicians in the United States each year are because of infections, and more than half of these are virus infections. This does not include the undoubtedly larger number of virus infections not seen by a physician or those diseases not recognized as virus infections. Two thirds of the acute illnesses in persons under 17 years of age are infections; half of these are viral. The annual economic impact of virus infections is measured in many billions of dollars; the cost of influenza alone in the United States in a pandemic year is estimated to be $5 billion to $6 billion, with an excess mortality of up to 70,000. Even in an interpandemic year, influenza results in costs of hundreds of millions of dollars and is a leading cause of excess mortality. Worldwide, infections remain the most common cause of illness and death, and viruses play a major role in mortality. Each year, for example, several billion cases of diarrheal disease occur throughout the world, causing 5 million to 10 million deaths, and a large number of these cases are the consequence of viral infections.

More than 400 different viruses are known to infect humans, and still more are being isolated. They cause diseases ranging from mild respiratory infections to acute and chronic neurologic diseases and neoplasms. Viruses, in addition to their role in human diseases, play an important role as pathogens of lower animals, insects, plants, and bacteria.

Although this book focuses on human viral diseases, no discussion of the basic properties of viruses, however brief, should omit mention of the enormous role of viruses in the explosion of knowledge in cellular and molecular biology during the past two decades. Much of what we know about the mechanisms of nucleic acid replication, messenger RNA synthesis and processing, protein synthesis, and membrane structure and assembly has been learned from viral systems. Virology has taken its place as a major contributor to the current biological revolution.

Historical Notes and Definition

Viruses have long been recognized as the causes of disease, but it was not until 1892 and 1898 that Iwanowski and Beijerinck, respectively, demonstrated a plant virus, tobacco mosaic virus (TMV), to be a transmissible infectious agent. Defining viruses as infectious agents distinct from bacteria was first based on the ability of a virus to pass through a bacteria-tight filter. The first animal and human viruses to be discovered were foot and mouth disease virus by Loeffler and Frosh (1898) and yellow fever virus by Reed (1902). Viruses causing malignant diseases in chickens—leukemia (Ellerman and Bang, 1908) and a solid malignant tumor (Rous, 1911)—were among the earliest isolated. Since then, virus-induced tumors have been demonstrated in innumerable animal species, from reptiles to primates. Only recently have viruses begun to emerge as possible agents for human cancers, although a virus has been known to cause the human wart, a benign tumor, since 1907.

The first book on viruses, titled *Filterable Viruses,* was written by Rivers in 1928. It drew together all the available knowledge on these agents and can be said to mark the beginning of virology as a science. Among the important points emphasized by Rivers was that virus replication absolutely required a living cell. He also pointed out at that early date that the primary pathologic changes in virus diseases were hyperplasia alone, hyperplasia followed by necrosis, or necrosis alone, and that inflammation was usually a secondary response resulting largely from damage to virus-infected cells. These observations over half a century ago were prescient.

Because filterability and submicroscopic size were early criteria for classifying an agent, emphasis was placed on smallness as a characteristic of viruses. However, eventually it emerged that there was little difference in size between the largest of the viruses and the smallest of the rickettsia or the agents now known as *Chlamydiae;* thus, size is not a defining factor. Arriving at an adequate definition to encompass all the various kinds of viruses has been difficult. Lwoff in 1957 was among the first to do so. He defined viruses as strictly intracellular and potentially pathogenic entities with an infectious phase, and only one type of nucleic acid, which multiply in the form of their genetic material, are unable to undergo binary fission, and are devoid of the enzymes needed to produce energy. Such a definition distinguishes viruses from cells and all other forms of infectious agents and emphasizes the intracellular nature and dependence on host-cell synthetic machinery. However, that definition was formulated before investigators found that certain viruses (e.g., the negative-strand viruses) could not replicate from their nucleic acid alone. Luria and Darnell (1978) provided an updated operational definition: "Viruses are entities whose genomes are elements of nucleic acid that replicate inside living cells using the cellular synthetic machinery and causing the synthesis of specialized elements that can transfer the viral genome to the other cells." It is a testimony to the diversity of viruses that even this excellent and carefully worded definition does not cover all instances, if the "viroids" are included as viruses. Viroids are plant pathogens that consist only of small circular RNA molecules; they have no protein coat or any elements involved in the transfer of the nucleic acid to other cells.

One important property of viruses is that their nucleic acids represent genetic information that, on entry into an appropriate host cell, redirects the synthetic machinery of the cell toward ends specified by the virus. The possible outcomes of such redirection include (1) cell death and acute disease; (2) stimulation of the cell to uncontrolled growth, that is, malignant transformation; and (3) neither cell death nor transformation but persistent infection with synthesis of some or all of the components of the virus, in some cases leading eventually to cellular malfunction and chronic disease. Examples of each of these outcomes' of a virus/cell encounter are provided in later chapters.

The fact that viruses are the ultimate intracellular parasites, using the cell's synthetic apparatus for their replication, often including cellular enzymes, underlies the problem of antiviral therapy. How does one intervene in the process of viral replication without injuring the cell? Fortunately, through extensive research on the cellular and molecular biology of viruses, unique steps in virus replication are being elucidated; the chemotherapy of virus infections, once despaired of, is now becoming a reality in certain instances, and reason exists for continued, though cautious, optimism.

VIRUSES: STRUCTURE, COMPOSITION, CLASSIFICATION

The modern era for studying the structure and chemistry of viruses began with the crystallization of the tobacco mosaic virus (TMV) by Stanley in 1935, followed in 1940 by the first purification and chemical analysis of an animal virus (vaccinia) by Hoagland, Smadel, and Rivers and by the introduction of the electron microscope for the examination of viruses. Early attempts to classify viruses on the basis of disease entity, host range, or organ tropism proved unsuccessful, because different viruses cause similar diseases, and similar viruses may infect several species or organ systems and cause different diseases. That a rational system of classification must be based on the structural and chemical

TABLE 1.1   Properties of Viruses Used in Virus Classification

Type of nucleic acid: DNA or RNA
Strandedness of nucleic acid: single or double
Polarity of nucleic acid: positive or negative
Capsid symmetry: icosahedral, helical, complex
Site of capsid assembly: nucleus or cytoplasm
Envelope: presence or absence
Site of envelopment: surface membrane, intracytoplasmic membrane, nuclear
   membrane
Virion: diameter
Capsid: diameter
Capsomers: number
Nucleic acid in virion: molecular weight

properties of the virus particles (virions) became apparent. Such a scheme was made possible by the rapid advances in the morphologic and chemical characterization of viruses in the late 1950s and early 1960s, including the introduction of the negative staining technique for electron microscopy and the development of methods for the large-scale production, purification, and chemical analysis of virions.

Properties of Virions

The amount of information on the physical and chemical properties of viruses has grown so explosively that this introductory chapter cannot consider in detail any one virus family, let alone all of the human viruses. However, of great importance here are the general structural principles that can be applied to all viruses, and on which the current classification of viruses is based. Viruses can be subdivided, according to their primary host, into animal viruses, plant viruses, and bacterial viruses, although some overlaps occur because some plant viruses may multiply in insect hosts. However, the general principles that have been developed for classification on structural grounds apply to all the viruses, regardless of the primary host (Table 1.1).

Viral Nucleic Acids

SIZE

   Viruses contain either ribonucleic acid (RNA) or deoxyribonucleic acid (DNA) as their genetic material. Thus, they can be divided into two large categories, the RNA and the DNA viruses. The nucleic acid content of the virus may range from as little as 1% of the mass (e.g., in parainfluenza virus) to as much as 25% (e.g., in poliovirus). Nucleic acid genomes range in size from those of the picornaviruses (RNA) and the parvoviruses (DNA), with molecular weights of $1.5-3 \times 10^6$, to those of the poxviruses and herpesviruses (DNA) with molecular weights of 100 to 200×10^6. In terms of genetic information, these viral genomes can code from about three or four average-sized proteins to a few hundred.

STRUCTURE

Viral nucleic acids, whether RNA or DNA, may be either single stranded or double stranded. In the case of some RNA viruses, e.g., influenza and reoviruses, the genome may be segmented rather than covalently linked. The segmented nature of the virus genome facilitates the easy exchange of RNA segments, and thus of genetic information, in cells infected with two virus strains. This process is called *genetic reassortment,* to distinguish it from true *genetic recombination,* which involves the exchange of regions of nucleic acid that are covalently linked in the genome. Genetic reassortment is particularly frequent with influenza viruses, and it is thought to be the mechanism for the periodic major antigenic shifts that occur with this virus and lead to pandemics (see Chapter 5).

Several other structural characteristics are found in some, but not all, viral nucleic acids. Some of these will be discussed in more detail in later chapters dealing with specific viruses. The significance of all structural features of the nucleic acids is not altogether clear, but some of them are involved in nucleic acid replication. Although most viral nucleic acids are linear, some (e.g., the papovaviruses) are covalently closed circles that are supercoiled. The nucleic acids of other viruses (e.g., retroviruses) are *terminally redundant,* i.e., nucleotide sequences are the same at the two ends of the strand, whereby a double-stranded DNA molecule can be represented as

$$5'\ \underline{A\ B\ C} \qquad\qquad \underline{A\ B\ C}\ 3'$$
$$3'\ \overline{A'B'C'} \qquad\qquad \overline{A'B'C'}\ 5'$$

From this arrangement it can be seen that if a short segment of the 5' ends of each strand were digested away, the remaining single-stranded 3' ends would be complementary to each other (cohesive ends), and the molecule could circularize. Such circularization can play a role in DNA replication or integration into the host-cell genome.

In some DNA viruses (e.g., adenovirus), the terminal repeats are inverted, and thus a single DNA strand could form circles. In herpes simplex virus, a unique situation exists in which two sections of unique sequences of unequal length are bracketed by inverted repeated sequences. This arrangement is discussed in Chapter 11. The single-stranded DNA parvoviruses contain a terminal palindrome, which enables the DNA to form a hairpin turn at the end by self-hybridization. The poxviruses are unique in having the two strands of their DNA cross-linked at the ends, so that if the hydrogen bonds between the two strands are melted, a single-stranded circle is generated whose circumference is twice the length of the annealed linear molecule. The bunyaviruses and arenaviruses, single-stranded RNA viruses, have cohesive complementary sequences at their termini. This arrangement enables the individual segments of their RNAs to form circles.

Some viral nucleic acids are covalently linked to protein. The 5' end of poliovirus RNA is linked to a protein having 22 amino acids, and both of the 5' ends of the double-stranded DNA of adenoviruses are linked to a protein having a molecular weight of approximately 55,000 daltons. These proteins are thought to play a role in the replication of the viral nucleic acid.

POLARITY

The genomes of single-stranded RNA viruses can be defined as either positive or negative strand. The convention of these designations is that the positive strand serves as messenger RNA (mRNA) in protein synthesis, and the negative strand is complementary to the mRNA. The genome RNA (i.e., the RNA present in the virion) in picornaviruses, togaviruses, coronaviruses, caliciviruses, and retroviruses can serve as mRNAs; thus, viral proteins can be translated directly from them. They are therefore designated *positive-strand* viruses. In contrast, the genome RNAs of myxoviruses, paramyxoviruses, rhabdoviruses, arenaviruses, and bunyaviruses cannot serve as mRNA; RNA complementary to the genome RNA is the mRNA. These viruses are designated *negative-strand* viruses.

Because they function as mRNAs, the positive-strand RNA viruses usually have a structure similar to that of cellular mRNA (i.e., polyadenylic acid [poly A]) at the 3' end, and a "cap" structure (a 5'-5' linkage of a 7-methyl guanylate residue to the next base) at the 5' end. This cap plays a role in the binding of the mRNA to the ribosome and in the efficiency of translation. Poliovirus RNA is an exception in that it has a covalently linked protein at its 5' end, as mentioned previously. The genome RNA of negative-strand viruses does not have a cap or poly A; the 5' terminus is a nucleoside triphosphate. The mRNAs transcribed from these genome RNAs in infected cells are capped and polyadenylated.

Morphology

THE CAPSID

The nucleic acids of viruses are enclosed in a shell of protein called a capsid. This capsid is composed of multiple identical units of one or a few proteins arranged in precise geometrical patterns. Crick and Watson predicted in 1957 that virus capsids would be composed of multiple identical subunits, because the limited amount of genetic information present in most viruses leaves them unable to code for large

TABLE 1.2   DNA Viruses with Icosahedral Symmetry, Nonenveloped, Capsid Assembled in Nucleus

Family Name	Genus Name (Examples)	Molecular Weight ($\times 10^6$) and Strandedness of DNA[a]	Diameter of Virion (nm)	No. of Capsomers
Adenoviridae	Mastadenovirus (human adenovirus) Aviadenovirus (avian adenovirus)	20–30, DS	70–90	252
Papovaviridae	Papillomavirus (wart virus)	5, DS	55	72
	Miopapovirus (SV40, polyoma, JC)	3, DS	45	
Parvoviridae	Parvovirus (Aleutian mink disease) Dependovirus (adeno-associated virus) Densovirus (densonucleosis virus of arthropods)	1.5–2.0, SS	18–26	32

[a] DS = double stranded; SS = single stranded.

numbers of different proteins. The use of identical subunits also enables the efficient self-assembly of stable, geometrically regular structures by virtue of the physical properties of the protein subunits. Viral capsids exhibit two basic types of symmetry, icosahedral and helical. The capsids with icosahedral symmetry have their protein shells assembled in the form of an icosahedron (a structure with 20 triangular faces, 30 edges, and 12 vertices).

Many viruses have such symmetry, including adenoviruses, papovaviruses, picornaviruses, and reoviruses. The number of protein subunits that can be arranged in such a regular icosahedral array varies with the properties of the subunits; however, only certain numbers of subunits are possible according to the geometrical rules for construction of such structures. All icosahedra must be composed of 60 equivalent parts, but each of these parts is in turn composed of one or more polypeptides called *protomers*. The protomers may cluster together to form morphologic units, called *capsomers*, which are usually resolved by means of the electron microscope (Tables 1.2 through 1.8). The minimum number of capsomers in an icosahedral virus is 12; each capsomer then forms a vertex of the icosahedron. Some of the other possible numbers found in viruses are 32, 72, 92, 162, and 252. The numbers of capsomers in different viruses are given in Tables 1.2, 1.3, 1.5, and 1.6.

TABLE 1.3 DNA Viruses with Icosahedral Symmetry, Enveloped

Family Name	Subfamily Name	Genus Name (Examples)	Molecular Weight ($\times 10^6$) and Strandedness of DNA	Diameter of Capsid (nm)	No. of Capsomers	Site of Envelopment	Site of Capsid Assembly
Herpesviridae	Alphaherpesvirinae	Simplexvirus (herpes simplex) Polkilovirus (pseudorabies)	80–150, DS	100	162	Nuclear membrane	Nucleus
	Betaherpesvirinae	Cytomegalovirus (human cytomegalovirus) Muromegalovirus (murine cytomegalovirus)					
	Gammaherpesvirinae	Lymphocryptovirus (Epstein Barr) Thetalymphocryptovirus (Marek's disease) Radinovirus (herpesvirus saimiri)					
Iridoviridae		African swine fever virus Ranavirus (frog virus 3) Iridovirus (iridescence viruses of arthropods) Chloridovirus (180-nm insect viruses) Lymphocystis virus	100–250, DS	125–300	1500	Cytoplasmic membrane	Cytoplasm

TABLE 1.4   DNA Viruses with Complex Structure Assembled in the Cytoplasm

Family Name	Subfamily Name	Genus Name (Examples)	Dimensions of Virion (nm)	Molecular Weight ($\times 10^6$) and Strandedness of DNA
Poxviridae	Chordopoxvirinae	Orthopoxvirus (smallpox, vaccinia) Avipoxvirus (fowl pox) Capripoxvirus (sheep pox) Leporipoxvirus (rabbit myxoma) Parapoxvirus (pseudocowpox, milker's nodules) Suipoxvirus (swine pox) Ungrouped viruses (molluscum contagiosum)	225×300 (except parapox, 150×200)	125–185, DS
	Entomopoxvirinae	Insect poxviruses		

The size, shape, and morphologic appearance of icosahedral viruses is inherent in the properties of the protomers. The protomers within a capsomer are held together by noncovalent bonds; these bonds are usually somewhat stronger than the bonds between protomers in different capsomers. (For a thorough discussion of the structure and design principles of virus particles, see the chapter by Caspar in *Viral and Rickettsial Diseases of Man*, edited by Horsfall and Tamm. Other discussions of virus structure, including extensive illustrations, can be found in the chapters dealing with virus morphology in the texts by Joklik, Luria et al, and Davis et al, listed in the Recommended Readings.)

ADENOVIRUSES

One type of arrangement of capsomers in an icosahedral capsid is well illustrated by the adenovirus virion. The icosahedral capsid of adenoviruses contains 252 capsomers, of which 240 are located on the faces and the edges of the icosahedron; each of these has six neighboring capsomers and is called a *hexon*. The remaining 12 capsomers are located at the 12 vertices; each of these, called a *penton*, has 5 neighboring capsomers. Adenovirus pentons have an unusual structural feature, a long fiber with a terminal knob projecting from the penton base. This structure is thought to be involved in attachment of the virus to the cell.

The protein shell of icosahedral viruses protects the viral genome, for example, from nucleases; however, it also frequently serves as the means for spreading the infection to other cells by providing the mechanism

TABLE 1.5 RNA Viruses with Icosahedral Symmetry, Nonenveloped, Capsid Assembled in Cytoplasm

Family Name	Genus Name	Molecular Weight ($\times 10^6$) and Strandedness, and Polarity of RNA[a]	Diameter of Virion (nm)	No. of Capsomers
Picornaviridae	Enterovirus (polio, Coxsackie, echo virus) Rhinovirus (human rhinoviruses) Cardiovirus (EMC virus of mice) Aphthovirus (foot and mouth disease)	2–3, SS, +	24–30	32
Caliciviridae	Calicivirus (vesicular exanthem of swine, ? human calivirus, Norwalk agent)	2–3, SS, +	34–39	32
Reoviridae	Orthoreovirus (mammalian reoviruses)	15, DS, segmented	75	92
	Orbivirus (Colorado tick fever) Rotavirus (human and calf rotaviruses)	12, DS, segmented	60–80	32

[a] + = positive strand; – = negative strand; SS = single strand; DS = double strand.

for attachment to, and in some cases penetration of, the host cell. The capsid, together with the enclosed nucleic acid, is termed the *nucleocapsid*. In viruses with icosahedral nucleocapsids, the viral nucleic acid is tightly packed within the shell and in some cases makes contact at specific points on the inner surface of the protein shell. In some icosahedral DNA viruses, the DNA is packed in association with histonelike proteins, forming structures resembling cellular chromatin. With many viruses, capsids that do not contain nucleic acid can be isolated, indicating that the nucleic acid is not required for assembly of the protein shell. However, such empty capsids are usually more fragile than the normal nucleic acid–filled capsids; thus, the viral genome can contribute to capsid stability.

HELICAL CAPSIDS

Viral capsids with helical symmetry are rod shaped, consisting of identical protein subunits (protomers) associated with nucleic acid in a

TABLE 1.6 RNA Viruses with Icosahedral Symmetry, Enveloped, Capsid Assembled in Cytoplasm

Family Name	Genus Name	Molecular Weight ($\times 10^6$) and Strandedness, and Polarity of RNA[a]	Diameter of Virion (nm)	No. of Capsomers
	Alphavirus (eastern and western equine encephalitis)	4–5, SS, +	50–70	32 or 42
	Rubivirus (rubella)			
Togaviridae	Flavivirus (yellow fever, dengue, St. Louis encephalitis)	3–4, SS, +	40–80	?
	Pestivirus (hog cholera virus)			

[a] + = positive strand; SS = single strand.

ribbonlike arrangement, which is wound into a helix, forming a cylindrical structure with a hollow central region. A line down the center of this region provides a single axis of rotational symmetry; that is, rotation around this axis to any extent does not change the observed pattern of the capsid structure. As with viruses having icosahedral symmetry, a helical virus capsid's diameter and its stability are determined by the properties of the protomers. However, in addition, the length of a helical nucleocapsid is determined by the length of the viral nucleic acid with which the protomers are associated. The protomers in each turn of the helix interact side-to-side with protomers in the same turn. Each protomer, however, also forms quasi-equivalent bonds with those in the turns of the helix above and below it. When these bonds are strong, the result is a stable, rigid, rodlike structure. The prototype of a helical virus is the plant virus TMV. From x-ray crystallographic, electron microscopic, and chemical studies more is known about the precise details of the structure of TMV than of other viruses, including morphology, protein–protein interactions, protein–RNA interactions, and the mechanism of self-assembly. The TMV virion consists of only the nucleocapsid and is a rigid rod 300 nm long.

In contrast to viruses with naked nucleocapsids, such as TMV, the animal viruses with helical nucleocapsids (e.g., orthomyxoviruses, paramyxoviruses, and rhabdoviruses) are enveloped, and, because the length of the nucleocapsid exceeds the diameter of the virus, these nucleocapsids must be flexible to be coiled within the envelope during virus assembly. For example, the nucleocapsids of paramyxoviruses are 1000 nm long and must fit into envelopes with diameters of 150 to 450 nm.

TABLE 1.7   RNA Viruses with Helical Symmetry, Enveloped, Capsid Assembled in Cytoplasm[a]

Family Name	Genus Name	Molecular Weight ($\times 10^6$) and Strandedness, and Polarity of RNA[b]	Diameter of Capsid (nm)	Diameter of Virion (nm)	Site of Envelopment
Orthomyxoviridae	Influenza virus (influenza viruses of humans and animals)	4–5, SS, − segmented	9–15	80–120	Surface membrane
Paramyxoviridae	Paramyxovirus (mumps, parainfluenza) Morbillivirus (measles, canine distemper) Pneumovirus (respiratory syncytial)	5–6, SS, −	17–20 17–20 12–15	150–300	Surface membrane
Rhabdoviridae	Vesiculovirus (vesicular stomatitis, chandipura) Lyssavirus (rabies)	3–4, SS, −	18	75×180 oblong	Surface membrane Cytoplasmic or surface membrane
Coronaviridae	Coronavirus (infectious bronchitis, mouse hepatitis)	5–6, SS, +	11–13	80–130	Cytoplasmic or surface membrane
Bunyaviridae	Bunyavirus (bunyawera, California encephalitis) Phlebovirus (sandfly fever) Nairovirus (Crimean-Congo hemorrhagic fever) Uukuvirus (uukunimi virus)	6–7, SS, − segmented	10–12	80–110	Cytoplasmic membrane

[a] In the case of influenza virus, the site of assembly of capsid is not certain.

[b] + = positive strand; − = negative strand; SS = single strand.

The nucleocapsids of these viruses are flexible and, under conditions of isotonic salt concentrations, are loosely coiled, although in some cases they become more rigid at high salt concentrations. In such flexible helical nucleocapsids, the bonds between adjacent protomers in the same

TABLE 1.8   RNA Viruses Assembled in Cytoplasm with Capsids Whose Symmetry Is Not Clear, Enveloped

Family Name	Subfamily Name	Genus Name (Examples)	Molecular Weight ($\times 10^6$) and Strandedness, and Polarity of DNA[a]	Diameter of Virion (nm)
	Oncovirinae	Oncornavirus C (avian and murine sarcoma and leukemia viruses) Oncornavirus B (mouse mammary tumor virus)		
Retroviridae		Oncornavirus D (Mason-Pfizer monkey virus)	6–7, SS, +, 2 copies of RNA in each virion	100
	Spumavirinae (human foamy virus) Lentivirinae (visna, Maedi viruses)			
Arenaviridae		Arenavirus (lymphocytic choriomeningitis, Lassa fever)	3–5, SS, −, segmented	80–130

[a] + = positive strand; − = negative strand; SS = single strand.

turn of the helix are stronger than those between protomers in adjacent turns of the helix.

In TMV the viral RNA lies in a groove in the protomers between the turns of the helix, and is thus extended rather than condensed as it is within the icosahedral capsid. The precise location of the RNA in helical animal virus nucleocapsids is not yet known. In rigid helical nucleocapsids such as TMV, the RNA is protected from ribonuclease by the protomers. In some flexible nucleocapsids (paramyxoviruses and rhabdoviruses), the RNA in the nucleocapsid is also resistant to ribonuclease, but in the orthomyxovirus (influenza virus) nucleocapsid, the RNA is ribonuclease-sensitive and the viral envelope serves to protect the genome from nucleases. The structure of the influenza virus nucleocapsid is unusual; it appears to be a double helix formed by the ribbonlike strand of RNA with associated protomers being folded back on itself and then twisted into a double helix.

ENVELOPED VIRUSES

The envelope enclosing the nucleocapsids of some viruses consists of a membrane containing a lipid bilayer with a few species of associated proteins. The nucleocapsid is enveloped during assembly of the virion. With many viruses (e.g., orthomyxoviruses, paramyxoviruses, some togaviruses, and rhabdoviruses), this envelopment occurs by a budding process at the plasma membrane of the host cell. With other viruses, however (e.g., herpesvirus and some rhabdoviruses), envelopment may occur at the nuclear or endoplasmic reticulum membranes. The lipids of the viral membrane are derived from the host cell; however, with few exceptions (such as cellular actin), the viral membrane proteins are virus-specific. Those on the outer surface of the membrane are glycoproteins, which appear as spikelike projections (peplomers) from the membrane. These proteins are involved in virus attachment to receptors, and, in some cases, in receptor destruction by enzymatic action. In some viruses (e.g., orthomyxoviruses and paramyxoviruses), viral membrane glycoproteins also mediate penetration of the virus into the cell. In orthomyxoviruses, paramyxoviruses, and rhabdoviruses, a nonglycosylated protein is associated with the inner surface of the viral membrane and is involved in virus assembly. Icosahedral nucleocapsids (e.g., herpesviruses and togaviruses) may also be enveloped.

COMPLEX VIRIONS

Some viruses have a capsid structure more complex than a simple helix or icosahedron. In animal viruses, this complexity is exemplified by the poxviruses, which do not have a well-defined capsid structure, but rather consist of a DNA-containing nucleoprotein core surrounded by protein that is enclosed, together with two lateral bodies of unknown structure and function, within a lipoprotein coat covered with a layer of tubules or filaments. Extracellular poxviruses are also surrounded by a membrane acquired in the cytoplasm and containing several virus-specific proteins.

UNCLASSIFIED VIRUSES

The structure of some viruses is not sufficiently well understood to permit assigning them to families on the basis of structural criteria. The most important human virus in this category is hepatitis B virus. The virion is a complex structure 45 nm in diameter with a nucleoprotein core 28 nm in diameter, surrounded by a lipoprotein shell. Within the core is a unique double-stranded DNA consisting of one circular, complete strand and a second, incomplete strand. The core contains a DNA-dependent DNA polymerase that is capable of completing the unfinished strand in vitro.

Classification of Viruses—Principles and Criteria

A comprehensive discussion of virus classification, or even a listing of all the rules used or all the classified viruses, is beyond the scope of this chapter; the Fourth Report of the International Committee on Taxonomy of Viruses, published in 1982, was 200 pages long. The important criteria that have made possible the classification of viruses into 17 major groups or families are given in Table 1.1. Tables 1.2 through 1.8 list these 17 families and some of their important properties and show their subclassification into genera and, in some cases, subfamilies. Examples of each are given. These tables are not all-inclusive; only one or a few virus species are listed under each genus, and unclassified groups of viruses are not shown.

ASSAY OF VIRUSES

Viruses can be quantitated by a variety of methods based on either infectivity or chemical, physical, or immunologic properties. These methods vary considerably in their precision, in ease and expense of application, and in time required for their completion. The choice of assay depends on the virus (not all viruses can be quantitated by certain methods) and on the facilities in a given laboratory.

Assay of Infective Units

Two general types of assays are based on the ability of the virus to infect cells. The first, exemplified by the plaque assay, depends on quantitation of individual, localized areas of infection in a layer of cells. The second, the end-point dilution method, depends on the detection of virus infection in a group of cell cultures or animals inoculated with serial dilutions of the virus.

PLAQUE ASSAY
This is the most precise, efficient, and economical assay of virus infectivity. The introduction of the plaque assay (long used to quantitate bacterial viruses) to animal virology in 1952 by Dulbecco was a major advance and played an important role in the development of virology as a quantitative science. Not only is the plaque assay accurate and highly reproducible, but also the number of plaques produced is proportional to the concentration of the virus inoculated, that is, the dose-response curve is linear. This proportionality indicates that one virus particle is responsible for the formation of each plaque. Thus, not only is plaque formation a simple and precise assay, but it also has made possible the important demonstration that one virus particle is sufficient to infect a cell, one of the cornerstones of quantitative virology. Because each plaque results from the infection of a cell by a single virus particle, the popu-

lation of virus in the plaque represents a clone, and one may isolate a clonal line by obtaining the virus from a plaque. In practice, this isolation is usually done by selecting the virus from an individual plaque two or three times in succession to avoid possible contamination with virus in the overlay from adjacent plaques. By this method, one can obtain genetically pure strains of virus even though mixed populations might have been present in the initial virus stock.

Plaque assays are done by inoculating monolayers of cells with equal volumes of serial dilutions of virus. After a period for adsorption of the virions to the cells, a semisolid overlay (usually agar) containing the appropriate maintenance medium for the cells, is added to minimize the spread of virus by dispersion, which would occur in liquid medium. A single virus particle infects a cell, and progeny virus produced by that cell spread to adjacent cells. Repetition of this process leads to an enlarging area of infection that, after a period that varies with the virus-cell system, results in a localized area of infection and cell damage (a plaque) that can be visually identified and counted. In some virus-cell systems, the plaques can be seen without any visual aids; however, a vital stain such as neutral red is frequently added to a second overlay to provide contrast between the living, uninfected cells and the damaged, infected cells. The amount of virus quantitated by the plaque assay method is expressed as plaque-forming units (PFU).

Some viruses do not cause sufficient cell damage to be seen as plaques, but their presence can be detected by special means. One such variation of the plaque assay, which is used with certain parainfluenza viruses, involves flooding the monolayer of cells with erythrocytes (RBCs) that will adsorb to the infected cells, and localized areas of RBCs can be seen. These areas, known as *hemadsorption plaques*, can then be counted. Another variation of the plaque assay involves the staining of the infected cells with fluorescein-labeled antiviral antibodies and counting the fluorescent foci with the microscope. This method has the disadvantage of requiring the immunofluorescence procedure, but the foci can usually be counted very rapidly, for example, within 24 hours, because they are detected microscopically. Although such an assay has its place, an end-point dilution method is more often used for viruses that do not cause sufficient cytopathic effects to result in detectable plaques.

FOCUS ASSAY

A variation of the plaque assay applicable to the quantitation of some tumor viruses that do not cause cell death is based on the ability of these viruses to cause cell proliferation. The same general procedure of inoculating monolayers of cells as for plaque assay is used, but the virus induces the formation of a localized area of transformed, proliferating cells called a *focus*. As with plaques, each focus is initiated by a single virus particle, and the units are expressed as focus-forming units (FFU).

POCK ASSAY

Just as viruses can cause localized lesions in monolayers of cells that are recognized as plaques or foci, some viruses can induce localized lesions on the chorioallantoic membrane of the chick embryo. Indeed, this method of assay has been in use since the 1930s. These lesions, called *pocks*, are produced by such viruses as poxviruses and herpesviruses. They may be purely degenerative, or they may be hemorrhagic and appear as red pocks. Certain tumor viruses, such as Rous sarcoma virus, will produce proliferative lesions (small tumors) on the chorioallantoic membrane. As with plaques, each pock is formed by a single virion. Pock assays are limited to those viruses that are released inefficiently or slowly, so that secondary pocks do not occur to complicate the assay. These assays also have the disadvantage that different embryos may vary in susceptibility to the virus, and thus accuracy and reproducibility are not as great as in a standard plaque assay done in a homogeneous population of cultured cells. Because the viruses that produce pocks or foci on the allantoic membrane usually also produce plaques in the cultured cells, pock assays are used infrequently today.

END-POINT DILUTION ASSAY

This type of assay is still used for certain viruses that do not cause sufficient cytopathic effects to be recognized as plaques but whose presence can be detected in cultured cells, or for viruses that do not replicate in tissue culture but do cause disease in embryos or adult animals. Before the advent of plaque assays, end-point dilution was the major method for assaying the infectivity of viruses. In this procedure, serial dilutions are made, and equal aliquots are used to inoculate replicate cell cultures, chick embryos, or animals. At a fixed time, which is long enough for infection to be manifest even if the inoculum contained only a single infectious virus particle, the inoculated cell cultures, embryos, or animals are examined for the presence of virus. This examination may be done in several ways. In cell culture, virus may be detected by the presence of a characteristic cytopathic effect (degeneration, cell rounding, syncytia formation) or by the physical presence of virus particles using the hemagglutination reaction (see below). In animals, virus may be recognized by disease or death or by lesions in specific organs (consolidation of the lungs, foci in the spleen). In chick embryos, the virus may be detected by death or by the presence of virus in the allantoic fluid, as demonstrated by hemagglutination.

In such titrations the end-point is the last dilution at which virus can be detected in an inoculated culture or animal; that dilution, by definition, contains at least one infective unit in the volume inoculated. Usually a formula is used to calculate the dilution at which 50% of the inoculated persons would be infected, and the concentration of virus is expressed as 50% infective doses (ID_{50}). In animals, death may be used as an end-point and a 50% lethal dose (LD_{50}) calculated. This dose may be the same as the ID_{50}, but with some viruses death of the animal may require a larger inoculum than infection, which can be manifest without death. If sufficient numbers are inoculated at each dilution, the end-point dilution method can be made reasonably accurate, but even under these conditions it does not have the precision of the plaque assay. Furthermore, cloning of virus by the end-point dilution method is inherently more difficult than by plaque selection because of the greater danger of more than a single particle being present in the inoculum. Thus, a plaque assay, if feasible, is the preferred method for quantitating viruses and for isolating genetically pure virus strains.

Assays Based on Physical or Chemical Properties

HEMAGGLUTINATION ASSAY

Some viruses have the ability to adsorb to RBCs and agglutinate them. This phenomenon provided the first workable assay for viruses as physical entities, independent of their infectivity. The discovery of hemagglutination (HA) in 1941 by Hirst and independently by Mc-Clelland and Hare represented an enormous advance in virology; it provided for the first time an accurate, rapid, and inexpensive method for assaying certain viruses as well as their antibodies. The HA test and the plaque assay, introduced 11 years later, are the two greatest contributions to the quantitation of viruses. Before the discovery of HA, assay of influenza viruses was done routinely by an end-point dilution method in mice, an expensive and time-consuming procedure that took many animals and days to complete a single titration, and required mouse-adapted strains of virus. HA made detection and quantitation possible within an hour, using any strain of influenza virus. (As was shown later, many other viruses also hemagglutinate, and the technique has been applied to them.)

The HA assay is based on a virus particle's ability to adsorb to one RBC and then to another, bridging the two cells; by successive bridging, an aggregate of RBCs is built up that can be easily detected with the unaided eye. The virus preparation to be assayed is serially diluted, and a constant amount of a standard RBC suspension is added; the titer obtained is expressed in hemagglutinating units (HAU) as the reciprocal of the last dilution showing a standard pattern of agglutination. The accuracy of this method is generally taken to be a factor of 2; however, techniques using fractional dilution series in steps less than twofold can increase the accuracy. In careful studies in the early 1950s, Horsfall, using influenza virus, showed that in the most sensitive and well-controlled system the ratio of virus particles to RBCs at the end-point approached one. However, this number will depend on the virus strain, the temperature of assay, and the type and concentration of RBCs used.

Although the HA assay quantitates physical particles, with a standardized virus-RBC system it is possible to establish a ratio between HA units and infective units. Thus, by doing an HA assay, one can obtain the relative concentration of infective units, within the limits of accuracy of that assay. This is valid only in a standardized system in which there has been no unusual inactivation of infectivity, or the production of defective virus particles. Hemagglutination is also useful in the direct assay of infective units, such as in an end-point dilution assay of influenza virus or other hemagglutinating viruses. The presence of virus in inoculated chick embryos or cell cultures is frequently detected by the HA reaction. Because many strains of influenza virus do not form plaques efficiently, the end-point dilution method is still widely used with certain strains.

By utilizing the property that specific antiviral antibodies inhibit HA, antiviral antibodies can be readily quantitated; serial dilutions of an antiserum are made, followed by the addition of virus and RBCs, and a hemagglutination-inhibition (HI) titer is obtained. In addition, the HA reaction is used in neutralization tests, that is, assay of the ability of antibodies to neutralize virus infectivity. After inoculation of cell cultures of chick embryos with dilutions of virus-antibody mixtures, the presence of progeny virus is detected by the HA reaction.

HEMAGGLUTINATION AS A MODEL FOR VIRUS ADSORPTION

The importance of the HA reaction goes beyond its role in the assay of virus or antiviral antibodies. The adsorption of the virus particle to the RBC has proved to be an excellent model for virus-receptor interactions. As mentioned below, we know the most about the receptor chemistry and specific sites on the viral protein that interact with the receptors of the orthomyxoviruses and paramyxoviruses. Much of what is known was learned using RBCs, or glycoproteins isolated from them, as the receptors.

HEMAGGLUTINATION BY DIFFERENT VIRUSES

A number of viruses other than orthomyxoviruses and paramyxoviruses hemagglutinate, although less is known about their receptors on the cell surface. A list of hemagglutinating viruses includes some members of these virus families: picornaviruses, reoviruses, togaviruses, rhabdoviruses, bunyaviruses, coronaviruses, adenoviruses, papovaviruses, and poxviruses. Although orthomyxoviruses will agglutinate a wide variety of RBCs (e.g., chicken, human, guinea pig) over a wide pH range, other viruses require specific cell types, and the reaction may be exquisitely sensitive to pH and temperature. For example, measles virus will agglutinate only monkey RBCs at 37°C; certain rhabdoviruses will agglutinate only goose RBCs at 4°C; many togaviruses will agglutinate one-day-old chick RBCs and are pH sensitive; echoviruses and Coxsackie viruses will agglutinate only human cells.

DIRECT COUNTING OF VIRUS PARTICLES

Virus particles can be counted in the electron microscope, thus directly determining the number of physical particles. This method does not distinguish between infective and noninfective particles. Viruses can be counted after concentration of virus particles onto electron microscope grids. By extrapolation from the starting volume of virus suspension, the total number of virus particles can be calculated. Another technique involves mixing the virus suspension with a given volume of a suspension of latex spheres of known concentration, and then applying the mixture to electron microscope grids and counting the spheres and virus particles. From the known concentration of latex spheres and the number of spheres and virus particles counted, the absolute number of virus particles can be calculated.

RATIO OF PHYSICAL PARTICLES TO INFECTIVE VIRUS PARTICLES

With bacterial viruses, the ratio of physical to infective particles is frequently one. However, with most animal viruses, this ratio is usually significantly higher. Under certain optimum conditions, ratios of physical to infective particles of from one to four have been reported with some viruses; however, in preparations of most animal viruses this ratio varies from 10 to as high as 1000. Several factors are involved in such high ratios. There may be noninfective virus particles in the population, because of defective or incomplete genomes, thermal inactivation, or physical disruption. However, in virus stocks prepared under conditions that minimize such factors, the ratio is frequently still high. This results from the fact that in a given population of virions inoculated, most do not initiate infection, even though they have the potential to do so. Factors responsible for this inefficiency of infection include variation in susceptibility of the host cells to infection, inefficient virus adsorption, elution of adsorbed virus from the cell before penetration, degradation of phagocytosed viruses by the host cell, and inefficient spread of the virus from cell to cell, resulting in failure to undergo the multiple cycles of infection that are necessary to register as an infectious event in a plaque or end-point assay.

ASSAYS BASED ON IMMUNOLOGIC PROPERTIES OF THE VIRUS

The fact that most viruses are good antigens can be used as the basis for assays. Now such assays are usually used for viral diagnosis, for examining antigenic relatedness of different strains, or for quantitating individual viral proteins. However, they can be used to estimate the number of virus particles present, if appropriate known standards are used. Historically, the complement-fixation test was the immunologic procedure used to detect and quantitate viral antigens; however, this test is relatively cumbersome and insensitive. In recent years, radioimmune assays (RIA) and enzyme-linked immunosorbent assays (ELISA) have been used extensively to quantitate viral antigens, and these procedures have largely replaced the complement-fixation test except for certain specialized purposes. RIA and ELISA are highly sensitive assays capable of detecting much smaller quantities of antigen than a complement-fixation test.

Several variations of each of these procedures have been used, but each basically involves measuring the amount of radioactivity or enzymatic activity in an antigen-antibody complex. The RIA may be done with either the viral antigen or antibody labeled with an isotope, usually ^{125}I, and the reaction may take place on a solid phase, with one of the reactants fixed to a surface, or in a liquid phase. With labeled antibodies, either a direct or indirect assay may be used. In the former, the antiviral antibodies are labeled, and in the latter, antiglobulin antibodies against the antiviral antibody are labeled. A competition assay in which the ability of a virus preparation to compete for antibody with a labeled, standard virus preparation can also be used. This method is a particularly sensitive method for examining antigenic relatedness of different virus strains.

In the ELISA, the virus or a viral antigen is immobilized on a plastic surface, and an antibody linked to an enzyme whose reaction product can be readily measured (e.g., alkaline phosphatase) is added, followed by washing and addition of a substrate for the enzyme. The reaction product is then measured. Either a direct assay, in which the antiviral antibody is linked to the enzyme, or an indirect assay, in which an antiglobulin antibody against the viral antibody carries the enzyme, may be used.

ASSAYS BASED ON VIRAL ENZYMES

Under certain conditions, viruses can be quantitated on the basis of virus-specific enzymes that are components of the virus particle. Thus, the neuraminidase activity of influenza virus or parainfluenza viruses can be assayed. The reverse transcriptase of retroviruses—that is, the viral enzyme that catalyzes the transcription of DNA from RNA—can be used as an assay for retrovirus particles. A case in which a viral enzyme is useful as an assay for virus particles because no cell culture system is yet available for the virus is hepatitis B virus. The virus particle contains a virus-specific DNA-dependent polymerase, which can be used to detect the presence of the virus.

ASSAY OF VIRUS PARTICLES BY OTHER PHYSICAL PROPERTIES

A highly purified and concentrated preparation of virus particles can be assayed by standard physical and chemical means, such as optical density measurement or chemical analysis for nucleic acid or protein.

Such techniques require, in addition to reproducibly pure preparations of virus, standardization of the procedure by another technique, such as electron microscopy.

STEPS IN THE VIRUS REPLICATION CYCLE

Members of the different families of animal viruses frequently have features of their replication cycles that characterize and distinguish them from members of other families. Nevertheless, there are some general steps in the replication cycle that all viruses undergo—adsorption, penetration, uncoating, synthesis of viral components, assembly, and release. As discussed in the following sections, with some viruses, two of these steps (e.g., penetration and uncoating, or assembly and release) may be combined.

Adsorption

The first step in virus infection and therefore the beginning of the replicative cycle is the adsorption of the virus particle to the surface of a susceptible cell. This process is begun by a random collision between the virus particle and the surface of the cell. Because the surfaces of both the virus particle and the cell are negatively charged at physiologic pH, the process of initial contact between them is facilitated by the presence of cations in the medium in which the virus-cell interaction is taking place.

Adsorption involves the interaction between specific sites on the virus surface with specific receptors on the surface of a susceptible cell. In few virus-cell systems is the chemical nature of either of these specific sites known in detail. On enveloped viruses, the reactive sites reside on the glycoproteins that form projections from the surface of the virion. On viruses with naked capsids, the reactive sites reside on one of the proteins of the capsid (e.g., the fiber protein projecting from the penton of adenoviruses) or, in some viruses, possibly a combination of different capsid proteins. Virus receptors on the cell surface have been biochemically characterized in detail only for orthomyxoviruses and paramyxoviruses. The receptors for these viruses are neuraminic acid-containing glycoproteins or, under certain conditions, glycolipids (gangliosides). Specific regions of the hemagglutinin glycoprotein of orthomyxoviruses or the hemagglutinin-neuraminidase glycoprotein of paramyxoviruses react with these receptors. Not only is the nature of the receptors for these viruses understood in detail, but also, in the case of the influenza virus hemagglutinin protein, the three-dimensional structure of the protein has been determined by x-ray crystallography. Because antigenic variation in this protein is the major factor in recurring epidemics of influenza, correlations are now being made between biologic function and structure at high resolution.

Members of both the orthomyxovirus and paramyxovirus families have an enzyme, neuraminidase, that is capable of destroying the receptors for the virus. Although this enzyme is not required for virus adsorption, it does play a role in the spread of the virus within the host. It does so by enabling the virus to free itself from neuraminic acid–containing glycoproteins, such as those found in mucous secretions in the respiratory tract, which might act as virus receptors and thus compete with the cellular receptors for the virus. Indeed, the virus particles themselves contain glycoproteins, and the carbohydrate moieties of these proteins are assembled by host-cell glycosyl transferases. Thus, if the viruses that react with neuraminic acid–containing receptors did not have a neuraminidase to remove the neuraminic acid residues from their own surface glycoproteins, the virus particles would adsorb to each other and the resulting aggregation would inhibit the spread of infection.

Although little is known about the chemistry of the receptors for most other groups of viruses, it is likely that glycoproteins, which are present in great variety on cell surfaces, serve as receptors for many types of viruses. Proteolytic enzymes, and enzymes and chemicals that attack carbohydrates, have been shown to abolish receptor activity for several viruses. In some cases (e.g., picornaviruses), lipid solvents destroy receptor activity. This destruction could indicate that a lipoprotein is the receptor, or it could result from disruption of a protein complex embedded in the lipid bilayer of the cell membrane.

In several virus-cell systems, the adsorption step can be subdivided into two phases: a reversible phase, in which a loose virus-cell complex is formed that can be easily dissociated (e.g., by elevating the temperature or by means of a chelating agent), and an irreversible phase, in which tight binding occurs. This tight binding probably involves multiple-point attachment between virus and cell. Elution of virus after reversible attachment occurs frequently with some viruses, and this elution is one of the reasons for a high virus-particle-to-infectivity ratio. As mentioned earlier, orthomyxoviruses and paramyxoviruses have the enzyme neuraminidase, which can destroy receptors and facilitate elution. Thus, in these viruses the receptor-binding and receptor-destroying activities of the virus compete, and the competition may be tilted either way by environmental conditions such as temperature, halide-ion concentration, and pH. For example, the neuraminidase activity of several paramyxoviruses is optimal at acid pH and diminished by chloride-ion concentrations found in extracellular fluids; such a condition could favor adsorption over neuraminidase activity in an extracellular environment and neuraminidase activity in a more acidic intracellular environment.

SPECIFIC RECEPTORS AS DETERMINANTS OF VIRUS-HOST RANGE AND TISSUE TROPISM

Because adsorption is a prerequisite for infection, a cell must have receptors for a given virus on its surface to be susceptible to infection by that virus. Thus, in some cases, the adsorption step is the major determinant of host range and tissue tropism of viruses, and consequently of the nature of the diseases that they produce. This fact is strikingly illustrated by the picornaviruses. Poliovirus will adsorb only to human or other primate cells, and, within the intact host, only the cells of the central nervous system (CNS), the oropharnyx, and the intestinal epithelium express receptors. Many other types of primate cells (e.g., renal epithelial cells) express receptors after being cultured in vitro for a short time. On the other hand, many types of nonprimate cells can be infected with poliovirus RNA extracted from the virus particle, thus establishing that the barrier to infection is at the level of virus capsid-receptor interactions. Other picornaviruses can adsorb to receptors present in a wider range of tissues; encephalomyocarditis virus adsorbs to heart muscle and brain, a tropism that is reflected in the name of the virus. Receptors may also be present at only certain stages of the development of an animal, which explains the pathogenicity of certain viruses for the newborn. Coxsackie B viruses can adsorb to receptors found only in newborn mice, not adult mice, and cause disease only in the newborn.

The exquisite specificity of the receptors for picornaviruses is also illustrated by the fact that the receptors for polioviruses, echo viruses, Coxsackie A and Coxsackie B viruses, and human rhinoviruses—all members of the picornavirus family—differ from each other. These receptors can be differentially inactivated, or the receptors on a cell for one virus can be saturated whereas another virus can still adsorb. On the other hand, there is some unexpected sharing of receptors between viruses of different families, for example, between Coxsackie B3 and human adenovirus 2.

In contrast to the situation with polioviruses, echo viruses, and Coxsackie viruses (the receptors for which are highly specific and found only on certain cells in the body), many different neuraminic acid–containing glycoproteins can act as receptors for orthomyxoviruses and paramyxoviruses. Because such sialoglycoproteins are ubiquitous on the surface of vertebrate cells, virus adsorption cannot be the step that determines host range and tropism. With these viruses, host range and tropism are determined at a later step.

The total number of receptors on the surface of a single cell has been estimated for several viruses. That number is difficult to determine with great accuracy; it may vary over a wide range with different viruses (e.g., 10^4 receptor sites for poliovirus and as many as 10^5 receptor sites for Sindbis virus, a togavirus).

Penetration

After the specific binding of a virus particle to receptors on the cell surface, the next step in infection is penetration, the process by which the virus or its genome enters the cytoplasm of the cell. This entry is accomplished in a variety of ways by different viruses, depending on their physical and biologic properties. In certain cases, such as some nonenveloped icosahedral viruses, the entire virion gains access to the cytoplasm; in others, such as enveloped RNA viruses, only the nucleocapsid enters the cytoplasm. Some viruses breach the plasma membrane at the cell surface, whereas with others the virus particle is first taken into an endocytic vesicle (a process known as *viropexis*) and then it or its nucleocapsid traverses the cell membrane (which is now the vesicle membrane) to enter the cytoplasm. One must recognize that a virus within an endocytic vesicle still has not gained access to the cytoplasmic matrix, and thus true penetration of the cell has not yet occurred, even though the virus is no longer present on the cell surface and is not accessible to antibody added to the medium bathing the cell. Resistance to neutralization by such antibody has been considered to be an operational definition of virus penetration, but this definition is not valid in cases in which the virus is first endocytosed, because the cell membrane still separates the virus from the cytoplasm. Examples of different modes of penetration of viruses with different physical properties are discussed briefly in the following paragraphs.

ENVELOPED VIRUSES

A large number of electron microscopic studies of early virus-cell interactions done in the 1960s and early 1970s yielded conflicting results as to the mechanism of penetration of viruses, particularly enveloped viruses. Two mechanisms were proposed for penetration: (1) endocytosis followed by release of viral components within an endocytic vesicle and (2) fusion of the viral membrane with the plasma membrane, with entry of only the nucleocapsid into cytoplasm. Because these studies were, of necessity, done with virus preparations with physical-to-infective-particle ratios greater than one and with the inoculation of very high multiplicities of virus to observe the early stages of infection in the electron microscope, one could not be sure that any particle observed, be it fusing with the plasma membrane or present in an endocytic vesicle, was involved in the infectious process. When more was learned about the proteins involved in penetration, and investigators could use infectivity as an end-point rather than having to rely on electron microscopy, the envelopes of some viruses were found to fuse with the plasma membrane at the cell surface. On the other hand, other viruses were found to be first taken into endocytic vesicles and their membranes subsequently to fuse with the membrane of the vesicle, enabling the nucleocapsid to enter the cytoplasm.

Unambiguous evidence for the penetration of an enveloped virus by fusion of the viral membrane with the plasma membrane, and for the direct involvement of a specific viral protein in the induction of membrane fusion, was obtained by Scheid and Choppin with paramyxoviruses. The F protein, a glycoprotein that forms one of the two types of spikelike projections on the surface of paramyxoviruses, mediates the fusion of the viral membrane with the plasma membrane of the host cell, and the helical nucleocapsid enters the cytoplasm as a result of this fusion process. Although fusion occurs over a broad pH range, from 5 to 9, it is enhanced at basic pH. The membrane-fusing function of the F protein is activated by a specific proteolytic cleavage by a host enzyme of an inactive precursor (F_0) to yield two disulfide-linked polypeptides (F_1 and F_2). The new N-terminus on F_1 generated by this cleavage appears to be directly involved in the fusion reaction. Fusion, and therefore infectivity, can be specifically inhibited by oligopeptides with amino acid sequences similar to that at the new N-terminus, providing a possible new approach to specific chemical inhibition of viral replication.

Different paramyxoviruses, or mutants of a given virus, require different proteases for activation, and not all cells have an appropriate protease that will activate the F protein. Because such activation is required for multiple cycle replication and spread in the host, the availability of a cellular protease to cleave and activate the F protein is a major determinant of host range, tropism, and virulence of paramyxoviruses. Thus, in this case, penetration rather than adsorption is the step at which tropism is determined. As mentioned previously, because the neuraminic acid–containing receptors for these viruses are ubiquitous on vertebrate cells, adsorption is not a step at which tropism is determined.

The penetration of orthomyxoviruses (influenza viruses) is mediated by the hemagglutinin glycoprotein and involves fusion of the viral membrane with the cell membrane. As with paramyxoviruses, influenza viruses' membrane-fusing activity is activated by a specific proteolytic cleavage by a host enzyme, and thus the penetration step is also a major determinant of pathoegenicity of influenza viruses. Unlike paramyxoviruses, however, influenza virus can only initiate membrane fusion at acidic pH. Furthermore, the available evidence suggests that, after adsorption, the virus is endocytosed, the endocytic vesicle fuses with lysosomes, and the resulting lowering of the pH in the secondary lysosomal vesicle activates the fusion of the viral membrane with the cell membrane to which it adsorbed, permitting the nucleocapsid to enter the cytoplasmic matrix.

A similar pathway has been demonstrated for an enveloped RNA virus with an icosahedral nucleocapsid, Semliki Forest virus (SFV), a togavirus. Adsorbed virus is endocytosed, and, in the presence of acid pH, in a secondary lysosome fuses with the cell membrane, and the icosahedral nucleocapsid enters the cytoplasmic matrix. The SFV virion can fuse with the plasma membrane at the surface of the cell without the requirement for endocytosis, if the cell with attached virus is exposed to a brief pulse of acid pH. Thus, with each of the above-mentioned enveloped viruses, penetration involves fusion of viral and cell membranes; however, the orthomyxoviruses, togaviruses, and rhabdoviruses require an acid pH for fusion. This acid pH is provided by endocytosis and fusion with lysosomes. In contrast, paramyxoviruses can fuse at physiologic pH, and thus fusion and penetration can be accomplished at the cell surface.

Although similar detailed studies have not been done with members of all families of enveloped viruses, it seems likely that penetration of enveloped viruses by fusion will be a general property of these viruses, occurring in some cases at the cell surface and in others within endocytic vesicles.

POXVIRUSES

The situation with the poxviruses, which have a complex structure, is more complicated. Detailed electron microscopic studies have been done with the so-called intracellular form of the virus obtained from disrupted infected cells. This form is infectious, and early electron microscopic evidence obtained by Dales suggested that it enters the cell by endocytosis and that the core of the virus is released in the vesicle and subsequently traverses the membrane and enters the cytoplasm. Later evidence suggested that the membrane of this intracellular form of the virus may fuse either with the membrane of an endocytic vesicle or with the plasma membrane at the cell surface. In addition to the intracellular form of poxvirus particles, there is also an extracellular form that has an outer membrane (not present on the intracellular form) acquired from intracytoplasmic membranes. This form of poxvirus particle, which has a higher specificity for infectivity, may enter by fusion of its outer membrane with the plasma membrane without the requirement for endocytosis.

NONENVELOPED VIRUSES

The penetration of animal viruses with naked icosahedral nucleocapsids is less well understood. Among the possible mechanisms are (1) endocytosis of the virion followed by disruption of the endocytic vesicle and release of the nucleocapsid into the cytoplasm, either intact or structurally altered; (2) the virion traversing the plasma membrane directly by an unknown mechanism, without being first endocytosed; and (3) the viral capsid being structurally altered by interaction with receptors on the plasma membrane, so that the viral nucleic acid is released into the cytoplasm, possibly without the capsid leaving the membrane. There is electron microscopic evidence that mechanisms 1 and 2 can occur with adenoviruses, but the relative roles of each mechanism in the initiation of infection is not clear. Mechanism 3 seems to occur with picornaviruses. It is thought that the picornavirus particle first binds to a receptor on the plasma membrane, and then additional receptors are recruited in the membrane to interact with viral capsid proteins, forming multiple sites of attachment. In this process, the cell membrane invaginates as it wraps around the virus particle. This interaction leads to a conformational change in the capsid, which may be partly intercalated into the membrane, and, as a result of this alteration in capsid structure, the viral nucleic acid is released into the cytoplasm.

Uncoating

The release of the viral genome from its outer shell is termed uncoating. Most viruses, at the uncoating step, cease to be recoverable from the cell as an infective entity. The virus thus enters a phase in the replication cycle known as the eclipse phase or latent period, a period that lasts until the first infectious progeny virus is produced. As with the penetration step, the uncoating process varies with different viruses, depending on their structural characteristics.

ENVELOPED VIRUSES

In the case of the enveloped viruses with helical nucleocapsids (i.e., orthomyxoviruses, paramyxoviruses, rhabdoviruses), the uncoating process occurs simultaneously with penetration. In the penetration step, the helical nucleocapsid leaves the viral envelope after fusion of viral and cell membranes, and no further uncoating is required to initiate biosynthesis, which begins with the transcription of RNA from the genome RNA. In these viruses this primary transcription occurs while the genome RNA is still associated with the nucleocapsid protein subunits.

Uncoating of enveloped viruses with icosahedral symmetry involves further steps, because the viral genome must be released from the nucleocapsid before transcription of the genome can begin. With togaviruses this occurs in the cytoplasm after fusion of the viral membrane with the endocytic vesicle membrane and entry of the nucleocapsid. With herpesviruses (enveloped DNA viruses), the icosahedral nucleocapsid enters the cytoplasm through fusion of viral and cell membranes. In the cytoplasm the capsid is removed from a DNA-protein core, which then enters the nucleus and initiates transcription.

NONENVELOPED VIRUSES

This uncoating also varies with the virus type; however, detailed information is available with only a few viruses, and even in those instances not all of the events are clear. With picornaviruses, as previously indicated, the interactions of the virus with receptors on the cell surface apparently induce a conformational change that leads to release of the viral nucleic acid into the cytoplasm, and the capsid may never leave the membrane. Thus, with these viruses also, uncoating seems to occur simultaneously with the penetration step.

DNA viruses with icosahedral nucleocapsids may be uncoated in the cytoplasm or the nucleus. The uncoating of adenovirus is a multistage process. The nucleocapsid appears to lose its penton, the capsomer at the vertex of the particle, soon after it enters the cytoplasm. Indeed, the fiber protein that projects from the pentons may remain associated with the cell membrane. The capsid minus the pentons has a somewhat more spherical appearance, and the DNA within it has become partially susceptible to deoxyribonuclease. In a subsequent step, the hexons of the capsid are lost, leaving a DNA-protein core, which may either enter the nucleus through a nuclear pore or deliver the nucleic acid into the nucleus at the nuclear membrane. The nucleocapsids of papopaviruses are transported to the nucleus rapidly after infection, and the uncoating process seems to occur entirely in the nucleus.

Biosynthesis of Viral Nucleic Acids and Proteins

After the viral genome has gained access to the appropriate region of the cell, the biosynthesis of the viral components begins. In this phase of the replicative cycle, great variety and complexity is found in the mechanisms for accomplishing virus reproduction, and it is in this area that the most rapid advances in our knowledge are occurring. Detailed consideration of the replication of each type of virus, or even of a few viruses, is beyond the scope of this chapter. Furthermore, new information is accumulating at such a pace that any such description would soon be dated. The brief discussion that follows will outline the basic strategies involved in the replication of different viruses, with specific examples cited where appropriate. Such a consideration of important general principles and special features of certain viruses, although not inclusive, should provide a cellular and molecular basis from which to consider, in later chapters, the discussions of disease processes and of attempts at immunologic and chemical intervention in virus infections.

A useful approach to the strategy of virus replication, originally outlined by Baltimore, is to consider the relationship of the mRNA of the virus to the nucleic acid present in the virion as the genome, using the convention that the mRNA is the positive or "plus" strand and nucleic acid complementary to the mRNA is the negative or "minus" strand. On this basis viruses can be placed into six different categories.

POSITIVE-STRAND RNA VIRUSES

The simplest case is represented by the positive-strand viruses, such as the picornaviruses and the togaviruses. With these viruses, the virion RNA can serve as mRNA and can initiate biosynthesis by coding for the virus-specific proteins, including the virus specific, RNA-dependent RNA polymerase. The virus must synthesize this enzyme to replicate the viral RNA; because the cell does not have such an enzyme, it must be encoded in the viral genome. Once this enzyme is available, the viral RNA can be reproduced and more proteins synthesized, and the replication cycle can continue. Because the genome RNA can function as mRNA, the nucleic acid of the positive-strand viruses is infectious.

NEGATIVE-STRAND RNA VIRUSES

In these viruses (e.g., orthomyxoviruses, paramyxoviruses, and rhab-doviruses), the RNA in the virion complements the mRNA; therefore, it cannot initiate protein synthesis. Because the cell cannot supply the RNA-dependent RNA polymerase needed to transcribe the first mRNA molecule and initiate the synthesis of viral proteins, the infecting virus must carry its own RNA polymerase (called a transcriptase) into the cell to start the process of virus replication. The helical nucleocapsids of the negative-strand viruses have such a transcriptase associated with them; in most viruses, two or three proteins are involved in this enzyme complex. It follows that the naked genome RNA of negative-strand viruses, which is deprived of its transcriptase, is not infectious. Because the genome is not mRNA, it cannot initiate the synthesis of the enzyme required to replicate itself.

DOUBLE-STRANDED RNA VIRUSES

The genome of these viruses (e.g., reoviruses) is segmented and contains both positive and negative strands. Because double-stranded RNA cannot serve as mRNA, and the cell cannot supply the appropriate RNA polymerase to transcribe mRNA from the double-stranded RNA genome, the infecting virion brings such a transcriptase into the cell. After partial removal of the outer capsid shell of the virion, this enzyme transcribes mRNA from each of the viral genome RNA segments, and these mRNAs are released into the cytoplasm to initiate protein synthesis. The unique replicative cycle of these viruses is discussed further below.

RETROVIRUSES

These viruses (e.g., RNA tumor viruses) represent a unique situation; the viral genome RNA, which is of the same polarity as mRNA, is first transcribed into DNA by a virus-specific enzyme, *reverse transcriptase*, contained in the virus particle. The DNA can then serve as a template for mRNA synthesis. The formation of a DNA intermediate, which can be integrated into cellular DNA, is fundamental to the role of these viruses in the malignant transformation of cells (as described in Chapter 12).

DOUBLE-STRANDED DNA VIRUSES

These viruses (e.g., poxviruses, adenoviruses, papovaviruses, herpes-viruses) synthesize their mRNA by using the same general strategy as the cell; that is, double-stranded DNA is used as a template for the transcription of mRNA. In most instances, the virus utilizes the DNA-dependent RNA polymerase of the host cell for this purpose; however, the poxviruses, whose replication occurs in the cytoplasm, have a virus-specific enzyme that is packaged in the virus core and transcribes the viral mRNA. For the replication of viral genome DNA, some viruses (e.g., adenoviruses and papovaviruses) rely on the cellular DNA polymerase, but herpesviruses and poxviruses utilize their own virus-specific DNA polymerases.

SINGLE-STRANDED DNA VIRUSES

Most of the *parvoviruses* contain single-stranded DNA of negative polarity; however, the adeno-associated viruses exhibit the unusual feature of incorporating equal numbers of both plus and minus strands into capsids, so that when the DNA is extracted from the two kinds of virions, it hybridizes. This hybridization originally led to the erroneous conclusion that these viruses contained double-stranded DNA. Cellular enzymes are used to transcribe the viral mRNA and replicate the DNA of these viruses.

Aspects of Biosynthesis of Selected RNA Viruses

POSITIVE-STRAND VIRUSES

As indicated above, the positive-strand viruses begin biosynthesis by the infecting RNA first serving as mRNA for the synthesis of viral proteins, including the RNA polymerase required to replicate the viral RNA. Positive-strand viruses can be subdivided into two groups that differ with respect to the details of protein and mRNA synthesis. In one group, exemplified by poliovirus, the genome RNA sets as a monocistronic mRNA, from which a single, large polyprotein is translated. The polyprotein is then subjected to a series of regulated proteolytic cleavages, some of which are due to a host-cell enzyme, but many of which are carried out by a virus-specified protease. These cleavages yield the four proteins that eventually become the capsid proteins, as well as nonstructural proteins, that is, the RNA polymerase and protease. In the process of cleavage, the precursor proteins and the final cleavage products may play roles in replication and morphogenesis. A special feature of the poliovirus RNA is the covalent linkage through a tyrosine residue of a small protein to the 5' end of the genome RNA found in virions. This protein is thought to be involved in the transcription of the RNA. Poliovirus mRNA is unusual in that, unlike cellular and other viral mRNAs, it does not have a cap structure at its 5' end. The absence of a cap may contribute to the ability of poliovirus mRNA to continue the synthesis of poliovirus proteins after cellular protein synthesis is inhibited. In the replication of the viral RNA, only full-length copies of the plus and minus strands are synthesized, in contrast to the situation with some other viruses, discussed later.

The togaviruses exemplify the other type of positive-strand RNA virus biosynthesis. For togaviruses, when the genome RNA infects a cell, a large protein is translated from about two thirds of the RNA, beginning at the 5' end. This protein is then cleaved into four nonstructural proteins, two of which represent the viral RNA polymerase. This enzyme copies the infecting positive strand to yield minus strands, which serve as templates for more plus strands. This system diverges from the poliovirus strategy in that two sizes of plus strands are synthesized: (1) a full-length copy can serve either as template for more minus strands or as mRNA or be incorporated into virions; (2) a shorter RNA serves as the mRNA for the structural proteins of the virus. As with poliovirus, this mRNA is translated as a polyprotein that later undergoes a series of proteolytic cleavages to yield the structural proteins of the viral capsid.

NEGATIVE-STRAND VIRUSES

These viruses can also be divided into two groups according to their mechanisms of RNA and protein synthesis. One group is represented by the rhabdoviruses and the paramyxoviruses. These viruses have a single, large, covalently linked strand of RNA as a genome. The virus-specific RNA transcriptase that is associated with the viral nucleocapsid transcribes the negative strand in two different ways: (1) into full-length plus strands that serve as templates for progeny minus strands, and (2) into five to seven subgenomic-sized RNAs, each of which represents a monocistronic mRNA for the synthesis of a viral protein. The exact mechanism by which these mRNAs are synthesized is not understood, although some investigators think it involves a stop-start effect of the transcriptase, rather than cleavage of a long transcript into pieces. Also not clear is how the transcriptase can transcribe full-length copies to be used as templates for minus strands versus the subgenomic-sized mRNAs. The transcription of the infecting RNA into plus strands is called "primary transcription," and, because it is done by the transcriptase carried into the cell with the RNA, it does not require protein synthesis. The copying of the plus strands into progeny minus strands, termed *replication,* does require protein synthesis, as does the secondary transcription from newly synthesized minus strands. The reason the continued protein synthesis is needed for replication is unknown but may be related to the encapsidation of minus strands by newly synthesized capsid proteins.

Orthomyxoviruses exhibit a unique strategy of replication. They have a segmented genome consisting (in influenza A and B viruses) of eight segments. As with other negative-strand viruses, an RNA polymerase complex (consisting of three proteins) exists in the virion. This polymerase complex is associated with each of the eight nucleocapsid segments. However, unlike the previously described situation, the transcription of influenza virus mRNAs from the parental minus strands takes place in the nucleus and requires the action of the cellular RNA polymerase II. This enzyme is required because the influenza virus mRNAs acquire the cap and the first few nucleotides from the 5' ends of newly synthesized cellular mRNAs, and the nucleocapsid-associated viral proteins that form the transcriptase complex have the necessary enzymatic activities to accomplish this transfer. These recent findings have explained the requirement of nuclear function for influenza virus replication; this requirement had been a mystery for many years.

The parental influenza RNA minus strands are transcribed in two ways: (1) into full-length plus strand copies, which do not have a cap at their 5' end, are not polyadenylated, and serve as templates for progeny minus strands; (2) into mRNAs that have a cap acquired from cellular mRNAs, are missing about 20 nucleotides from their 3' ends, and are polyadenylated. Exactly how the transcription of the two forms of RNA is regulated is not clear. As with paramyxoviruses, primary transcription of minus strands into plus strands does not require protein synthesis, but the replication of new minus strands does.

Formerly, each influenza RNA segment was believed to code for one protein and be transcribed into a single monocistronic mRNA. Recently, however, two of the influenza virus RNA segments have been shown to code for two proteins. The genes for these two proteins overlap on the genome RNA segments. The two proteins share a few nucleotides at their amino termini, but the remainder of the proteins are transisted in different reading frames. One of the mRNAs is essentially a copy of the genome segment, whereas the other is derived by a splicing mechanism, the first demonstration of mRNA splicing in an RNA virus.

As mentioned before, the segmented nature of the influenza virus genome leads to extensive genetic reassortment among strains, and this reassortment plays an important role in the epidemiology of influenza. Advantage has been taken of this genetic reassortment to establish the functions of the genome segments of the virus and to introduce desired viral genes for specific purposes, such as vaccine production. For example, a new influenza virus strain that has the surface antigens needed for a new vaccine, but that multiplies poorly, has been recombined with an old strain that multiplies to high titer. Genetic reassortment is also being used extensively in attempts to produce suitable live-virus vaccine strains. Another consequence of the segmented genome is the problem of getting at least one copy of all eight segments of RNA into the viral envelope. How the virus solves this problem is not clear. One school of thought is that the segments incorporate randomly into the envelope. This action assumes that, given the fact that more than eight segments can be incorporated into a viral envelope, enough virions will have copies of all eight segments for the virus to be successfully transmitted. The alternative is the existence of some kind of organizing principle that will ensure the selection of the right eight segments into a virion; however, no such mechanism has as yet been identified.

DOUBLE-STRANDED RNA VIRUSES

These viruses (reoviruses and rotaviruses), with their unique double-stranded RNA genome, exhibit a strategy of replication that differs from that of other viruses. The virion is taken into the cell in an endocytic vesicle, where lysosomal enzymes remove some of the proteins from the outer capsid, creating a structure called a *subviral particle*. This particle enters the cytoplasm of the cell, where the virion-associated transcriptase transcribes each of the ten segments of the segmented genome while they remain within the particle. Indeed, the double-stranded genome RNA segments are never free in the cytoplasm; they are found only in subviral particles, in newly formed immature particles, or in mature virions. The newly synthesized mRNAs are capped by several additional enzymatic activities that are associated with the virion, but they are not polyadenylated. The mRNAs enter the cytoplasm and each codes for a viral protein. Some of these proteins associate with the viral mRNAs to form complexes within which the mRNA strands are transcribed once only into minus strands, with which they remain hybridized, creating double-stranded RNA segments within a protein complex. These immature virus particles then serve as templates for the conservative replication of a great excess of plus (mRNA) strands. The plus strands in turn produce more proteins that later associate with the plus strands to form new immature virus particles, within which they are transcribed once into minus strands to form the double-stranded genome RNA segments. These immature virus particles eventually associate with the other structural proteins of the virus to form mature virions. This method of replication differs not only from that of single-stranded RNA viruses, but also from that of double-stranded DNA viruses, in that the replication is conservative and the synthesis of plus and minus strands is widely separated in time.

Because the genomes of double-stranded RNA viruses are segmented, the question of how the correct ten segments are packaged into the virion arises, as it did with influenza virus. The mechanism is not clear, although recent evidence suggests that a nonstructural protein may act as an organizing principle, recognizing the single appropriate plus strands and bringing them together into the complex, where they are transcribed to form the double strands. As with influenza virus, genetic reassortment occurs as a consequence of the segmented genome. This phenomenon has been utilized not only to study the molecular biology and genetics of the virus, but also, following the lead of the influenza virus work, for the practical purpose of creating a hybrid rotavirus that will grow well in cultured cells, by virtue of genes from an animal virus, but that also has the desired human rotavirus proteins important in inducing neutralizing antibodies. This approach has obvious value for vaccine production.

DOUBLE-STRANDED DNA VIRUSES

In recent years the molecular biology of several double-stranded DNA viruses (i.e., adenoviruses, herpesviruses, papoviruses, and poxviruses) has progressed enormously. Genomes have been mapped (and in some cases completely sequenced), the programs for the transcription and processing of viral messenger RNAs elucidated, control mechanisms at the level of transcription and translation identified, functions of viral proteins defined, and considerable progress made in understanding the replication of the viral DNAs. In addition to providing exciting new information concerning virus replication and gene expression in eukaryotic cells in general, the advances in recent years have identified specific targets for potential antiviral chemotherapy, some of which are already being exploited. In this section, the features of biosynthesis common to the double-stranded DNA viruses will be discussed, and some important points relating to specific viruses will be described.

With the exception of the poxviruses, transcription of the viral mRNAs occurs in the nucleus and uses the host cell RNA polymerase II. As with cellular mRNAs, the primary transcripts are larger than the RNAs that appear in the cytoplasm and function as mRNAs. These transcripts are processed to remove intervening sequences and are capped at their 5' ends and polyadenylated, by cellular enzymes in most cases. The poxviruses provide their own enzymes to accomplish this processing, and the primary transcripts of poxviruses are capped and polyadenylated by enzymes contained in the viral core before they are extruded into the cytoplasm. Each of the viral mRNAs functions as a monocistronic message, coding for a single protein.

Transcription is temporally regulated and is divided into early (before DNA replication) and late (after DNA replication) phases. With the exception of herpesviruses, only a small portion of the viral genome is transcribed early. Transcription is also regulated spatially, that is, different regions of the genome are transcribed at different times. This process has been worked out in elegant detail for the adenovirus system. Early transcription is scattered over several regions of both strands of the DNA, so that it is occurring in both directions at the same time. Late transcription also occurs in different regions of the genome, though mostly from one (the rightward reading) strand. These late mRNAs derive from identical large transcripts of this strand, which are then processed by cleavage at different sites into five groups of mRNAs, each containing several species of mRNAs of different lengths but having the same 3' terminus. Each of these mRNAs also has the same untranslated leader sequence at its 5' end, derived from three short, noncontiguous regions on the genome that are joined by splicing out the intervening sequences. The mechanisms that determine the cleavage sites on the primary transcripts and the reason that some mRNAs are processed more often than others remain to be determined.

The control of the transcription of the various mRNAs in turn controls the synthesis of the various proteins. Although in most viruses, this control seems to be at the transcriptional level, in herpes simplex virus the temporal regulation of protein synthesis appears to be at the posttranscriptional level. Three classes of mRNAs are synthesized in herpes-simplex-virus-infected cells, designated immediate early, delayed early, and late, which code for three classes of proteins—alpha, beta, and gamma, respectively. The synthesis of each of the subsequent sets depends on the previous synthesis of the preceding set. Positive feedback regulation of early mRNA protomers by the alpha proteins produces the mRNAs for the beta proteins. Alpha protein synthesis is inhibited by beta proteins, which also regulate the processing of transcripts for gamma proteins; the gamma proteins in turn are the structural proteins of the virus.

One feature of the expression of the papovavirus genome deserves special mention—the coding regions for the viral proteins overlap, so that some of the nucleotide sequences are used twice and others three times to code for different proteins. In some cases, the same reading frame is used, so that the proteins share some, or many, amino acid sequences; in other cases, a different reading frame is used for translation of the different proteins. This use of the small genome of these viruses to code for the required numbers of different proteins is extremely economical.

DNA REPLICATION

Elucidation of the precise mechanisms of the replication of the DNA of animal viruses has not proceeded as rapidly as the work on transcription; nonetheless, models for the DNA replication of each virus have been developed. Replication is semiconservative, but the patterns differ because the structures of the DNAs vary.

Adenovirus DNA replicates asymmetrically, initiating synthesis at either of the 3' ends of the molecule, with displacement of the unreplicated strand occurring as replication proceeds.

The DNA of papovaviruses is a supercoiled closed circle, and its replication seems to be symmetrical and bidirectional, with the unwinding of the coil carried out by a "swivel enzyme."

The DNA of the poxviruses is cross-linked at the ends, and the exact mechanism of replication is not yet certain. It may begin at one end, proceeding bidirectionally to produce a loop at that end, with an endonuclease cutting the terminal cross-link at the appropriate time to permit separation of newly made double strands.

The structure of the DNA of herpesviruses is exceedingly complex, with unique sequences being bracketed by inverted repeats and different configurations being found. This complex structure will be discussed further in the chapter on herpesviruses (Chapter 10). The mechanism of replication of these molecules is not clear but must involve frequent recombinations in order to obtain the variety of structures found.

The DNA polymerase utilized to replicate the viral DNA may be cellular (for adenoviruses and papovaviruses) or virus-specific (for poxviruses and herpesviruses).

VIRUS REPLICATION AND TRANSFORMATION OF CELLS

The preceding discussion has been concerned with the replication of viruses during production cycles, which usually result in cell death. In addition to such lytic cycles, several of the DNA viruses, including papovaviruses, adenoviruses, and herpesviruses, can cause the transformation of cells to the malignant state. This transformation involves integration of viral DNA into cellular DNA, and mature virions are usually not produced. The role of viruses in transformation is discussed in a later chapter (see Chapter 12).

DEFECTIVE-INTERFERING VIRUS PARTICLES

Many viruses, particularly the RNA viruses, can, under certain conditions, induce the synthesis of virus particles having defective genomes. These particles are usually produced under conditions of repeated, high-multiplicity passage of the virus. They contain only a part of the viral genome, usually only a small part, but they contain all of the viral proteins. They cannot replicate alone but can do so in the presence of a homologous helper virus. They specifically interfere with the homologous virus and, in some cases at least, this inhibition occurs at the level of replication of the genome RNA of the virus, through competition for the viral RNA polymerase. These defective, interfering particles are interesting biologic entities, and studies of them have shed light on several aspects of virus replication. They are mentioned here because, by virtue of their ability to interfere with homologous virus replication, they have been implicated in the causation of some persistent infections. In the laboratory, conditions can be produced in which the maintenance of a persistent infection occurs to depend on the presence of the defective interfering particles. Whether these particles play any role in the initiation or maintenance of persistent infections in humans or animals under natural conditions, however, remains to be determined.

Assembly of Virions

Late in the replication cycle, after the biosynthesis of the viral nucleic acids and proteins, the viral components are assembled into mature virions. A brief discussion of the general principles of virion assembly can be divided into two parts: the assembly of the nucleocapsid, and the envelopment of the nucleocapsid of enveloped viruses.

NUCLEOCAPSID

As discussed in the section on virus structure, the assembly of viral capsids, whether icosahedral or helical, is a spontaneously occurring self-assembly process; the capsids are minimum energy structures. The structural features required for the assembly of the protein shell reside in the protomers that are the subunits of which the structure is built.

The assembly, size, and dimensions of icosahedral capsids are determined by the nature of the protomers and do not depend on the presence of the viral nucleic acid. With many viruses, some empty capsids
are formed during viral replication; however, as mentioned, the nucleic
acid may stabilize the structure of the capsid. In helical nucleocapsids,
the diameter and all other structural features except length are determined by the properties of protomers. The self-assembly process of the
helical plant virus TMV has been studied in exquisite detail, and investigators have been able to assemble the helical structure in vitro both
with and without RNA. However, the length of a helical nucleocapsid
is determined by the nucleic acid that it contains.

The site of assembly of the capsid varies with the virus type. These
sites for the different virus families are given in Tables 1.2 through 1.8.
In general, the capsids of DNA viruses are assembled in the nucleus,
except for those of poxviruses and iridoviruses, and the capsids of RNA
viruses are assembled in the cytoplasm, with the possible exception of
influenza virus, in which transcription occurs in the nucleus and nucleocapsid proteins are present in the nucleus; thus, at least some association of nucleocapsid protomers with RNA probably occurs there.

As discussed briefly in the section on virus structure, some proteins
associated with the nucleic acids of some viruses are not part of the capsid structure. This finding is most notable in DNA viruses such as adenoviruses and papovaviruses, in which histonelike proteins are associated with the DNA so that the nucleoprotein complex has a structure
resembling cellular chromatin. The association of such proteins with the
viral nucleic acid precedes the encapsidation of the nucleic acid and does
not play a role in capsid assembly.

ENVELOPED VIRIONS

The presence of an envelope adds another phase to the assembly of
the mature virion. Depending on the virus, envelopment may occur at
the nuclear membrane, at intracytoplasmic membranes, or at the plasma
membrane. Variation in the site of envelopment may occur among different members of the same virus family, or even between different
strains of the same virus. In the togavirus family, the alphaviruses (e.g.,
Semliki Forest virus) bud from the plasma membranes, whereas the fla-
viviruses (yellow fever virus) bud into intracytoplasmic vesicles. Depending on the virus strain, rabies virus may bud from either intracytoplasmic membranes or the plasma membrane.

Although the sites of envelopment may differ, one general feature is
common to all enveloped viruses studied thus far. The membrane in
which the capsid is enveloped contains largely, if not entirely, virus-
specific proteins. The proteins on the outer surface of the virus membrane are usually glycoproteins, and in some, but not all, viruses, a
nonglycosylated protein is associated with the inner surface of the
membrane. The lipid composition of the viral membrane usually re-

flects closely that of the cell membrane from which the virus buds. The carbohydrate side chains of the viral glycoproteins are synthesized by cellular enzymes; they thus resemble in structure the carbohydrate moieties of the cell membrane of origin. A striking exception to this is the absence of neuraminic acid residues from orthomyxoviruses and paramyxoviruses due to the action of the viral neuraminidase.

RNA VIRUSES WITH HELICAL NUCLEOCAPSIDS

A considerable body of information has accumulated on the assembly of these viruses, particularly the orthomyxoviruses, paramyxoviruses, and rhabdoviruses. Indeed, the assembly of the membranes of these viruses has been used as a model for the biogenesis of the plasma membrane of the cell. Although different aspects of the assembly process have been elucidated to a greater or lesser extent with different viruses, in general the process is similar, and the assembly of these viruses will be discussed together, with a few differences pointed out as they arise.

The orthomyxoviruses and paramyxoviruses each have two glycoproteins that form spikelike projections from the viral membrane (i.e., the hemagglutinin and neuraminidase, and the hemagglutinin-neuraminidase and fusion proteins, respectively). The rhabdoviruses have only a single-spike glycoprotein. All three viruses have a single, nonglycosylated membrane protein associated with the inner surface of the lipid bilayer of the viral membrane. The glycoproteins are synthesized on membrane-bound polysomes, with a signal sequence at the N-terminus of the polypeptide chain being inserted into the membrane of an endoplasmic reticulum (ER) vesicle soon after its synthesis. As the synthesis of the polypeptide chain proceeds, it is translocated into the lumen of the vesicle.

Glycosylation of the protein begins in the ER vesicles, with completion of the carbohydrate chains occurring in Golgi vesicles. The complete glycoproteins are transported from the Golgi to the plasma by fusion of the vesicle membrane with that of the plasma membrane. The viral glycoproteins can now migrate in the plasma membrane, until they contact and interact specifically with other viral proteins. Each of the viral glycoproteins has a hydrophobic region near one terminus that spans the bilayer of the membrane, and a short segment of the polypeptide chain is exposed on the inner surface of the bilayer. With the exception of the neuraminidase protein of influenza virus, all the viral membrane glycoproteins that have been examined have their C-terminal region associated with the membrane and their N-terminal region on the external surface of the virion, whereas the N-terminal region of the neuraminidase protein is embedded in the membrane.

The viral membrane or matrix (M) protein, which is synthesized on free polysomes, interacts with the segment of the glycoproteins that are exposed on the cytoplasmic surface of the membrane. This interaction leads to the formation of a patch of membrane containing virus-specific proteins, from which cell membrane proteins are excluded. The viral nucleocapsid protein subunits are synthesized on free polysomes and associate with the viral genome RNA, forming the helical nucleocapsid, which interacts with the M protein at the inner surface of the plasma membrane. Recent evidence with paramyxoviruses has shown that the M protein is present at the inner membrane surface in a highly ordered array, and the nucleocapsid associates with these arrays in a regular pattern. There then occurs a cooperative interaction among the proteins, the details of which are unclear, resulting in the initiation of the budding process and the envelopment of the nucleocapsid in the membrane and release of the complete virion.

From this description, it is clear that the M protein of these viruses plays a key role in the assembly of these viruses, interacting with the exposed ends of the viral membrane glycoproteins, on the one hand, and providing a recognition site for the nucleocapsid on the other. Interestingly, in subacute sclerosing panencephalitis, a chronic neurologic disease caused by persistent infection by measles virus, there is failure to express the viral M protein in brain cells, resulting in an abortive infection in which no mature virions are produced and nucleocapsids accumulate in the cells. Because all of the other viral proteins continue to be produced, extremely high levels of antibodies to all the viral proteins except M are found in the serum and cerebrospinal fluid of these patients.

RNA VIRUSES WITH ICOSAHEDRAL NUCLEOCAPSIDS

This process, as studied with togaviruses such as Sindbis and Semliki Forest viruses, follows a scheme generally similar to that of the helical RNA viruses, except that there is no M protein. Therefore, the nucleocapsid protein subunits interact specifically with the segments of the glycoproteins that are exposed on the cytoplasmic side of the membrane.

DNA VIRUSES

The envelopment of herpesviruses, DNA viruses with icosahedral symmetry, differs from the process with RNA viruses in that the capsid is assembled in the nucleus and then enveloped by budding from the inner nuclear membrane. This envelopment places the virion within the system of the endoplasmic reticulum, through which it passes to reach the surface of the cell.

The poxviruses exhibit a unique method of envelopment, being complex DNA viruses that are synthesized in the cytoplasm. The intracellular form of the virus has a membranelike, lipid-containing structure that is synthesized de novo in the cytoplasm. Such forms of the virus are released from the cell only after the disintegration or mechanical disruption of the cell. Most studies of poxviruses have been done on this type of virus particle. However, an extracellular form of the virion exists that acquires an envelope in the cytoplasm by being wrapped by vesicles resembling those of the endoplasmic reticulum, acquiring a double membrane in the process. These double-membrane-enclosed particles migrate to the cell surface, where the outer membrane presumably fuses with the plasma membrane, and the virion with the inner membrane is released from the cell.

Release of Virions

The final step in the virus replication cycle is the release of mature progeny virions from the cell. As one would expect from the different modes and sites of virus replication and assembly, the process of release varies with virus type. Again this difference in release can be considered in terms of viruses with naked-virus-enveloped nucleocapsids.

NONENVELOPED NUCLEOCAPSIDS

In general, the mechanism for the release of viruses with naked nucleocapsids is simply cell lysis resulting from the cytopathic effects of virus infection or, in the laboratory, from mechanical disruption by the experimenter. Depending on the rate of replication of the virus and the rapidity and severity of the damage to the cell, disintegration of the cell may occur relatively rapidly, as with poliovirus, or relatively slowly, as with papovaviruses. Because there is no active mechanism for release, mature virions accumulate in the cell until it lyses. Thus, if one quantitates both released extracellular virus and intracellular virus, the accumulation of the former in the medium lags significantly behind the amount of intracellular virus, until late in the replication cycle, when all the cells are degenerating.

ENVELOPED VIRUSES

Because the maturation of enveloped viruses is not complete until the nucleocapsid is enveloped, the processes of release and assembly are one and the same with many enveloped viruses, specifically those viruses assembled by budding from the plasma membrane. With those viruses, there are no "intracellular virions," only extracellular virions. The term *cell-associated* virus is sometimes used with reference to these viruses to designate virions still found with cells rather than free in the medium; however, these represent virions that have budded from the cell but are still associated with it, that is, adsorbed to adjacent receptors on the plasma membrane. For those viruses that are enveloped at the nuclear or intracytoplasmic membranes, there are intracellular virions, that is, virions within intracytoplasmic vesicles. Such virions either make their way through the endoplasmic reticulum or are carried to the surface in vesicles that release them by a process resembling exocytosis or are released after degeneration of the cell. Depending on the virus, the release by one of these mechanisms may be delayed; therefore, a consid-

44 VIROLOGY IN MEDICINE

erable amount of intracellular virus may be present until late in the replicative cycle. The release of poxviruses represents a unique situation that has been discussed.

CONCLUSIONS

Probably no field of biology is progressing more rapidly at this time than virology. This rapid progress augurs well for the future, not only for a greater understanding of how these important agents of diseases of humans, lower animals, and plants replicate, damage cells, and cause disease, but also for a greater understanding of cellular functions, such as gene expression, control mechanisms, cellular proliferation, and membrane biology.

As an example of how understanding of the basic mechanisms of virus replication has already been translated into effective therapy in humans, the fact that herpesvirus uses its own DNA polymerase has made possible the design of specific selective antiviral agents. Indeed, the variety of different enzymes now known to be coded for by viruses is one reason for optimism regarding the prospects for antiviral chemotherapy. Two decades ago, only one animal virus-specific enzyme was known, the influenza virus neuraminidase. In recent years, dozens of virus-specific enzymes have been identified, and individual viruses (e.g., poxviruses, reoviruses) have been found to have multiple enzymatic activities.

Virus-specific enzymes are not the only point of attack on viruses. As we learn more about the functions of other viral proteins, we should be able to design means for interfering with their activity chemically as well as immunologically. One can be confident that future volumes such as this will contain much new and exciting information.

Recommended Readings

Andrewes CA, Pereira HG, Wildy P: *Viruses of Vertebrates*, 4th Ed. London, Balliere Tindall, 1978.
Ball LA, Wetz GW: VSV RNA synthesis: How can you be positive? *Cell* 25:143–144, 1981.
Beers RF Jr, Bassett EG (eds): *Cell Membrane Receptors for Viruses, Antigens and Antibodies, Polypeptide Hormones, and Small Molecules.* New York, Raven Press, 1976.
Baltimore D: Expression of animal virus genomes. *Bacteriol Rev* 35:235–241, 1971.
Butler PJB, Klug A: The assembly of a virus. *Scientific American* 239:62–69, Nov. 1978.
Caspar DLD: Design principles in virus particle construction. In Horsfall FL Jr, Tamm I (eds): *Viral and Rickettsial Infections of Man*, 4th Ed. Philadelphia, Lippincott, 1965.
Choppin, PW: Measles virus and chronic neurological diseases. *Ann Neurol* 9:17–20, 1981.
Choppin PW, Scheid A: The role of viral glycoproteins in adsorption, penetration, and pathogenicity of viruses. *Rev Infect Dis* 2:40–61, 1980.

Crowell R: Comparative generic characteristics of picornavirus receptor interactions. In Beers RF Jr, Bassett EG (eds): *Cell Membrane Receptors for Viruses, Antigens and Antibodies, Polypeptide Hormones, and Small Molecules.* New York, Raven Press, 1976.

Davis BD, Dulbecco R, Eisen HN, Ginsberg HS (eds): *Microbiology.* New York, Harper & Row, 1980.

Dulbecco R: The nature of viruses; and Ginsberg HS, Dulbecco, R: Multiplication and genetics of animal viruses. In Davis BD, Dulbecco R, Eisen HN, Ginsberg HS (eds): *Microbiology.* New York, Harper & Row, 1980.

Fields BN (ed): *Virology.* New York, Raven Press, 1985.

Flint SJ, Broker TR: Lytic infection by adenoviruses. In *The Molecular Biology of Tumor Viruses,* 2nd ed, part 2. New York, Cold Spring Harbor Laboratory, 1981.

Fraenkel-Conrat H, Wagner RR (eds): *Comprehensive Virology,* vols 2–17. New York, London, Plenum Press, 1974–1981.

Joklik WK: *Principles of Animal Virology.* New York, Appleton-Century-Crofts, 1980.

Joklik WK: Structure and function of the reovirus genome. *Microbiol Rev* 45:483–501, 1981.

Kitamura N, Adler CJ, Wimmer E: Structure and expression of the picornavirus genome. *Ann NY Acad Sci* 354:183–201, 1980.

Krug RM, Broni BA, Bouloy M: Are the 5' ends of influenza viral mRNAs synthesized *in vivo* donated by host mRNAs? *Cell* 18:329–334, 1979.

Lamb, RA, Breidis DJ, Lai CJ, Choppin PW: Multiple mRNAs and coding regions derived from influenza A and B virus RNA segments. In Nayak DP (ed): *Genetic Variation Among Influenza Viruses,* ICN-UCLA Symposia on Molecular and Cellular Biology, vol 22. New York, Academic Press, 1981.

Lonberg-Holm K, Philipson L (eds): *Virus Receptors, Part 2 Animal Viruses, Receptors and Recognition,* series B, vol 8. London, New York, Chapman and Hall, 1981.

Luria SE, Darnell JE Jr, Baltimore D, Campbell A: *General Virology,* 3rd ed. New York, Wiley, 1978.

Mathews REF: Classification and nomenclature of viruses. Fourth Report of the International Committee on Taxonomy of Viruses. *Intervirology* 17:1–200, 1982.

Mosa B: Poxviruses, in Nayak DP (ed): *The Molecular Biology of Animal Viruses,* vol 2. New York, Marcel Dekker, 1978.

Richarson CD, Scheid A, Choppin PW: Specific inhibition of paramyxovirus and myxovirus replication by oligopeptides with amino acid sequences similar to those at the N-termini of the F_1 and F_2 or HA_2 viral polypeptides. *Virology* 105:205–222, 1980.

Roizman BR: Herpes simplex viruses. In Tooze J (ed): *The Molecular Biology of Tumor Viruses,* 2nd ed, part 2. New York, Cold Spring Harbor Laboratory, 1981.

Wilson I, Skehel JJ, Wiley DC: Structure of the haemagglutinin membrane glycoprotein of influenza virus at 3 Å resolution. *Nature* 289:366–372, 1981.

2

Host Response to Viral Infection

JERRY W. SMITH and CALDERON HOWE

Because of major developments in antibiotic therapy during and after World War II, many bacterial infections can now be treated effectively. Chemotherapy, in contrast, has only recently emerged as a real possibility for treatment of viral infections, signaled by the discovery that DNA analogs are effective against various manifestations of herpesvirus infection. With few exceptions, one must still rely on the natural ability of the host to control and eliminate viruses when they are recognized as the cause of acute disease. Recent advances in active and passive immunoprophylaxis and their impact on the incidence, morbidity, and mortality of viral infections underscore the importance of immune responses as the critical factor in recovery and prevention. Smallpox has been eradicated, and the incidence of paralytic poliomyelitis and measles has been reduced to the vanishing point, largely as a result of the application of effective immunoprophylactic measures.

IMMUNE MECHANISMS

Host resistance is a complex phenomenon involving multiple components. Each component has a different degree of importance, depending on the virus, the portal of entry, the organs involved, the patterns of viral pathogenesis, and the age, physiologic conditions, and genetic constitution of the host. For example, viral infection often affects neonates, young infants, and the elderly more severely than young healthy adults. At both ends of the life span, specific as well as nonspecific immune host responses are of major importance and, along with other factors, govern the final outcome.

Besides the immune mechanisms, a number of passive barriers limit the capability of viruses to damage cells and disturb host physiology.

The skin and mucous membranes impede contact between virus and susceptible cells. Few viruses can penetrate skin or mucosal epithelium unless those tissues are damaged by physical trauma such as burns, cuts, abrasions, insect or animal bites, or by preexisting disease. In addition, viruses may encounter secretions that compete for virus receptor sites or tissue enzymes that degrade viruses and render them noninfectious. Once barriers have been breached and cells have been infected, virus-specific proteins are produced and, being foreign, evoke inflammatory and specific immune responses. The mechanisms of host resistance then shift to a level of even greater complexity.

VIRUSES AS ANTIGENS

Intracellular parasites depend on host-cell machinery for synthesis of their components. Viruses code for a surprising number of specific "new" proteins, depending on the size of the viral nucleic acid. Generally virus-coded proteins can be classified as ultimately forming either a part of the mature virion (structural) or a nonvirion (nonstructural) protein. Nonvirion proteins are, for the most part, enzymes directly involved in various steps of viral synthesis. Virion proteins may initiate an immune response, not only as components of complete virions but also as viral subunits, such as capsid protein and glycoprotein complexes in virion envelopes, which are synthesized during viral morphogenesis and liberated when the cell dies.

Proteins at the surface of the mature virion, and the same proteins when they are found in infected cells, are generally those that, by eliciting an immune response, ultimately protect the host from further onslaught by the same virus. These antigens are therefore referred to as protective antigens. Not all viral proteins are important as protective antigens, because they may be situated deep within the mature virion and hence are not accessible to immunologic attack, even though antibodies are formed against them. The most specific virion components, ones that usually vary from strain to strain within a group (e.g., different serotypes of group A influenza viruses), are the surface antigens (i.e., type-specific antigens). In enveloped viruses, these comprise proteins, glycoproteins, and glycoprotein complexes that cover the external aspect of the envelope. In nonenveloped viruses, externally exposed capsomers comprise the protective antigens.

In contrast, internal antigens usually share antigenic determinants that define a group (e.g., the nucleocapsid antigens of group A influenza viruses) and are therefore referred to as group-specific antigens. Both group- and type-specific antigens provide the basis for immunologic classification of viruses. As indicated previously, nonvirion antigens are produced during the course of viral replication, usually as early antigens that shut off host-protein synthesis, aid in the takeover of host-cell

machinery, or constitute enzymes for subsequent viral nucleic acid synthesis. Early antigens elicit both humoral and cellular immune responses, none of which, however, is known to have any role in eliminating infectious virus. If viral replication does not complete the full
cycle of production of viral progeny, as happens in virus-induced transformation of cells, detection of an immune response to these early nonvirion protein antigens offers a possible approach to the recognition of
virus-induced neoplastic change. Detection of immune responses to
nonvirion proteins of herpes simplex virus has helped to link that agent
to cervical cancer.

At the cellular level, the picture of the potential magnitude and diversity of antigens becomes clear. One tends to think of viruses, and
the immune response to them, strictly in terms of antigens comprising
the fully formed infectious viral particle. The range and magnitude of
antiviral antibodies and cellular immune response produced under
conditions in which viruses infect and multiply intracellularly differ
greatly from the response to viruses that are noninfectious. In the first
instance, viral replication results in great amplification of the antigenic
mass initially introduced, as with live, attenuated measles virus vaccine.
In the second instance, for example with inactivated influenza viral vaccine, only antibodies to surface components form, and the extent of the
response is dose-dependent. These are important considerations in assessing the relative efficacy of live-virus vaccines on the one hand and
inactivated-virus vaccines on the other.

When a virus multiplies, not only do the progeny viral particles act
as antigens but also virus-coded nonvirion proteins elicit immune responses in the host. Perhaps of even greater importance to the character of the immune response is the manner in which a given infectious
virus is produced. With most viruses, production of progeny is a highly
inefficient process whereby, at the time of cell death, excess antigen,
unassembled and partially assembled viral components, and defective,
coreless particles may be present in greater quantity than fully infectious virions. These viral components and particles constitute an amplification of the antigenic load. Consequently, the immune system is called
on to process a wide range of molecular forms—from free polypeptide
antigens to viral subunits and complete viral structures. It follows that,
if a virus replicates in a highly efficient manner, after infecting a limited number or range of target cells, and without being acutely cytocidal, the immune response to that virus will be measurably less pronounced than the response to a virus that is invasive, widely disseminated,
and destructive of cells, provided that the host survives the initial infection. Unfortunately, viral replication and the responses to its products
depend on too many variables to be entirely predictable. Viruses that
generally undergo incomplete nonlytic cycles of infection, however, such
as those that transform or latently infect cells, also produce relatively
little viral antigen. This characteristic limits the overall immune re-

sponse and complicates containment and elimination of virus from the host.

Through any of several different mechanisms, many viruses cause the appearance of "new" antigens, that is, antigens foreign to the host, on the membranes of infected cells. With viruses that reach maturity by budding from the cell membrane (e.g., myxoviruses, RNA tumor viruses), viral envelope constituents are found to be integrated into the cell membrane before they take their places in the fully developed viral particle. In cells transformed by oncogenic viruses, virus-coded proteins may remain as cell-membrane constituents and, in some instances, be coupled to carbohydrate residues by host-cell mechanisms (glycosylation). In some instances, new antigens are induced by viral infection and result from cryptic unmasking of existing cellular proteins or from derepression of host genes governing synthesis of host antigens not normally produced (e.g., embryonic antigens). Other viruses may induce formation of antigens that function as receptors to bind the constant region of the immunoglobulin molecule (F_c) (e.g., herpes simplex viruses). These new antigens, whatever the mechanism by which they are produced, provide targets for immunologic attack by the immunocompetent host and play important roles in the capacity of the host to recognize and eliminate infected cells.

IMMUNE RESPONSES

Both cell-mediated and humoral responses are important in resistance to viral infection. Historically, antibodies appearing during the course of viral infection were the first to be studied extensively, because of the relative ease of in vitro neutralization assays. These methods include inhibition of cytopathology in cell culture or of hemagglutination, or modulation of disease in laboratory animals, including embryonated eggs. Long ago clinicians observed that patients with hypogammaglobulinemia recovered from acute viral infections and, conversely, that diminished numbers of functional T cells, occurring naturally or as a result of immunosuppressive measures, enhanced susceptibility to viral infection. Such early observations have underscored the relative importance of cell-mediated immunity in viral infection.

A major impediment to investigating cellular mechanisms is the necessity for using living cells in vitro to detect and measure characteristic responses and specific functions. Nonetheless, despite these restrictive conditions—which only imperfectly represent the situation in vivo— knowledge concerning B-cell and T-cell diversity and function has been expanding rapidly. Consequently, the definition of cell types and culture conditions and the recognition of normal functions (and hence of their derangement by disease) have become exceedingly complex. A wide range of variables must be controlled and accounted for in any experimental or diagnostic procedure designed to investigate any single function or group of functions.

T cells play a central role in the immune response, directly through their interaction with other cell types and indirectly through the effects of lymphokines. For example (Table 2.1), helper T (T_h) cells are required for an effective antibody response to viruses by B cells, resulting in antibody production. The initiation and amplification of the total immune response, and the switch from gamma-M to gamma-G globulin (IgM to IgG) synthesis, also depend on T_h cells. Another important regulator, the suppressor T (T_s) cell, likewise seems to function in both T-cell and B-cell responses. Effector cells, that is, those cells that actually mediate the effects, beneficial or otherwise, on the host tissues, include those responsible for delayed (tuberculin)-type hypersensitivity (T_{dth} cells) and cytotoxic T lymphocytes (T_c cells). In response to appropriate stimuli, each kind of T cell produces lymphokines that are nonspecific for any antigen or other mitogen. With the development of hybridoma techniques, monoclonal antibodies (OKT series) are now available for finely discriminating these interrelationships among cells, and for defining subsets (e.g., T inducer/helper, OKT4+; T cytotoxic/suppressor, OKT8+).

Most viral infections induce a classic humoral immune response, with production of circulating IgM, IgG, and small amounts of IgA antibody. IgM antibodies are the first to be produced during primary infection, followed by IgG and IgA. The presence of IgM antibody is indicative of recent infection. IgM levels increase rapidly and reach a peak at 3 to 5 weeks after the initial stimulus and then, within a few months, decline to preinfection levels. IgM antibody levels are seldom as high after a secondary antigenic challenge with the same or related virus or viruses. In contrast, IgG antibodies develop later and persist at significant levels for longer periods than IgM antibodies—sometimes for life. IgG antibody accounts for most of the antiviral activity in serum and helps to block the hematogenous spread of virus to target organs. Secondary challenge with the same or related antigens results in an enhanced (anamnestic) response attributable to the production by an expanded clone of B lymphocytes of larger amounts of antibody than occurs after primary infection. In general, IgG serum antibody levels, when measurable, correlate well with resistance to infection. They are also critically important in evaluating the efficacy of vaccines and in determining the incidence of previous infection in epidemiologic surveys of defined populations.

Small amounts of IgG and IgM may also be synthesized locally by lymphocytes in tissues that contain secretory epithelium, for example, those from the lung or gastrointestinal (GI) tract. The large amount of antibody frequently found on mucosal surfaces is usually secretory IgA, which may become detectable a few weeks after primary infection. Although many factors concerning the production and role of IgA in viral

TABLE 2.1 Immune Responses to Viral Infection

Cell Type	Commonly Used Designations	Characteristics and Functions
Lymphocytes		
T lymphocytes		
Cytotoxic T lymphocytes	T_c cells, CTL	T-cell surface markers. Specific recognition and lysis of virus-infected cells by direct interaction with virus-induced surface antigens. Restricted requirement for common MHC[a] gene products. Antigen-stimulated development. Proliferation (in response to interlukin 2, IL-2) and maturation modulated by T helper T suppressor cells. Clonal expansion and memory.
Delayed-type hypersensitive cells	T_{dth} cells	Specific antibody-independent interaction with virus-infected cell; release macrophage-arming factor (MAF).
T helper/inducer cells, T suppressor/cyto-toxic cells	T_h cells (OKT4+) T_s cells (OKT8+)	Surface markers identified with monoclonal antibodies (OKT series). Interact with T and B cells to diminish or enhance total immune responses.
Killer cells	K cells, null cells	Lack T-cell surface markers. F_c and C_3 receptors present. Mediate ADCC.[b] No clonal expansion or MHC[a] restriction.
Natural killer cells	NK cells, large granular lymphocytes, LGL	Surface markers vary and may overlap with those of other cell types. F_c receptors present. Direct antibody-independent cytotoxicity. Numbers and function amplified by interferon, IL-2.
B lymphocytes	B cells	Surface IG of same class and specificity as that produced by plasma cell derived from same clone. Blastogenesis, clonal expansion and differentiation to plasma cells. Each clone produces single class of antibody of uniform specificity. Memory cells. Response modulated by T_h and T_s cells, which, along with B lymphocytes interact with macrophages in primary immune response.
Phagocytes		
Macrophages	Mϕ, blood monocytes	F_c and C_3 receptors present. Mediate ADCC.[b] Central to primary immune responses by B and T lymphocytes. Variable ability to support viral infection. Activated by lymphokines (MAF, IFN-Y) from immune T cells produce interleukin 1, IL-1.
Polymorphonuclear leukocytes	Granulocytes, PMNL, neutrophils	F_c and C_3 receptors present. Early inflammatory reaction and phagocytosis. Mediate ADCC.[b]

[a] Major histocompatibility complex.
[b] Antibody-dependent cellular cytotoxicity for IgC-sensitized virus-infected cells.

disease have yet to be resolved, secretory IgA, when present, seems to control virus infection, primarily by limiting spread from cell to cell. In certain situations, therefore, the IgA antibody level may be a better index of immunity than is the serum IgG level. For example, high levels of secretory IgA antibody in the oropharynx and tracheobronchial tree correlate with immunity to respiratory viruses more closely than do serum IgG antibody levels. Immunization by mouth (e.g., the Sabin poliomyelitis vaccine) or intranasally (experimental live influenza virus vaccine) stimulates formation of specific secretory IgA antibody and, secondarily, IgG serum antibody. Initially, the mucosal antibody is the chief barrier to local spread. Parenteral immunization, whether with live-virus (measles) or killed-virus (influenza or rabies) vaccines, generally does not elicit secretory IgA responses, although serum IgA antibody levels are induced that may be comparable to those attained by oral immunization.

Traditionally, descriptions of cell-mediated immunity have been restricted to the biologic functions of T lymphocytes. We must again emphasize, however, that host immune response to viruses represents complex interactions among T and B lymphocytes and macrophages and their products, as well as many other specific and nonspecific serum and tissue components. Besides T_c cells, which are produced in response to virus-induced, cell-surface antigens, other types of effector lymphoid cells can lyse or kill infected cells. These include natural killer (NK) and killer (K) or null cells, as well as activated macrophages.

Delayed hypersensitivity in previously infected persons who receive an intradermal injection of the same, but inactivated, virus shows that cellular immunity has developed. T_{dth} cells cause a slowly progressive (24- to 48-hour) inflammatory (tuberculin type) reaction at the site of injection and systemically mediate resistance to infection. The extent to which both virion and nonvirion antigens stimulate specifically sensitized T_{dth} cells governs the intensity of the hypersensitivity tissue response and the degree of resistance. Sensitized T_c lymphocytes likely also respond to virion and nonvirion proteins other than those present on plasma membranes of viable infected cells, although with presently available in vitro techniques clear distinction is not possible.

ELIMINATION OF INFECTIOUS VIRUS

The ability of antibody to bind to infectious virus and thereby block infectivity is called *neutralization*. Antibodies of three classes of immunoglobulin (IgM, IgG, and IgA) are capable of neutralizing infectious virus with varying degrees of efficiency. Effectiveness depends on many factors, such as molecular heterogeneity, antigenic determinants to which the antibody molecules are directed, binding affinity, and complement-binding characteristics. In early textbooks, neutralization was explained simply as the ability of antibody to prevent viruses from attaching to cells. That view is too simplistic to satisfy current concepts of neutralization. Indeed, no single model, or combination of models, has emerged that can explain the range of experimental observations concerned with mechanisms of neutralization. Clearly, simple combination with antibody does not necessarily result in neutralization. In some instances viruses are neutralized after encountering a single molecule of antibody, whereas in other instances virus-antibody complexes remain infectious, and the antibody is said to be nonneutralizing. The resulting complexes are not always readily cleared from the body and may be deposited in arterioles and in the renal glomerular basement membrane, leading to chronic immune complex disease.

Interaction between viruses and antibodies involves some special features not generally found in nonviral immune reactions. Because of the multimeric nature of the capsid or envelope, antigenic sites tend to be repeated and closely arrayed on viral surfaces and to interact with antibodies in ways that would be impossible with more widely dispersed or isolated antigenic sites or with antigens in solution. Viruses themselves are not irreversibly altered by reaction with antibody and may be recovered in infectious form after artificial dissociation from complexes. The multimeric nature of the viral surface allows a single antibody molecule to bind to two adjacent combining sites, thereby forming an extremely stable complex that cannot readily be separated. During the early stages of interaction, however, when only one antibody antigen-binding site (Fab) per antibody molecule may be bound, the complex is freely reversible by simple dilution. The components may then establish equilibrium with the complex.

Figure 2.1 illustrates several of the basic mechanisms by which viruses are neutralized. The first depends on interference with attachment and penetration. Adsorption of virus to host-cell receptors may be prevented through simple steric hindrance by antibody in high concentration. Alternatively, by formation of immune aggregates, virus may be prevented from establishing infection. The size of the aggregates depends on the class of antibody and its reaction with virus. Viral aggregation effectively reduces the number of infectious particles free to initiate infection, and promotes phagocytic engulfment and rapid clearing of the complexes. Certain enveloped viruses, such as herpes simplex, rubella, and influenza viruses, may be inactivated by complement-dependent immune lysis, with the production of lesions in the antibody-sensitized viral envelope similar to those that occur in membranes of erythrocytes and animal cells after interaction of antibody and complement.

In a second basic mechanism, virus with antibody attached to it binds and penetrates into the host cell. The reaction, exemplified by poliovirus and antibody, follows first-order kinetics, one antibody molecule inactivating one virus particle. Antibody apparently prevents infection

PREVENTION OF ATTACHMENT AND/OR PENETRATION

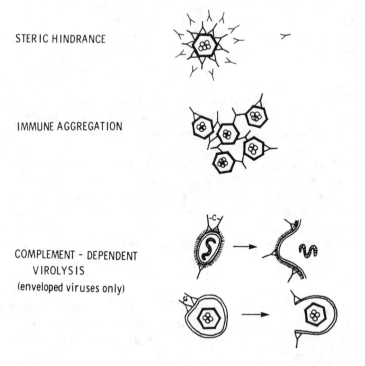

STERIC HINDRANCE

IMMUNE AGGREGATION

COMPLEMENT – DEPENDENT
VIROLYSIS
(enveloped viruses only)

NEUTRALIZATION AFTER PENETRATION

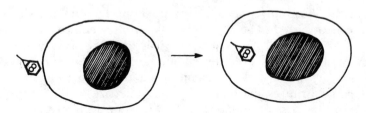

FIG. 2.1. Basic mechanisms of virus neutralization. **Prevention of attachment and/or penetration:** Antibody can combine with externally located viral antigens (attachment proteins or envelope proteins), which are thereby prevented by *steric hindrance* from interaction with specific cellular receptors. *Immune aggregation,* whereby virions are likewise prevented from attachment, also reduces the number of infectious viral particles free to initiate infection. If attachment should occur via unneutralized aggregated viral surface proteins, internalization may still be inhibited because of the size of the immune aggregates. Enveloped viruses, sensitized by antibody, are rendered noninfectious through *complement-dependent virolysis,* whereby the integrity of essential lipid-containing envelopes is lost. **Neutralization after penetration:** Some naked viruses, even though combined with antibody, may be internalized by pinocytosis; but they will resist uncoating (i.e., be noninfectious) because of alterations or loss of essential capsid proteins, or structural shift of capsomers during internalization.

by causing a shift in capsomer conformation, thereby preventing the virus from being uncoated. The virions are degraded in phagolysosomes, and their RNA is thus prevented from causing infection.

The ability of antibody to neutralize virus may be greatly enhanced by complement. Antibody in low concentrations or of low avidity may bind to and sensitize viruses, which are then neutralized by complement, especially by the polyvalent complement component C_3. With herpes simplex virus, for example, IgM and IgG antibodies produced early in infection possess in themselves little or no neutralizing activity, although they are reactive with virion-envelope glycoproteins. Addition of complement greatly enhances neutralization of infectivity. IgG antibody produced late in infection and inherently having strong neutralizing activity is less potentiated by complement.

The exact mechanism by which complement enhances neutralization remains unclear. Some of the proposed models are (1) stabilization of single molecules of antibody bound to critical site or sites, (2) cytolysis of virion envelopes when these are present, (3) envelopment of antibody-sensitized virions, with resulting steric hindrance and complete blockage of the virion proteins involved in attachment to cell receptors, and (4) enhanced formation of immunoaggregates.

As already mentioned, exposure of infectious virus to antibody and the combination of antibody with virion elements does not necessarily result in neutralization. In many virus-antibody reactions, a small amount of infectious virus remains unneutralized and has therefore been termed the *persistent fraction*. Aside from simple dissociation of unstable virus-antibody complexes, the genesis of the persistent fraction is variously thought to stem from the (1) breakup of aggregated viruses in which internally located particles have remained inaccessible and hence unaffected by antibody, (2) formation and dissociation of immune aggregates in which some virions remain unneutralized, and (3) interaction of virus with antibodies that themselves do not neutralize but, by their combination with virus, block neutralization by other antibodies. Often the persistent fraction can be decreased by adding reactants that stabilize the complexes, or that more effectively coat surface components, such as antibody reactive with additional antigenic determinants and complement.

ELIMINATION OF INFECTED CELLS

Basically, the mechanisms of specific immunity by which host defenses directly limit the spread of infectious virus are two: neutralization, as already discussed, and immune cytolysis of infected cells. Many viruses can elicit the early formation of new antigens on the surface of viable infected cells. All enveloped viruses code for glycoproteins that are incorporated into plasma membranes and ultimately into mature viral

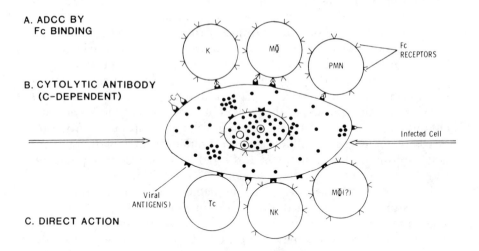

ANTIBODY–DEPENDENT LYSIS

A. ADCC BY
   Fc BINDING

B. CYTOLYTIC ANTIBODY
   (C–DEPENDENT)

C. DIRECT ACTION

ANTIBODY–INDEPENDENT LYSIS

FIG. 2.2. Destruction of infected cells by antibody-dependent and antibody-independent mechanisms. **A. Antibody-dependent cellular cytolysis** (ADCC) occurs when the F_c portion of antiviral antibody bound to cell surface virus-specific antigen(s) reacts with F_c receptors on killer (K) lymphocytes (present in nonimmune as well as immune persons), macrophages (M) and polymorphonuclear neutrophils (PMN). **B. Cytolytic antibody to virus** reacts specifically with cell surface virus-specific antigen(s), thereby sensitizing the cell to lysis by complement (C). **C. Antigen-specific cytotoxic T lymphocytes (T_c)** are evoked in the immune response to virus-induced glycoproteins, and, subject to histocompatibility restriction, react directly with the same or related antigens on the surface of the infected cell without the mediation of antibody. Natural killer (NK) lymphocytes, as well as macrophages "activated" by T lymphocyte-derived lymphokines, also recognize and react directly with virus-specific cell surface "foreign" antigens, and thus contribute to the destruction of the cell.

envelopes. A few nonenveloped viruses (DNA tumor viruses, Coxsackie B virus) induce surface antigens that allow immunologic recognition of infected cells. Of obvious importance in immune elimination are the characteristics of surface antigens thus expressed and their number, variability, and distribution density. Immune cytolysis has the greatest effect when it occurs early in infection, before the formation of mature virions.

The effector mechanisms involved in immune cytolysis may or may not depend on antibody (Fig. 2.2). Antibody may participate in cell lysis either through the mediation of complement (cytolytic antibody) or in antibody-dependent cellular cytotoxicity, in which antibody serves as a bridge between effector lymphocytes or phagocytes and infected target cells.

Antibody-Dependent Cellular Cytotoxicity

In addition to T and B lymphocytes, a third population of lymphocytes in the peripheral blood can lyse infected cells that bear surface antigens. These are referred to as killer (K) or null cells, because they lack T-cell markers or surface Ig characteristic of B cells, and carry membrane receptors for C_3 and the F_c portion of Ig. K cells themselves are not virus-specific and in vitro recognize only target cells that have been sensitized by Ig from immune donors. Thus, the cell-attached Ig itself confers specificity. The Fab portions are bound to surface antigens, and the distally exposed F_c portion binds to K cells (Fig. 2.2A). K cells are not produced as part of the immune response and accordingly have no immunologic memory. They are present in both immune and nonimmune hosts and do not seem to change appreciably in number or activity during the course of viral infection.

Macrophages and polymorphonuclear leukocytes are also capable of mediating antibody-dependent cellular cytotoxicity (Fig. 2.2A). They, like K cells, bear F_c receptors, do not require immunologic memory, and are present in nonimmune hosts. For the most part, the ability of these effector cells to bring about antibody-dependent cellular cytotoxicity has been demonstrated in vitro, so that the importance of this mechanism in eliminating infected cells in vivo remains to be determined, as does the nature of the cell-killing process.

Cytolytic Antibody

Cytolysis may occur with either IgG or IgM antibodies. IgM is quantitatively the more potent, as a single IgM molecule is capable of initiating a complement-mediated lesion. Two IgG molecules, on the other hand, bound by the Fab ends to antigens in close proximity to one another, are required for complement lysis (Fig. 2.2B). Complement is abundant in the serum and binds to the F_c end of the immunoglobulin molecule, which has been distorted through interaction of the Fab portions with cell surface antigens. Fc binding initiates the classic complement pathway that results in plasma membrane lesions and eventual lysis of the cell.

Antibody-Independent Mechanisms: Direct Cytotoxicity

Three cell types probably kill by direct cytotoxicity: cytotoxic T (T_c) lymphocytes, natural killer (NK) cells, and activated macrophages (Fig. 2.2C). Of these three, only T_c lymphocytes have virus-specific memory and are usually found in the blood, spleen, and sometimes lymph nodes draining the sites of infection. They reach maximal levels 6 to 12 days after primary infection and slowly subside, persisting in small numbers as memory lymphocytes. On renewed contact with virus-infected target cells bearing surface viral antigens, secondary T_c lymphocytes are generated in large numbers as a result of clonal expansion. The induction of T_c lymphocytes specific for virus-induced plasma membrane protein as well as their ability to lyse infected target cells require that the T_c lymphocyte and target cell have in common at least part of the major histocompatibility (HLA) region. Two models have been suggested to account for this dual requirement by T_c lymphocytes for both the viral antigens and HLA gene products. The theory of dual receptors postulated that T_c lymphocytes have independent receptors, one for viral antigens and a separate receptor for HLA products. The altered-self hypothesis suggests that viral antigens and HLA products are recognized by a single receptor on the T cell. It has in fact recently been established that the receptor on T lymphocytes for foreign cell surface antigens and HLA consists of a heterodimeric structure comprising two different protein chains (alpha and beta), both structurally similar to and having homology with immunoglobulins. Both have molecular weights of about 33,000, and amino acid sequences that are identical up to about 30%, but very different throughout the remainder of the respective chains. Each contains a short region extending into the cytoplasm, a transmembranal segment, just distal to which is an SS bond, and a large extracellular segment. The latter contains a constant region, which is the same for all chains of the same type, and a variable region, which differs with each clone and is involved in specific reactivity with antigens. It is thought that the alpha chain may recognize primarily histocompatibility gene products and that the beta chain may react primarily with foreign antigen(s) (e.g., virus- or tumor-specific surface antigens).

Natural killer cells lack the markers characteristic of T and B cells, and, like K cells, have F_c receptors. Their mode of killing, which does not require antibody, apparently is by direct cytotoxicity for the target cell without activation by antigen or mitogen. NK cells are not restricted in activity (as are T_c lymphocytes or K cells) and will lyse and attach to various cell types, both virus-infected and uninfected. They are present in the peripheral blood of both the immune and nonimmune host. Their numbers and cytotoxic activity seem to be amplified by the action of interferon produced by virus-infected target cells.

The third category of antibody-independent effector cells, and probably the least understood of the cellular cytotoxic mechanisms responsive to viral infection, is the activated macrophage. Effector T lymphocytes (T_{dth}) interact with viral antigens and release lymphokine(s) (MAF, IFN-Y) that cause macrophages to differentiate into activated or "angry" macrophages containing increased numbers of lysosomes and displaying enhanced ability to ingest and destroy bacteria. Similarly, activated macrophages show an increased ability to lyse tumor cells and presumably to kill cells bearing virus-induced tumor-specific surface antigens (Fig. 2.2C). However, not enough data have accumulated, particularly in human systems, to determine whether or not activated macrophages are important for cell-mediated destruction of cells infected with non-tumorigenic viruses.

Most evidence points to the T_c lymphocyte as the most important and effective cell type in immune cytolysis of virus-infected cells. Other cell types mainly have accessory roles, depending on disease conditions and the time after infection. For example, in an individual exposed for a second time to a given virus specific antibody and limited numbers of memory T_c lymphocytes may already be present. NK cells and nonspecific cells involved in antibody-dependent cellular cytotoxicity (K cells, PMN leukocytes, macrophages) may then provide immediate defense and thus may serve to limit the spread of infection until adequate numbers of specific T_c lymphocytes have been marshaled to the attack.

IMMUNE CLEARANCE

When viruses are neutralized by antibody, or when cells with surface antigens are destroyed by immune cytolysis, what happens to the Ag-Ab complexes and the residua of infection? This question, almost entirely overlooked by researchers, is one that has critical potential impact on the outcome of disease. With the great amount of evidence available concerning clearance of bacteria, virologists have assumed, but usually have not attempted to prove, that similar patterns of opsonization, phagocytosis, and phagocytic degradation occur with the products of viral infection. For the sake of discussion, we will present phagocytosis and intracellular degradation in terms of what is known regarding bacterial diseases, adding what is known of viral disease when appropriate.

One hallmark of acute tissue inflammation is the infiltration of many PMN leukocytes at the site of the lesion. When incited by viral infection, the infiltrate is also found to contain many macrophages. We have already mentioned that PMN leukocytes can mediate antibody-dependent cellular cytotoxicity in the presence of specific antibody and thus serve as early effectors of immune cytolysis. Whether this mediation occurs in the circulation or only at sites of inflammation is a moot question. Even less is known of the phagocytic capabilities of PMN leukocytes in viral infection. Because they carry both F_c and C_3 receptors, PMN leukocytes likely ingest virus-antibody complexes, immunoaggregates, and other antigenic materials to which opsonins such as antibody and complement have bound.

In contrast to macrophages, PMN leukocytes apparently do not support replication of viruses. This is not too surprising, because the half-life of the mature PMN leukocyte does not exceed 6 or 7 hours. However, interaction with virus might still affect these cells adversely. For example, PMN leukocytes exposed in vitro to influenza or Newcastle disease viruses (the latter a paramyxovirus) show diminished capacity to

phagocytize zymosan, a yeast-cell component used to monitor phagocytic activity in vitro. On lysis of the cells, a complex array of enzymes
and toxic products is released, with the result that tissue necrosis, including destruction of normal cells, supervenes. In some tissues that are
capable of little or no regenerative repair, such as the cornea, permanent scarring and impaired function result. In the course of viral diseases in which there may be some degree of localization, PMN leukocyte infiltrates are relatively rapidly replaced by a mononuclear reaction
comprising largely macrophages, lymphocytes, and plasma cells. This
change in the character of cellular response is characteristic of viral
meningitis, in which the cellular content of the cerebrospinal fluid may,
initially, consist mainly of "pus cells" (as in bacterial infection). The acute
reaction soon gives way to a predominantly "round cell" or lymphocytic
infiltrate.

Macrophages comprise a diverse group of cells, ranging from monocytes (immature macrophages) in the peripheral blood to fixed phagocytic cells of the reticuloendothelial system, such as the Kupffer cells
in the liver, and large-tissue macrophages found in the lung, spleen and
lymph nodes, and gastrointestinal tract. In these areas, macrophages help
to clear viruses from the lymphatic system and bloodstream, and thus
limit the extent of viral dissemination. Like PMN leukocytes, macrophages seem able to digest and degrade viruses and virus-antibody
complexes. The effectiveness of macrophages in limiting the spread of
infection, however, may well be dampened by their demonstrable ability to support viral replication. This phenomenon may be a two-edged
sword, the virus-infected macrophage becoming a vehicle for transport
of virus, inaccessible to antibody, to distant sites. Moreover, by supporting viral infection, other cellular functions may become impaired,
particularly those related to the immune response. Studies have shown,
for example, that after infection with poliovirus or influenza virus,
macrophages prevent the blastogenesis of lymphocytes exposed to phy-
tohemagglutinin. There may also be profound effects on antigen processing and the initiation of primary specific lymphocyte responses. Finally, cell death resulting from viral infection would abrogate the
production of antiviral macrophage functions such as antibody-
dependent cellular cytotoxicity and IL-1 production.

Although many viruses have been shown to have immunosuppressive
effects, few data are available that define the exact pathway(s) through
which the immune response may be affected. When immune complexes escape phagocytosis, as with lymphocytic choriomeningitis lactic
dehydrogenase viruses in mice, complexes are disseminated and collect
in the glomerular basement membrane and in the choroid plexus. Once
localized in tissues, the complexes bind complement, and an inflammatory reaction ensues. In a similar manner, virus-specific immune
complexes have been detected in human infection with Epstein-Barr virus, respiratory syncytial virus, and hepatitis B virus. Immune complex
glomerulonephritis occurs in chronic active hepatitis.

TABLE 2.2 Interferons

| | NOMENCLATURE | | |
Current[a]	Previous	Inducers	Producer Cells[b]
Alpha(HuIFN-α)	Leukocyte, type I	Foreign or transformed cells, viruses	Lymphocytes, others?
Beta(HuIFN-β)	Fibroblast, type I	Viruses, nucleic acids, synthetic polynucleotides	Fibroblasts, epithelial cells, macrophages
Gamma(HuIFN-γ)	Immune, type II	Microbial antigens, mitogens	Immune lymphocytes

[a]Each interferon is identified by species of origin, such as human (HuIFN), murine (MuIFN), bovine (BovIFN).

[b]Cells most commonly considered to be sources of IFN. Some cell types may be capable of producing more than one kind of interferon.

INTERFERON

More than a quarter century has passed since the discovery of interferon, and enthusiasm for its therapeutic potential has alternately waxed and waned. Today, with the advent of genetic engineering techniques for synthesizing interferons in cloned bacteria, the analytical approaches of unsurpassed precision provided by monoclonal antibodies, and the suggestion that interferon may alter the course of certain forms of cancer, interest in interferon waxes anew.

The major difficulty in interpreting results of relevant research is that interferon takes several forms, each difficult to purify. To add to the difficulty, the nomenclature of interferon has been changing as methods for characterization have improved. Table 2.2 summarizes the current nomenclature.

Interferons are a heterogeneous group of low-molecular-weight glycoproteins (about 20,000 daltons) released by lymphocytes, macrophages, fibroblasts, and epithelial cells in response to viral infections and other stimuli, not all of which are of infectious origin. Interferons have been classified into three basic types: alpha, beta, and gamma. They are produced within hours after initiation of infection and, on entry into uninfected cells, prevent viral replication by mechanisms not fully understood. Maximum serum titers are usually reached 2 or 3 days after infection.

Designing experiments that would determine the importance of interferon in preventing disease or promoting recovery is difficult, because no diseases are known to exist in humans in which interferon production is inhibited or deficient. If interferon does help to limit vi-

rus infection, meaningful protection likely occurs only among cells close
to those infected. If protection were more widespread, one would expect that natural infections or immunization would confer resistance to
later challenge with an unrelated virus, because interferon is host-specific and not virus-specific. Although high serum interferon levels may
be achieved, sustained resistance due solely to interferon cannot be attained.

For many years interferon has been intensively investigated as a
chemotherapeutic agent. Interferon is a natural by-product of virus infections, has high biologic activity, and is nontoxic, only weakly antigenic, and active against a wide spectrum of viruses. It would seem to
be the ideal antiviral drug, a potential for which universities and phar-
maceutical companies have spent untold hours and huge sums of money.
Basically two approaches to interferon therapy have been tried: (1) administration of preformed interferon to patients and (2) induction of
interferon.

A major problem arises with administration of interferon. In order
to evaluate adequately the ability of interferon to prevent or alter the
course of disease, large quantities are needed, because it is rapidly cleared
from the blood. Until recently, alpha interferon obtained from leukocytes exposed to viruses was used in most investigations. To do even
the simplest experiments has usually required many liters of human
blood, making all but a few selected experiments impractical. With the
advent of genetic engineering, it is now possible to produce large quantities in bacterial culture.

This achievement, however important in itself, brings us up against
a second major problem with administration of interferon, namely, lack
of knowledge as to how it is induced and its mode of action. Too little
is known about the range of interferon types, the basic properties of
each, and their roles in the biology of the host. Gamma interferon (IFN-
γ), for example, is produced by lymphocytes as a result of mitogen or
specific antigen stimulation and also functions as a lymphokine with
significant roles in immunoregulation. Many studies are underway to
evaluate the potential of the new technology in determining whether
there are active forms or combinations of interferons that can be of
practical therapeutic use in cancer and viral disease. Results thus far
have varied, some disappointing and some encouraging. The greatest
interest is being focused on those types of interferon produced by leukocytes (alpha and gamma).

The second approach to therapy concerns the possibility of inducing
the host to manufacture endogenous interferon. Although live attenuated viruses have been considered, most of the effort has been directed toward developing synthetic polynucleotides as inducers. As with
administration of interferon, prophylaxis is much more successful than
therapy. However, systemic administration of inducers is deemed impractical in most cases because of unacceptable side effects on hemo-

poiesis and liver function. Current research centers on synthesizing less toxic agents with higher interferon-inducing activities.

VIRUS-INDUCED IMMUNOSUPPRESSION

Viruses are unique in their interactions with elements of the immune system. Whereas some microbial agents may have defenses, such as a capsule that might limit the ability of the host to process or eliminate them, viruses are unique in that they may actually infect cells of the immune system, altering their function and using them as a source for further production of viral progeny. There is evidence of virus inter-action with immune elements ranging from stem cells to mature effec-tor cells. Observation of the anergy (e.g., conversion of positive to neg-ative tuberculin reaction) that accompanies measles is long-established and well-substantiated; however, it is not fully explicable as either a re-sult of actual infection or some more subtle influence imposed by the virus or lymphocytic cells. Leukemia viruses in experimental animals infect lymphoid tissue and thereby enhance or depress humoral or cell-mediated immunity. Acquired immune deficiency syndrome (AIDS) is caused by a retrovirus, human T-cell lymphotropic (leukemia) virus (HTLV-III) that specifically infects and destroys T-helper/inducer (OKT4+) cells, thereby significantly altering the ratio of T-helper to T-suppressor/cytotoxic (OKT8+) lymphocytes; this accounts for the heightened susceptibility of patients with AIDS to overwhelming op-portunistic infections (parasitic, fungal, and viral) and neoplastic dis-ease. Epstein-Barr virus and cytomegalovirus can infect and become la-tent in lymphocytes. Many other viruses can also infect lymphocytes, particularly after blastogenesis, and, as mentioned earlier, macro-phages may be permissive for infection.

CONCLUSION

It should be evident from the foregoing discussion that immune re-sponses to viral infection are at once complex and in many instances incompletely or poorly understood in terms of molecular events in-volved in resistance and recovery. During the past quarter century, the discovery of hundreds of new viral agents, many of which have been dissected at the molecular level, has been matched in extent and com-plexity by equally startling discoveries in immunobiology. Coupled with new biochemical and biophysical technology, these developments promise that significant advances in treating and preventing human disease caused by viruses are not too far in the future.

Recommended Readings

SPECIALIZED ARTICLES

Kohl S, Loo LS, Piekering LK: Protection of neonatal mice against herpes simplex viral infection by human antibody and leukocytes from adult, but not neonatal humans. *J Immunol* 127:1273–1275, 1981.

Notkins AL: Viral infections: Mechanisms of immunologic defense and injury. *Hosp Pract* (September); 65ff, 1974.

Oldstone MBA: Immunopathology of persistent viral infections. *Hosp Pract* (December); 61ff, 1982.

Oldstone MBA, Sinha YN, Blount P, et al: Virus-induced alterations in homeostasis: Alterations in differentiated functions of infected cells in vivo. *Science* 218:1125–1127, 1982.

Reichman RC, Pons VC, Murphy BR, Caplan EA, Dolin R: Cell-mediated cytotoxicity following influenza infection and vaccination in humans. *J Med Virol* 4:1–14, 1979.

Roder JC, Pross HF: The biology of the human natural killer cell. *J Clin Immunol* 2:249–263, 1982.

Smith JW, Sheppard AM: Activity of rabbit monocytes, macrophages, and neutrophils in ADCC of HSV-infected corneal cells. *Infect Immunity* 36:685–690, 1982.

BOOKS AND REVIEWS

Baron S, Dianzani F, Stanton GJ (eds): The interferon system: A review to 1982. *Texas Reports on Biology and Medicine* vol 41, 1982.

Belshe RB: *Textbook of Human Virology*, Littleton, Massachusetts, PSG Publishing Company, 1984.

Benacerraf B, Unanue ER: *Textbook of Immunology*. Baltimore, Williams & Wilkins, 1979.

Evans AS: *Viral Infections of Humans*. New York, Plenum Medical Book Co, 1983.

Fenner FJ, White DO: *Medical Virology*, 2nd ed. New York, Academic Press, 1976.

Hoeprich PD: *Infectious Diseases*, 3rd ed. Philadelphia, Harper & Row, 1983.

Howe C, Coward JE, Fenger TW: Viral invasion: Morphological biochemical and biophysical aspects. *Comprehensive Virology* 16:1–71, 1980.

Lonberg-Holm K, Philpson L (eds): Virus receptors, part 2, animal viruses, in Cuatrecasas P, Greaves MF (eds): *Receptors and Recognition*. New York, Chapman and Hall, 1980.

Mandell GL, Douglas RG, Bennett JE: *Principles and Practice of Infectious Diseases*. New York, Wiley, 1979.

Notkins AL, Oldstone BA: *Concepts in Viral Pathogenesis*. New York, Springer-Verlag, 1984.

3

Pathology of Virus Infection

MURRAY B. GARDNER, WILLIAM F. SULLIVAN, and STEVEN H. HINRICHS

Knowledge of the natural history and pathology of virus infection will help the physician select the proper specimens for rapid diagnosis, interpret the laboratory results, begin appropriate treatment and public health measures, and predict the clinical outcome. We begin this chapter, therefore, with a general discussion of viral pathogenesis, followed by an explanation of the general morphologic and immunopathologic findings that may be observed in virus-infected human tissues at the light and electron microscopic levels. We will explain the morphologic basis for the diagnosis of virus infection through isolation of the agent in cell cultures, and then conclude by briefly discussing newer techniques.

PATHOGENESIS OF VIRAL INFECTION

The clinical features of a viral illness can be correlated with the host response and the defense mechanisms involved. In most infections, the virus enters the bloodstream through lymphatic channels and circulates as free or leukocyte-associated virions. During the incubation period, which lasts about 1 week, this primary viremia disseminates virus asymptomatically throughout the body. Although macrophages may remove some virions, other virus particles reproduce in various target organs, such as liver, spleen, lymph nodes, or bone marrow, from which they again enter the bloodstream. Eventually they accumulate in other target organs, such as the skin or central nervous system (CNS), multiply in certain cell types to which they have an affinity, and thus establish localized infections. This cytotropism is accounted for, in part, by the presence of specific receptor molecules on the cell surface that recognize peptides on the surface of the virus particles.

Toward the end of the incubation period, the immune system is called into action and the clinical illness begins. Destruction of virus particles and virus-infected cells and formation of viral antigen-antibody complexes contribute to the characteristic systemic manifestations of viral illness. The peak level of virus shedding occurs just before the onset of symptoms and decreases as the immune response strengthens, so that by seven to eight days after onset of infection, only rarely can the virus be recovered by laboratory techniques. Macrophages that have ingested virus particles present virus antigens to T lymphocytes, which then help stimulate B lymphocytes to secrete virus-specific immunoglobulins (Ig). Immunoglobulin M (IgM) antibodies are first produced by the primary infection, but a switch to IgG antibodies occurs within several weeks; subsequent reinfection triggers primarily the IgG antibody class. Specific stimulation of other immunoglobulin classes in virus infections does occur. A localized production of IgA antibodies is responsible for protecting mucosal surfaces from invasion. The tissue response to respiratory syncytial virus by IgE-triggered release of histamine from mast cells may result in symptoms such as wheezing and difficulty in breathing.

In addition to stimulating B-cell immunity, T lymphocytes are involved in host responses to viruses. T killer cells destroy virus-infected cells but not free virus particles. The killer cells lyse only those infected target cells with which they share a human leukocyte antigen (HLA) determinant, a form of recognition that allows precise discrimination of virus-infected cells from normal target cells. Natural-killer-cell-mediated cytotoxicity, antibody-dependent cell cytotoxicity, and interferon production are other major, but less specific, host defenses against viral infection. Although of primary importance in protecting against cell-associated viruses, these measures of cellular immunity are not yet routinely assayed in the clinical laboratory. To examine viral pathogenesis in more detail, we will discuss representatives of the two main morphologic categories, the enveloped and nonenveloped viruses.

Enveloped Viruses

Because most of the virion's antigenic mass is hidden within a lipoprotein envelope derived from the host cell's cytoplasmic membranes, the virus is recognized as "self" and is temporarily afforded protection from the host's immune response. The only viral antigens exposed to the immune system are the low-molecular-weight polypeptides (peplomers) in the outer envelope—a vulnerability the virus must allow for further cellular attachment to occur. Therefore, the host's initial immune reaction is against these exposed viral proteins. Only after infection has progressed, and clearance of virus and infected cells begun, are the viral nucleoprotein core antigens recognized.

As measured in the laboratory, the hemagglutination-inhibition or

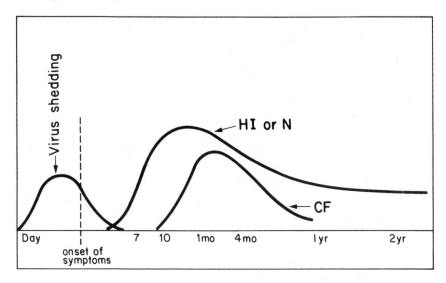

FIG. 3.1. The antibody response in paramyxoviral infection represented by three serologic procedures; the hemaglutination-inhibition (HI), neutralization (N), and complement-fixation (CF) tests. The peak level of virus shedding occurs before the onset of symptoms. The HI or N antibody titers are the first to become elevated as a response to the viral envelope peplomers (glycoproteins), followed later by the immune response to core (nucleocapsid) antigens as measured by the CF test.

neutralization antibody response (Fig. 3.1) is primarily a response to the envelope peplomers, and the complement-fixation antibody is mainly a response to the nucleoprotein antigens. Comparison of the responses indicates that the complement-fixation antibody begins to diminish about four months postinfection and disappears within one year. The hemagglutination-inhibition or neutralization antibody curve, on the other hand, usually remains at a detectable level for the lifetime of the host. The disappearance of the complement-fixation antibody is logically explained. Virus in the acute infection is eliminated within the first seven days after the onset of symptoms. Therefore, the level of antibody produced to that virus will diminish with time. Why then does the level of hemagglutination-inhibition or neutralization antibody remain elevated so long? The answer rests with the critical difference between two terms often used interchangeably: *infection* and *disease*. We will illustrate the difference by comparing the pathogenesis of two enveloped viruses, namely, rubella and influenza.

RUBELLA VIRUS

Rubella has an incubation period that ranges from 7 to 21 days. During that period the virus initiates infection in the upper respiratory tract or conjunctiva. The virus spreads to the lymphatics, from which follows a primary viremia. After the virus further replicates in the liver and spleen, a secondary viremia results in spread of the virus to secondary targets, such as the capillary endothelial cells in the skin. By the time an immune response has begun, specific antibody and sensitized lymphocytes attack viral antigens in endothelial cells, resulting in capillary leakage and formation of a macular rash, the clinically recognized form of the disease in children. Note that the "infection" has existed some time before the disease is detected (Fig. 3.1). The site of disease production by rubella virus is thus distal from the site of first replication.

Because of the anamnestic response in the immunocompetent host, subsequent infections will not progress to the stage of a generalized infection. The nature of the rubella virus itself also contributes importantly to disease resistance after infection. Rubella virus is genetically stable and has long maintained its antigenic characteristics. Its recognition by the immunized host is immediate and its pathogenesis on reinfection will halt in the upper respiratory mucosa without causing any illness. The ubiquitous rubella virus continually challenges the individual throughout life. Therefore, the maintenance of hemagglutination-inhibition or neutralization antibody is the result of limited reinfection in the absence of disease.

INFLUENZA VIRUS

The incubation time for influenza-virus infection is dramatically shorter than that of rubella virus; the critical period between the first contact with the virus and the start of disease is estimated to be only 2 or 3 days. In uncomplicated cases, the site of initial replication, the upper respiratory mucosa, remains the site of disease. Primary or secondary viremia seldom occurs. However, infection may develop throughout the tracheobronchial tree and lead to pneumonia. The immune system must recognize the agent immediately and completely if disease is to be averted or modified. This recognition is complicated by the instability of the influenza-virus genome, which results in frequent genetic recombinations and subsequent antigenic variation of the peplomers of the envelope (antigenic drift) or viral nucleoproteins (antigenic shift), or both. When sufficient antigenic changes have occurred, the host's immune response is delayed and the influenza virus escapes destruction. Infection and disease will then follow, despite high levels of antibody to other influenza virus serotypes.

Nonenveloped Viruses

The replication of nonenveloped, or "naked," viruses is represented in this discussion by the picornaviruses and adenoviruses. Because these agents have no envelopes and their entire antigenic mass is exposed to the host's immunologic attack, they have evolved different strategies to temporarily evade these host defenses. The response to these different viruses illustrates the variability of the host defense. The adenovirus particle is a large (60–90 nm) cubic virion whose double-stranded DNA contains more than 35 genes; the picornavirus is a small (28 nm) cubic virion whose single-stranded RNA genome contains only seven or eight genes.

PICORNAVIRUS

Suppose that two identical cells were infected simultaneously, one with an adenovirus, the other with a picornavirus. Which would produce the greatest amount of progeny virus in the shortest time? Because of the lower demands on the cells' nucleotide/amino acid pools for the completion of a single virion, the picornavirus would certainly win this competition. Knowing, further, that its RNA genome is infectious and its whole growth cycle is confined to the cytoplasm, one can see why the replication of the picornavirus is extremely rapid and results in the release of an overwhelming number of infectious virions. The picornavirus shuts down the normal cellular machinery and subverts the cellular membranes to a goal of virus reproduction. Yet, at first, the cell's cytoplasmic organelles show no morphologic evidence that this cellular component has been dedicated to construction of thousands of infectious virions. When the cell is exhausted, its lysosomes release their enzymes internally, resulting in cell lysis and the sudden release of newly produced picornavirus virions. Imagine the dilemma of a phagocytic cell in proximity to a picornavirus-infected epithelial cell. The phagocytic cell must cope with overwhelming numbers of virions that are too small to easily phagocytize. In fact, viremia due to picornavirus infection represents free circulating virus. Fortunately, this sudden onslaught of virus represents a large antigenic mass, and the immune response is rapid and efficient. Antibody will aggregate the virions, thus facilitating phagocytosis. This picture corresponds to what is seen clinically in cases of aseptic meningitis caused by echo virus or Coxsackie virus. Despite the initial severity of the symptoms, the prognosis is excellent for a meningitis patient infected with one of these agents. However, infection of neonates and immunocompromised persons may be severe and is often fatal.

ADENOVIRUSES

If the adenoviruses, with their individually large antigenic mass and low numbers, were similarly released through cellular lysis, they would be readily eliminated by the host's immune system. How, then, can the adenovirus spread without eliciting an overwhelming immune response? The answer is that the virus remains cell-associated in the nucleus and spreads contiguously from cell to cell. Without making major antigenic changes on the cytoplasmic membrane of the infected cell, infection can proceed relatively unimpeded by phagocytic or antibody action. Sloughing of infected cells provides the vehicle for spread of the virus within the host, and from the host to the external environment.

The infected cell is literally a "bag of virus," protecting the agent from both the immune response and environmental stress.

It is no wonder that adenovirus infections can be among the most severe viral infections encountered. They are characterized by high infectivity and severe morbidity and mortality. Typical of these infections are outbreaks of epidemic keratoconjunctivitis in the hospital or institutional environment. The infection spreads rapidly throughout a population, and its source can be traced to such unlikely vectors as bars of soap or "disinfected" ophthalmologic instruments. Pneumonia is another major disease entity caused by adenovirus in institutional and pediatric populations. Because the humoral immune response to infections by the adenovirus is so limited, reinfection at the same anatomic site can occur. Fortunately, the genome of adenovirus is a strong inducer of interferon, a principal means of defense against adenovirus infection.

Interpretation of Virus Assays

From these examples of virus pathogenesis, several important clinical facts become apparent. By the time the patient is sick enough to seek medical help, the causative virus may be difficult or impossible to isolate because it has been largely eliminated by the immune system. Thus, the best opportunity to isolate or demonstrate virus is to obtain tissue or fluid from the sites of localized infection early in the clinical course. Unfortunately, the disease process is generally well established before a biopsy or culture is done.

Inability to demonstrate virus antigen or infectious virus at sites of localized infection may be due to inadequate sampling, neutralization of viral antigens by antibody, or clearance of virus-infected cells by phagocytic or cytotoxic activity. Of course, failure to isolate virus need not rule out a viral etiology. Sometimes immunologic reactivity or electron microscopy will show virus antigens or virus particles in the infected tissues although the virus itself is not recoverable. Common examples are the detection of hepatitis viruses and rotaviruses through immunochemical or electron microscopic assay. On the other hand, demonstration of infectious virus, viral antigen, or virus particles does not necessarily establish proof of their etiologic role. Viruses may be present that are not responsible for the disease in question (e.g., the isolation of herpes simplex from respiratory secretions in a case of pneumonia). However, the combined clinical picture, histopathology, and results of serologic tests usually confirm or establish the etiologic role of the virus. Serologic assays are excellent measures of immunity, but they can only be used with known viruses and generally establish or confirm the diagnosis in retrospect. Of particular note are the diagnoses of the generalized exanthemous infections of rubella virus and measles virus. These viruses are very difficult to grow in vitro, requiring prolonged incubation and complicated assays for the detection of virus. As illustrated earlier, because the rash of these diseases is a result of the patient's production of antibody, satisfactory serodiagnoses can be made.

In general, the clinician must have a high index of suspicion in requesting a specific viral assay. Because the immediate response to primary viral infection is a transient production of IgM antibody, detection of antiviral antibody of this Ig class may help the recognition of recent infection. By contrast, detection of serum IgG antibody alone (even at elevated levels) can generally not distinguish current from past infection. However, a significant (fourfold) rise in antibody titer is considered diagnostic. Thus, paired sera must be tested, the first collected as early as possible after onset of symptoms, the second collected 2 to 4 weeks later during the recovery phase.

The severity of disease will vary greatly, depending on the kind of virus, the exposure date, the level of preexisting immunity in the individual and community, the patient's general immunocompetence, and other host-resistance factors. Not surprisingly, severe, disseminated, and fatal virus infections often occur in immunosuppressed patients. Recent outbreaks of disseminated cytomegalovirus (CMV) have occurred in homosexual males as a result of immunosuppression associated with infection by a T-lymphotrophic virus.

LIGHT MICROSCOPIC CHANGES OF VIRUS-INFECTED CELLS

The morphologic changes that result from virus infection are caused by direct virus-induced lysis, fusion, or hyperplasia of infected cells and triggering of the host immune system against the virus and infected cells. The cytologic changes of virus infection are summarized in Table 3.1.

Cell Lysis and Viral Inclusions

As mentioned, viruses generally exhibit tropism for certain organs and tissues because they attach to, and enter, only certain cell types that bear on their surfaces the appropriate receptor molecules. For example, myocardial cells can be the target for group B Coxsackie virus. Because viruses reproduce inside the target cells to which they gain entry, the effect on these host cells is, not surprisingly, often that of damage and eventual death. This type of cytolytic virus-cell interaction is the basis of acute infectious virus diseases such as those mentioned in the previous section.

Early in infection the virus does not kill the host cell, but rather subverts the cell's protein-manufacturing mechanism to its own purpose— the making of more virus. The viral proteins or newly formed virus

TABLE 3.1 Pathology of Virus Infections Frequently Recognized Using Histopathologic Technique

Cytologic Changes	Virus Example
Cytoplasmic vacuolization and swelling	Hepatitis virus
Nuclear chromatin smudging	Adenovirus
Inclusions	
Intranuclear	Herpesvirus
	Adenovirus
	Papovavirus
Intracytoplasmic	Molluscum contagiosum virus
	Rabies virus
Syncytia (giant cells)	Herpesvirus
	Measles virus
	Influenza virus
	Respiratory syncytial virus
Cytomegaly (unique large cells)	Cytomegalovirus
Necrosis	Enterovirus
	Herpesvirus
	Togavirus
Hyperplasia or lymphoid depletion	Papovavirus
	Human T-cell lymphotrophic virus

particles may form crystalline aggregates, visible microscopically as solid "inclusions" in the cytoplasm or nucleus. In some instances viral inclusions represent alteration of internal cellular membranes rather than an accumulation of viral particles. Inclusions are generally spherical, about the size of a red blood cell, and, depending on the virus, are intranuclear or intracytoplasmic in location and acidophilic or basophilic in staining. In the nucleus, the viral inclusion is often surrounded by a clear halo because the chromatin becomes peripherally displaced. This halo helps distinguish the viral inclusion from a nucleolus. Examples (Table 3.1) are the cytoplasmic inclusion of vaccinia virus (Guarnieri body), the nuclear inclusion of herpes simplex virus (Cowdry Type A), the nuclear and cytoplasmic inclusions of strikingly enlarged cells infected with cytomegalovirus (Fig. 3.2), the nuclear inclusion of adenovirus, and the cytoplasmic inclusions of rabies virus (Negri body).

Later in infection, the progeny viral proteins cause a shutdown of both host and viral RNA and protein synthesis. Consequently, the normal cellular division and repair processes are disrupted and the cellular membranes become leaky. This effect allows extracellular fluid to enter the cell, resulting in swelling and intracellular edema. The pathologist describes the corresponding cytoplasmic appearance in the light microscope with terminology such as cloudy swelling, ballooning degeneration, vacuolization, poikilocytosis, and spongiform change. Hepatocytes chronically infected with hepatitis B virus show a "ground glass" ap-

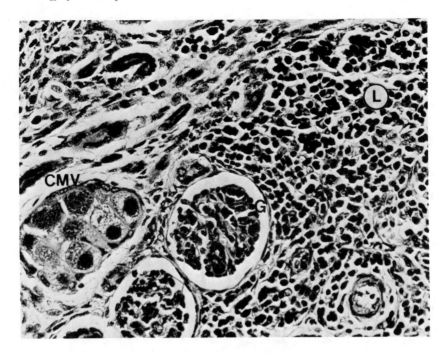

FIG. 3.2. Renal tubular epithelial cells infected with cytomegalovirus *(CMV)* show striking enlargement with both nuclear and cytoplasmic inclusions. A typical cellular infiltrate resulting from a viral infection consisting mainly of stimulated lymphocytes *(L)* is present near the renal glomeruli *(G)*. (H&E, ×365)

pearance in the cytoplasm as a result of accumulation of the virus surface antigen. Lysosomal enzymes escape inside the infected cell, causing further cytoplasmic degradation and structural alteration of nuclear chromatin, which the pathologist describes as clumping, pyknosis, smudging, and margination. Finally, autolytic digestion leads to cell dissolution or necrosis. The cytoplasm of necrotic cells often appears condensed and brightly eosinophilic. A moderate degree of cytoplasmic swelling is probably reversible, although cell function may be impaired. However, the nuclear changes, particularly fragmentation of chromatin, imply cell death.

Cell Fusion

A second type of virus-cell interaction causes infected cells to fuse. As a result, multiple nuclei share a common cytoplasm, forming so-called syncytial giant cells, or polykaryocytes. This cell fusion is brought about by synthesis of virus-coded glycoproteins (peplomers) in the plasma membrane and is a characteristic, but not an exclusive property, of certain enveloped RNA viruses that bud from the plasma membrane. Vi-

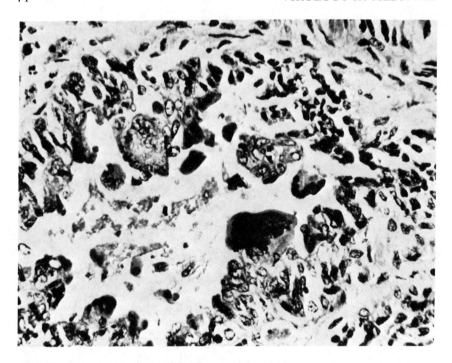

FIG. 3.3. Respiratory syncytial virus pneumonia in an immunosuppressed 2-year-old child. A syncytial giant cell is present in the bronchial epithelium. (H&E, ×640)

ruses of this type often cause a steady-state or persistent, latent infection in which virus particles are more or less continuously produced without appreciable effect on the metabolism of the host cell. Examples of clinically important viruses causing syncytial formation in vivo and in vitro are measles virus, respiratory syncytial virus (RSV), and the herpesviruses. The RSV is the most common cause of viral pneumonia in children less than 5 years of age (Fig. 3.3). In the proper clinical setting, such as pneumonia in the pediatric age group or cervicitis in sexually active women, the finding of syncytia will permit the pathologist/cytologist to make a tentative diagnosis of infection with RSV or herpes simplex, respectively.

Cell Hyperplasia

A third type of virus-cell interaction leads to cell replication and neoplastic transformation. Historically, these viruses were first described by Shope as causing papillomas or warts in rabbits and later associated with squamous-cell carcinoma. This important discovery illustrated how a virus caused neoplasia while avoiding detection by available diagnostic techniques. The papillomatous lesion is well characterized by histologic

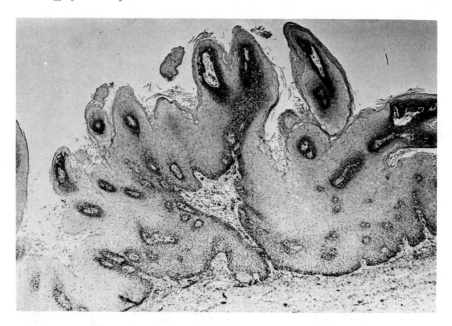

FIG. 3.4. The papillomatous lesion of virus-induced epithelial hyperplasia. (H&E, ×40)

criteria (Fig. 3.4), and the viral antigens can be demonstrated by im-munoperoxidase techniques (Fig. 3.5). Note, however, that the viral antigen can be demonstrated only in a few cells that are maturing and undergoing keratinization. In addition, intranuclear inclusions corre-sponding to crystalline assays of viral particles may sometimes be seen in these cells. The viral genome remains unintegrated, within the host-cell nuclei of the less-mature squamous cells, and therefore can persist without ready detection. Expression of only one or a few of the viral genes is apparently all that is required for starting and maintaining the transformed state. In humans, the papilloma viruses are the causative agents of warty growths or condylomas of the skin, cervix, vulva, penis, perineum, mouth, and larynx. The incidence of genital condylomas is increasing in epidemic proportions, and the identification of viral anti-gen in lesions considered to be premalignant (e.g., cervical dysplasia) has also implicated the virus as a possible causative factor in human cervical carcinoma.

Although human infection with retroviruses is rare, exogenous infec-tion has recently been associated with two separate disease entities: T-cell lymphoma or leukemia and acquired immunodeficiency syndrome (AIDS). Several different laboratories have recovered T-cell lympho-trophic retroviruses from patients with AIDS, including human T-cell lymphotrophic retrovirus III (HTLV III), lymphadenopathy associated virus (LAV), and AIDS-associated retrovirus (ARV). Additional evi-

FIG. 3.5. Nuclear viral antigen (human papilloma virus) stained with the immunoperox-idase technique. Virus antigen can be seen in only a few cells *(arrows)* that are maturing and undergoing keratinization. The virus may be present but not expressed in less mature squamous cells (×640)

dence to support a causal relationship between infection with a T-lymphotrophic retrovirus and acquired immunodeficiency has come from epidemiologic studies demonstrating the high prevalence of antibodies to HTLV III or LAV in AIDS and pre-AIDS patients. Antibodies to LAV and HTLV III have also been found in donors who gave blood to persons who subsequently developed AIDS. A similar disease entity occurring in rhesus monkeys and associated with a Simian retrovirus has been described at the California Primate Research Center. Further studies of acquired immunodeficiency may contribute to understanding the natural history of retrovirus transmission and the role these viruses play in human malignancies.

Microscopically, the histologic changes of cell lysis, inclusions, fusion, and hyperplasia are nonspecific in that identical features may result from causes other than virus infection. For example, intranuclear inclusions may be mimicked by giant nucleoli of rapidly growing cells, cytoplasmic inclusions may be mimicked by condensation of organelles occurring during cell degeneration (e.g., alcoholic hyaline in liver cells), and cell fusion may be caused by mycobacteria and other nonviral pathogens. Hyperplasia and neoplasia may certainly be caused by nonviral means, and most human cancer is not thought to be virus-induced. Nevertheless, in the proper clinical setting and absence of other explanations, a viral etiology for such characteristic findings as inclusions or syncytia must be strongly considered. Confirmation of a specific virus usually depends on identification by electron microscopy, labeled antibody, virus isolation, or serology.

CORRELATIONS BETWEEN PATHOLOGY AND VIRUS DETECTION IN THE LABORATORY

Many of the mechanisms behind the ability of viruses to cause infection and disease can be further explained by examining how viruses are detected in the laboratory. In essence, the goal is to detect the functional characteristics of the virus, using its ability to produce a specific protein, cause cell death, or alter cellular appearance and function. In many cases, the pathology of the infection caused in the patient can be correlated with the in vitro phenomenon observed in the laboratory. With newer techniques, however, the presence of the viral nucleic acid can be detected directly, bypassing the need to assay for the functional characteristics of the virus.

When a virus is suspected of causing morbidity in the proper clinical setting, a culture is taken and inoculated in the laboratory into cell cultures, embryonated eggs, or animals. Cell culture is the most commonly employed method for cultivation of viruses. The technologist can select from a wide range of host systems, but primary monkey kidney cells and human fetal lung cells offer adequate susceptibility for diagnostic purposes (Table 3.2). Inoculated cell cultures, growing as monolayers in glass or plastic tubes and plates, are observed for the presence of characteristic viral cytopathic effects (CPE) such as lysis, cell rounding, or syncytia formation or for demonstration of the physical presence of certain viral components using methods such as hemadsorption. The characteristic features of the infected cell suggest a diagnosis to the trained observer. A summary of these changes is found in Table 3.1. The laboratory provides a service to the clinician who is primarily concerned with establishing whether or not a virus is present. The ability of the laboratory to specifically identify the viral agent can provide the physician with information essential for therapy (avoidance of unnecessary antibacterial chemotherapy and perhaps even the initiation of antiviral chemotherapy), the determination of prognosis, and the modification of infection control regimen. See Chapter 13 for a summary of methods for virus identification (Table 13.3).

Many of the same principles involved in these qualitative assays can also be employed in quantitative assays. Although quantitative assays are not often employed in the clinical setting, their usefulness in research, in vaccine and reagent production, and as the assay method for moni-

TABLE 3.2 Viruses Frequently Encountered by the Clinician and Pathologist/Microbiologist[a]

				DIRECT DETECTION		VIRUS ISOLATION		
Virus	Disease or Organ System Infected	Specific Diseases[b]	Special Features	Electron Microscope	Immuno-Chemical	MKHEL		Other Host Systems
Adenoviridae Adenovirus	R, O. E, I	Pneumonia Epidemic keratoconjunctivitis (EKC). Fever of unknown origin (FUO)	Epidemic in military populations. Associated with severe acute pulmonary disease in children. Intranuclear inclusions	Naked virions, 60–90 nm in diameter, with fiber antigens projecting from pentomers pentons	FA	±	±	HEK, HEp-2
Herpesviridae Cytomegalovirus	O, C. E, R, H	CID FUO	Disseminated infections in immunocompromised host; chorioretinitis; associated with CMV pneumonia	Enveloped virions, 100 nm in diameter	FA, IFA, IP	−	+	HEK
Varicella-zoster (vz) virus	O, R. E	Chickenpox, shingles	Disseminated infection; VZ pneumonia			±	+	HEK
Herpes simplex virus types 1 & 2	O, C. G, N, E	Herpes genitalis, labialis; keratitis, aseptic meningitis encephalitis	CNS involvement more frequently with Type 1 and Type 2 Type 2 associated with uterine, cervical carcinoma		FA, IP, IFA, ELISA	±	+	HEK
Orthomyxoviridae Influenza A, B viruses	R	Influenza, pneumonia	Common agent of viral pneumonia in adults. Hemorrhagic pneumonia	Enveloped virions, helical symmetry 80–170 nm in diameter. Virions may appear spherical or pleomorphic	FA, IP	+	−	

			Hyperplastic	Morphology	Detection	Not cultivatable		Lab culture
Papovaviridae	G	Genital warts	Hyperplastic lesions. Probably sexually transmitted	Naked virions, cubic symmetry, 40–45 nm in diameter	FA, IP	+	–	
Paramyxoviridae								
Measles virus	R, N, E	Measles, pneumonia	Syncytia and giant cells	Enveloped virions, helical symmetry, may be 150–300 nm in diameter, usually spherical and highly pleomorphic with occasional filamentous forms	FA, IP	+	–	
Mumps virus	N, G	Aseptic meningitis, orchitis			FA, IP	+	±	HeLa
Respiratory syncytial virus	R	Bronchiolitis	Common viral bronchiolitis in children less than 5 years		FA, IFA, IP ELISA	+	+	HEp-2, HEK
Parainfluenza virus	R	Croup			FA, IFA	+	±	
Picornaviridae								
Poliovirus	N			Naked virions, cubic symmetry, 24–30 nm in diameter	None	+	±	HeLa, HEK
Echo virus	N, R, E, C, H, I	Aseptic meningitis				+	±	HeLa, HEK
Coxsackie A	O, R, N, E, C, H, I	Encephalitis, paralysis, hand-foot-mouth disease				–	±	RD, suckling mice
Coxsackie B	N, R, E, C, H, I	Aseptic meningitis				+	+	HEK, BGM
Enterovirus 68–71	R, O, N	Pneumonia				–	+	HeLa, HEK

[a]Key:
R = respiratory
O = ocular
C = congenital or neonatal
G = genitourinary
N = central nervous system
E = exanthem (vesicular/macular)
H = hepatic
I = gastrointestinal
[b]Lab cultures correlate with these.

ELISA = enzyme-linked immunosorbent assay
FA = direct fluorescent antibody
IFA = indirect fluorescent antibody
IP = immunoperoxidase-labeled antibody
MK = primary monkey kidney (usually rhesus, cynomolgus)
HEL = human embryonic lung
HEK = human embryonic kidney
MDCK, HEp-2, HeLa, RD and BGM = other available cell lines

toring antiviral chemotherapy is of growing importance. The plaque assay
is one method that not only establishes the presence of a virus but also
allows quantification of infectious particles. Plaque assays are performed by inoculating monolayers of cells on culture plates with serial
dilutions of virus. This method can be used to isolate genetically pure
strains of virus, since each plaque has theoretically been caused by infection from a single virus particle. By this method several different
strains of virus may be separated from the initial mixed population.

Some viruses do not cause sufficient cell damage to be seen as plaques
in culture plates or in tubes; however, their presence can be detected
by other means. For example, in cases in which a paramyxovirus is suspected, the monolayer of cells is flooded with erythrocytes that will ad-
here to infected cells in localized areas (hemadsorption). Another technique involves staining of the infected cells with fluorescein or peroxidase-
labeled antiviral antibodies to detect viral antigens in the cultured cells.
Both of these techniques can aid the laboratory technologist in making
a timely diagnosis because they can usually be performed within 24 hours.

Hemagglutination (HA) is a laboratory phenomenon in which virus
released into the cell culture medium is assayed. The basis of the HA
assay is that virus particles attach to one red blood cell and then another, eventually resulting in an aggregate of red blood cells that can
be easily detected with the unaided eye. Because hemagglutination requires free virus particles, their detection occurs some time after
hemadsorption can be detected on the cell surface. These are physical
properties of virus particles, and much of what we know regarding the
cell virus-receptor chemistry and specific viral protein receptors was
learned using glycoproteins isolated from red blood cells.

When antibodies are present in serum or a fluid medium, the antibodies can attach to viral antigenic components (peplomers or capsid
antigens) and prevent their adsorption to red blood cells. By utilizing
the property that specific antiviral antibodies inhibit HA, antiviral antibodies can be readily quantitated.

Virus particles can be detected by means of the electron microscope;
however, this method does not distinguish between infective and non-
infective/defective particles. This is an important point, which under-
scores the "supply and demand" limitations of viral replication. The intracellular assembly of complete virus particles occurs after the formation
of the component parts. The machinery of the infected cell may produce capsid proteins in different quantities from the viral nucleic acid,
often resulting in virions devoid of nucleic acid. Certain nucleotides may
have been depleted in the infected cells and the nucleic acid produced
may be either noninfectious or code for defective viral proteins. In some
situations the ratio of physical viral particles to infective viral particles
may be as high as 1000:1. In addition to the production of defective
viral particles, some of the factors that may account for this relative
inefficiency of infection are variations in susceptibility of the host cells,

elution of adsorbed virus from the cell before penetration, production of interferon by the host cell, and degradation of phagocytized virus by the host-cell enzymes.

If the identity of virus detected by laboratory methods correlates with the clinical picture, then a cause-and-effect correlation can be made. However, on occasion it may be difficult to correlate the laboratory data with the clinical picture or even establish the presence of a virus by laboratory methods in a clinical setting where a viral etiology is strongly suspected. Such situations can be detected and quantitated on the basis of virus-specific enzymes that are components of the viral particle. An example is in the diagnosis of hepatitis B virus infection. No cell culture system yet exists for this virus. The hepatitis B virus contains a virus-specific, DNA-dependent, DNA polymerase that, when detected, correlates with acute or chronic active hepatitis. (Likewise, the neuraminidase activity of influenza virus or parainfluenza virus can be detected and assayed.) Reverse transcriptase, the viral enzyme that catalyzes transcription of DNA from RNA in retroviruses, has also been used as an assay for detection of human T-cell lymphotrophic virus infection.

A new method that will have considerable application in the future is the DNA restriction enzyme analysis of viruses isolated from various body fluids or infected tissues. Using synthesized DNA probes that are homologous to viral DNA, very small amounts of the latter can be detected. Because the technique is highly specific, it is not necessary to wait for the development of characteristic viral cytopathic effects as the marker of viral infection.

Since the development of hybridoma methodologies, the production of monoclonal antibodies against viruses and viral components has recently begun to serve an important role in the rapid diagnosis of viral diseases and has contributed to our understanding of viral pathogenesis. We anticipate that current viral diagnostic techniques (cell/animal culture and serology) will soon be supplanted by the direct immunochemical (probably ELISA) detection of specific viruses in clinical specimens, using monoclonal viral antisera.

HISTOPATHOLOGY OF HOST IMMUNE RESPONSE TO VIRAL INFECTION

Viral infections, except some of those congenitally acquired, and the "slow viruses," generally induce a strong immune response in the host. Cell damage may actually result more from the immunologic response to the viral proteins on the cell surface than from direct viral cytotoxicity. This immune attack may even kill the virus-infected cell before virus particles are released. In some persistent viral infections, such as chronic active hepatitis, this mechanism may be the major cause of disease. Both B- and T-cellular arms of the immune response are activated by viral

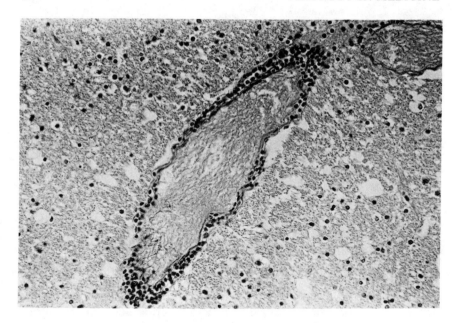

FIG. 3.6. In viral infections of the central nervous system, the lymphocytic infiltrates are often present surrounding blood vessels and are described as perivascular cuffing. (H&E, ×125)

infection, but B and T lymphocytes cannot be distinguished easily by routine histopathology. What is microscopically apparent to the pathologist in standard hematoxylin-and-eosin-stained sections is a cellular infiltrate consisting mainly of lymphocytes, macrophages, and, perhaps, a few plasma cells (Fig. 3.2). The lymphoid infiltrate localizes at sites of viral infection, and in the central nervous system surrounds blood vessels as characteristic perivascular "cuffs" (Fig. 3.6). Local or distant sites of injury may result from deposition of viral antigen-antibody complexes, for example, in kidney glomeruli and blood vessel walls.

Viral antigen-antibody complex deposition triggers complement activation and a secondary acute inflammatory reaction. This activation and reaction results in microvascular leakage and thrombosis, producing microscopically apparent changes such as edema, hemorrhage, infarction, and necrosis. Only in the presence of severe necrosis or secondary bacterial infection, however, does an infiltrate of neutrophils occur. Thus, mononuclear cells, lymphocytes, macrophages, and plasma cells rather than polymorphonuclear leukocytes characterize the immune cellular reaction induced by viral infection. Similarly, an increase in the blood of lymphocytes in relation to neutrophils usually occurs. Furthermore, some uninfected lymphocytes in the tissue or circulation may appear large and atypical because of their immunologic stimulation by viral antigens on other cells. Reactive or so-called atypical lymphocytes of this sort in a blood smear are characteristic of infectious mononucleosis caused by the Epstein-Barr virus. The debris from dead and dying parenchymal and inflammatory cells will be removed by macrophages, and the parenchymal cells may be renewed if the target tissue has regenerative capacity (e.g., epidermis or respiratory mucosa) and is not too severely damaged.

Areas of extensive damage or loss of such cells as skeletal muscle or neurons, which have little or no postnatal regenerative capacity, will be replaced by fibrous tissue. The end result of most viral infections is, therefore, either normal-appearing tissue or scar tissue. In chronic or persistent infection, a combination of inflammation and fibrosis will be evident microscopically. These histopathologic features are, generally, nonspecific and also follow other types of injury. Nevertheless, this pattern of reaction taken together with other suggestive microscopic features, such as inclusions, cytomegaly, or syncytia, and a characteristic clinical setting often suggest a specific viral diagnosis (e.g., hepatitis B and chronic active hepatitis or cytomegalovirus infection in newborn or immunosuppressed patients). In most instances, however, immunologic staining of tissue sections or virus-infected cell cultures is required to establish a specific diagnosis. These techniques are now readily available, and their important contribution to virus diagnosis is discussed next.

IMMUNOPATHOLOGY OF VIRAL INFECTION

Viral Antigen and Antibody

Fortunately, labeled antibodies are now readily available for detection of intracellular virus-specific antigen directly in clinical specimens (e.g., biopsy or autopsy tissues, vesicle scrapings, nasopharyngeal smears, respiratory tract aspirates, or viral isolates in tissue culture). Free virus-soluble antigen or antibody in body fluids is commonly detected by radioimmunoassay or enzyme-linked immunoassays (ELISA), as discussed elsewhere (Chapter 12). The immunologic techniques offer rapid, sensitive, and specific diagnoses, and they can be applied to agents such as hepatitis virus, which are not readily isolated in vitro. Each technique may be used to detect and identify viral antigen, provided that specific viral antibody is available, or to estimate the level of viral antibody in the serum, provided that antigen of known specificity is used.

Immunofluorescence and immunoperoxidase are the two most widely used methods to detect intracellular viral antigens in infected tissues (Fig. 3.7). For direct immunofluorescence, each virus-specific antiserum must be separately conjugated with fluorescein before application. Such conjugated viral immunoglobulins for the direct immunofluorescence test

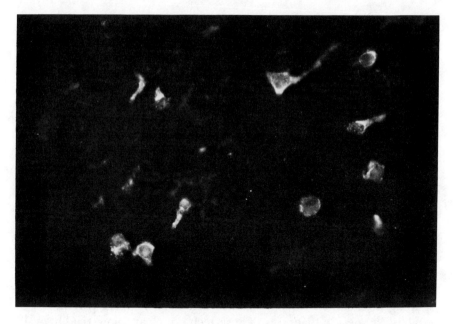

FIG. 3.7. Herpes simplex viral antigen is demonstrated by fluorescein-labeled antibody in the cytoplasm of cerebral neurons. The specific diagnosis of herpes encephalitis in this case was made 45 minutes after obtaining the biopsy. Subsequent treatment resulted in patient survival with minimal sequelae. (direct immunofluorescence stain, ×640)

are commercially available for most common viral infections, except enteroviruses and togaviruses. For indirect immunofluorescence, unconjugated virus-specific antiserum (goat or rabbit) is applied and incubated, and unattached antibody is washed away. A fluorescein-labeled antiserum, directed at the immunoglobulin of the animal species providing the virus antiserum, is then added and the specimen is examined with the fluorescence microscope. Viral reagents for use in the immunofluorescence test are commercially available for respiratory syncytial virus, herpes simplex virus, Epstein-Barr virus, and cytomegalovirus.

The direct immunofluorescence test is more commonly used because it is relatively quick, uncomplicated, and specific, but the indirect test has the advantages of greater sensitivity and fewer individual sera that need conjugation. The indirect test can also determine susceptibility to specific viruses or confirm the diagnosis by measuring the level of virus-specific antibody and class of immunoglobulin in the serum of individual patients. As with most laboratory procedures, the reliability of the fluorescent antibody technique is based largely on the quality of the specimen collected (see Chapter 13).

Immunofluorescence techniques, unfortunately, work only on unfixed cell imprints or frozen tissue sections; formalin fixation, routine in most pathology laboratories, ruins the fluorescence. By contrast, the immunoperoxidase technique can be applied to formalin-fixed, paraffin-embedded tissues and, therefore, is much more practical for biopsy and autopsy tissues. The immunoperoxidase procedures are as sensitive as immunofluorescence procedures and, in addition, do not require expensive fluorescent microscopic equipment. Immunoperoxidase-stained slides are stable and can be stored without loss of sensitivity. A steadily increasing number of labeled antisera suitable to the immunoperoxidase method are now available for detection of human virus pathogens (e.g., hepatitis B, herpes simplex, rubella, influenza viruses, cytomegalovirus, rotavirus, orthomyxo- and paramyxovirus, and papovavirus).

Inflammatory Response

In addition to looking for intracellular viral antigens in human tissue sections, one can analyze the inflammatory response with the immunofluorescence or immunoperoxidase techniques using highly specific or monoclonal antibodies to Ig determinants, complement (C_3), and T-cell subsets. One can determine which class of Ig (IgM, IgG, or IgA) is produced by the B lymphocytes and whether the lymphoid infiltrate expresses both kappa and lambda light chains, as expected in a polyclonal response to viral infection, or expresses only one light chain, which would suggest neoplastic monoclonal proliferation. This technique is particularly helpful in recognizing virus-induced lesions that mimic neoplasia, such as the lymphadenopathy of mononucleosis. Coating of infected target cells with Ig and C_3 at sites of tissue injury or necrosis helps to explain the inflammatory reaction and disease process. The ratio of helper T cells to suppressor T cells, as judged by staining with monoclonal antibodies to T-cell markers, may give new insight into the pathogenesis of viral infection or other complex immunopathologic entities (e.g., AIDS).

MOLECULAR DETECTION OF VIRAL GENOMES

It is now "state of the art" to detect virus DNA or RNA in serum, tissues, and cells by molecular hybridization with radiolabeled virus probes and autoradiography. Virus DNA, if plentiful ($> 10^5$ molecules/100 μl), can be identified in serum or urine "spotted" onto a nitrocellulose filter for hybridization. Virus genes in tissue can be identified in much lesser amount ($\sim$0.1 genome/cell) by DNA extraction, restriction enzyme digestion, and electrophoresis through a gel before transfer by Southern blotting to a nitrocellulose filter and hybridization. Of great promise, because of its safety and practicality, is the development of nonradioactive, biotinylated gene probes that can identify virus nucleotides in formalin-fixed paraffin-embedded tissue sections. Detection of the virus genes in this system depends on the same immunoperoxidase technique as used to detect viral antigens. Recognition of viral genes and proteins in the same cells in paraffin-embedded tissues is also feasible, by means of a combination of radiolabeled probes for in situ hybridization and immunocytochemistry. Techniques such as these are now being applied to further our understanding of the pathogenesis of various virus diseases, for example, hepatitis B in relation to chronic liver disease and HTLV III in relation to AIDS.

Recommended Readings

Evans AS (ed): *Viral Infection of Humans: Epidemiology and Control.* New York, Plenum Press, 1979.

Fenner F, McAusten BR, Sambrook J, White DO (eds): *The Biology of Animal Viruses,* 2nd ed. New York, Academic Press, 1979.

Gardner PS, McQuillin J (eds): *Rapid Virus Diagnosis—Application of Immunofluorescence.* Boston, Butterworths, Inc, 1980.

Hsiung GD, Green RH (eds): *Virology and Rickettsiology,* Vol 1, Parts 1 and 2. In Seligson D (ed): Section H, *Handbook Series in Clinical Laboratory Science.* West Palm Beach, CRC Press, 1978.

Joklik WK (ed): *Virology,* 2nd ed. Norwalk, CT, Appleton, 1985.

Lennette EH, Schmidt NJ (eds): *Diagnostic Procedures for Viral, Rickettsial and Chlamydial Infections,* 5th ed. Washington, DC, American Public Health Association, 1979.

Lennette DA, Spector S, Thompson KD (eds): *Diagnosis of Viral Infections, The Role of the Clinical Laboratory.* Baltimore, University Park Press, 1979.

4

The Epidemiology of Viral Diseases

NEAL A. HALSEY

Laboratory investigations are essential for a complete understanding of the pathogenesis of viral diseases. However, much of our knowledge of viral diseases has been acquired from epidemiologic studies. For example, detailed information regarding infectious hepatitis, including effective methods of preventing the disease, was known for more than two decades before hepatitis A virus was identified. Also, epidemiologic studies of AIDS identified the common modes of transmission long before the infectious agent was identified. This permitted public health officials to make the recommendation that populations at increased risk for AIDS should not donate blood. This early decision probably resulted in the prevention of many cases of AIDS. Moreover, knowledge of the epidemiology of common viral infections is an important tool for physicians in their everyday practice of medicine. Clinicians usually evaluate individual patients, and epidemiologists usually study disease patterns in large populations. However, accurate clinical data are the foundation for epidemiologic studies, and competent physicians must know and practice some epidemiology.

Table 4.1 presents definitions that are useful in understanding the epidemiology of viral diseases.

SURVEILLANCE

Surveillance is the cornerstone of epidemiology. If physicians and other interested persons did not report instances of diseases, no conclusions could be drawn regarding changing patterns of disease, relative differences in disease rates in various places, or any of the other factors discussed in this chapter. Some authorities have criticized surveillance data as inaccurate because of underreporting and inaccuracies in clinical di-

TABLE 4.1   Definitions of Epidemiologic Terms

Agent	The causative organism
Case	The single occurrence, in a person, of the disease in question
Cohort	A group of persons defined as identical for all of the variables in question
Contagious period	The time from the onset of the infectivity of the affected person (i.e., when virus first becomes detectable) until the danger of transmission to others is past (i.e., virus is no longer being shed). An infected person may be contagious before the appearance of symptoms
Control	A person known not to be affected by the disease or factor in question
Epidemic	The presence of a disease in numbers greater than expected
Immune	The state of being resistant to infection so that infection by the specific agent cannot occur
Incidence	The number of new cases of disease within a specified period in a defined population
Incidence rate	The number of new cases of disease divided by the population at risk in a specified period (i.e., the incidence for a given period, divided by the population)
Incubation period	The time from exposure to the agent to the onset of the first symptom
Latent period	The time from exposure to active disease up to the point at which infection is clinically evident or detectable by laboratory tests
Prevalence	The number of cases of a disease in a specified population, usually given as a "period prevalence," i.e., the number of persons affected during a specified period. This number is influenced by the incidence and duration of disease
Prevalence rate	The prevalence divided by the population
Prospective study	Groups (cohorts) of persons free of the disease in question who are identical except for varying degrees of exposure to one or more factors suspected of affecting the acquisition of the disease. The cohorts are followed for specified periods to determine the incidence rate of disease in each group
Susceptible	Subject to infection by a given agent, i.e., the absence of immunity. Susceptibility to reinfection or recrudescent infection (e.g., herpesviruses) represents special circumstances of specific immunity
Transmission	The transfer of an infectious agent from one person to another

agnoses. Nonetheless, as long as the diagnostic methods and case definitions of disease remain constant, changing patterns of disease may be detected. For some diseases, such as influenza or varicella, fewer than 10% of all cases are reported. Even with these limited data, changing patterns of disease can be accurately identified as long as the surveillance system remains relatively unchanged.

Every state maintains a list of reportable diseases. All physicians are strongly encouraged to obtain the list from the state health department and to report observations on the diseases so designated. The occurrence of some viral diseases is best reported through diagnostic laboratories by virus isolation or serologic evidence. Information from these sources can be useful in identifying which viruses are prevalent in the community at different times of the year. When a physician faces an unusual diagnostic problem, personnel in these laboratories and in state health departments can often assist with the diagnosis.

TRANSMISSION

A single viral agent may be transmitted by any of several modes; and several different terms may be used to describe the same mode of transmission. For example the "person-to-person" mode encompasses the terms *respiratory, direct contact, fecal-oral,* and several others. The transmission of viral agents is best described by defining the source or sources of an infectious agent, the vehicle of transmission, and the anatomic site or sites through which it is inoculated into the host.

Table 4.2 shows the source, vehicle, and portal of entry for transmission of viral infections. Most viral agents affecting humans are acquired from other humans who are infected and shedding virus. Respiratory and genital secretions, skin lesions, and excreta all represent potential sources from which infection may be readily spread by hand to eyes and mouth. Personal hygiene, particularly washing hands with soap or detergent, is an important means of limiting the transmission of viral and other infectious agents. Such measures are especially applicable to physicians and other health-care personnel.

The mode of transmission and site of acquisition may influence the severity of a viral disease. Susceptible young infants generally tolerate varicella-zoster virus infections well when they acquire the virus postnatally through inoculation of respiratory mucous membranes. However, if the virus is transmitted transplacentally from the mother less than 5 days before birth, then the infant's viral infection may be severe or even fatal. A similar phenomenon has been described for enteroviral infections transmitted transplacentally just before birth. Poliovirus is normally acquired by ingestion of the virus. However, when small amounts of live, wild-type poliovirus were inadvertently administered by injection, the rate of subsequent paralytic poliomyelitis was higher than would be expected had the virus been ingested.

DESCRIPTIVE EPIDEMIOLOGY

Time, place, and person are the traditional parameters used to describe the epidemiology of most infectious agents.

TABLE 4.2 Modes of Transmission of Viral Infections

Source[a]	Vehicle or Mechanism	Portal of Entry	Virus Examples[b]
Respiratory tract	Aerosol droplets, hands	Nose, mouth, eyes	Myxoviruses, paromyxoviruses, rhinoviruses, VZV
Oral secretions	Kissing, hands	Mouth	HSV, EBV, enteroviruses
Animal and human oral secretions	Bites, licks	Skin, open wounds	Rabies virus
Feces	Hands, food, water, sexual contact	Mouth	Enteroviruses, hepatitis A virus, rotavirus, adenoviruses
Urine (human)	Food (hands?)	Mouth	Hepatitis A virus, CMV
Urine (animals)	Water	Mouth	LCM, and Lassa fever viruses
	Dried aerosol	Respiratory tract	
Genital secretions	Sexual contact	Genital mucosa	HSV
	Hands	Mouth	HSV
	Transcervical spread	Fetal mucous membranes	HSV
	Parturition	Mucous membranes of infant	HSV
Skin	Hands	Mouth, nose, eyes	VZV
	Direct contact, hands	Skin	Papovaviruses
Eyes	Hands	Eyes	Adenoviruses, enteroviruses 70
Blood (human)	Needles	Blood by subcutaneous, intramuscular, or intravenous skin infection	Hepatitis B, nonA-nonB hepatitis, CMV, Ebola, Lassa fever, HTLV III
	Bites of infected mosquitoes, ticks	Blood	Yellow fever, dengue, Colorado tick fever
	Transplacental	Fetal blood	Rubella virus, CMV, enteroviruses, VZV
	Ingestion or aspiration during parturition	Blood of newborn	Hepatitis B virus, HTLV III
Blood (birds, mammals)	Bites of mosquitoes	Blood	Arboviruses

[a] Source is human unless otherwise specified.
[b] VZV = varicella-zoster virus, HSV = herpes simplex virus, EBV = Epstein-Barr virus, CMV = cytomegalovirus, LCM = lymphocytic choriomeningitis, HTLV III = human T lymphotropic virus type III.

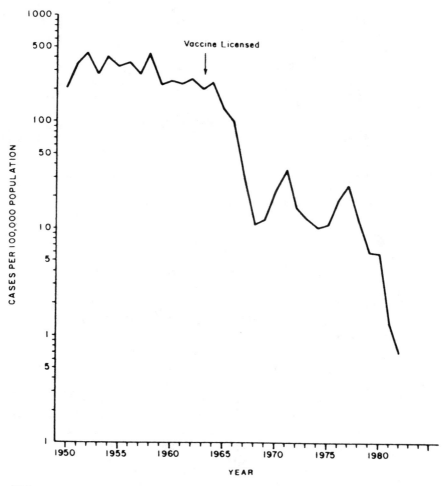

FIG. 4.1. Reported measles incidence—United States, 1950–1982. (Source: Morbidity and Mortality Weekly Report 32:49, 1983.)

Time

SECULAR TRENDS

To report disease trends, the annual incidence or incidence rate is usually plotted on an arithmetic or semilogarithmic scale. Figures such as these are useful for discerning long-term patterns of disease occurrence and the impact of preventive measures, as illustrated in the story of measles during the past three decades (Fig. 4.1). Some viral infections occur at relatively constant rates from year to year. Epidemics of others recur at variable intervals. For example, before the widespread use of vaccine, rubella epidemics occurred every 6 to 9 years (Fig. 4.2). Apparently, during the long interepidemic intervals, a gradual increase

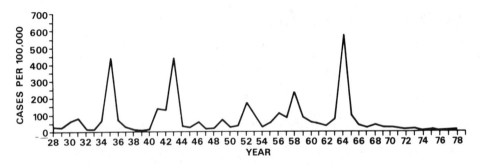

FIG. 4.2. Rubella incidence in 10 selected areas in the United States (Maine, Rhode Island, Connecticut, New York City, Ohio, Illinois, Wisconsin, Maryland, Washington, Massachusetts), 1928–1978. (Source: Center for Disease Control: Rubella surveillance January 1976–December 1978. Issued May 1980, p. 3.)

in the percentage of susceptible individuals reached a point where widespread transmission of disease could be maintained.

The impact of disease control programs or the association of disease with changing factors, such as economic status or development, can usually be discerned by reviewing secular trends. As noted previously, although only a fraction of all disease occurrences may be reported, the data are usually adequate for interpreting disease trends as long as the mechanisms and efficiency of reporting remain constant. When intensified efforts are made to find and report all cases, an apparent increase in disease incidence may be noted. This phenomenon was exemplified by the smallpox eradication program prior to the eradication of this disease. Intensified searches for smallpox cases in endemic countries usually resulted in 4- to 10-fold increases in the number of reported cases.

To emphasize a perspective, one may present epidemiologic data in any of several different ways. The impact of both inactivated (Salk) and live-attenuated (Sabin) poliomyelitis vaccines, for example, can be seen in Fig. 4.3A. The annual incidence of paralytic poliomyelitis is shown, plotted on a semilogarithmic scale. When the same data are plotted on an arithmetic scale (Fig. 4.3B), the impact of inactivated-virus vaccine appears more striking.

SEASONAL PATTERNS
The peak incidence of most viral respiratory disease occurs during the late winter and early spring months in the Northern Hemisphere, that is, January through May. In the Southern Hemisphere, the same diseases also peak in the winter and early spring months, which are August through November. In tropical countries, distinct seasonal patterns have been noted for some, but not all, respiratory viral infections.

In these climates, the level of disease activity frequently corresponds with cycles of rainfall.

Enteroviral illnesses usually peak in the late summer and early fall months (Fig. 4.4). Arbovirus infections (western equine encephalitis, St. Louis encephalitis, and eastern equine encephalitis viruses) also peak in the late summer, coinciding with the peak period of mosquito activity.

Some of the factors responsible for seasonal patterns of disease are known—for example, the availability of vectors. The peak in transmission of measles in tropical countries often coincides with times of low relative humidity. Some investigators have suggested that crowding and increased interpersonal contact during winter months are key factors in the transmission of respiratory viral infections. Although attendance at schools and day care centers influences outbreaks of viral infections, it does not fully explain the seasonal pattern of viral illnesses.

The reservoirs for several respiratory viruses during the "off" season have not been adequately identified. Intensive surveillance for influenza during the summer months has revealed that small numbers of persons develop clinical illness throughout the year. These persons probably provide the "seed" viruses for epidemics once the appropriate conditions have been established in the general population.

Some viral illnesses, for instance herpes simplex and hepatitis A and B, have not been associated with distinct seasonal patterns.

EPIDEMICS

Most viruses known to produce clinical illnesses in humans have been reported to occur in epidemics. Before one can declare an outbreak of disease to be an epidemic, however, care must be taken to ensure that the observed increase in disease activity is not due to expected seasonal patterns, increased surveillance, or incorrect diagnosis.

Several factors contribute to the occurrence of epidemics, including changing susceptibility of the population, unusual modes of transmission, increased numbers of vectors, and biologic variation of the infectious agent. Contamination of foodstuffs with hepatitis A virus or enteroviruses, for example, has resulted in epidemics (Fig. 4.5). Influenza viruses circulate throughout the world every year. Minor "drifts" in the antigenic structure of the virus ensure that some relatively susceptible persons are available for continued transmission. When major "shifts" occur every 10 to 20 years, the high percentage of susceptible persons in the population provides a fertile environment for rapid disease transmission. Similarly, the introduction of measles viruses into isolated populations has resulted in severe epidemics affecting over 99% of susceptible persons. Localized outbreaks of St. Louis encephalitis and western equine encephalitis occur every few years in the United States. These outbreaks or "miniepidemics" often occur after heavy rains or flooding has provided abundant breeding grounds for mosquitoes.

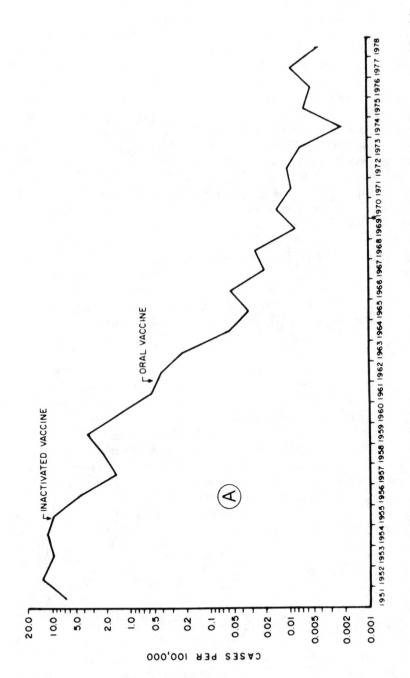

CASES PER 100,000

INACTIVATED VACCINE

ORAL VACCINE

Ⓐ

1951 1952 1953 1954 1955 1956 1957 1958 1959 1960 1961 1962 1963 1964 1965 1966 1967 1968 1969 1970 1971 1972 1973 1974 1975 1976 1977 1978

*PRIOR TO 1969 INCLUDED CASES REPORTED TO MMWR; SINCE 1969 INCLUDES ALL CASES REPORTED TO THE VIRAL DISEASES DIVISION, BUREAU OF EPIDEMIOLOGY

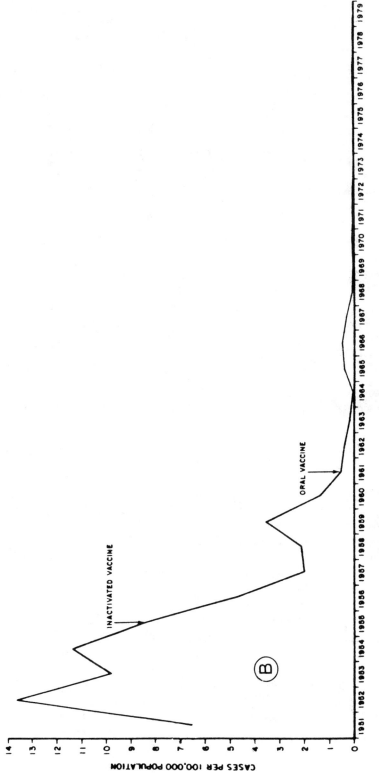

FIG. 4.3. Paralytic poliomyelitis, reported cases in the United States. A, logarithmic scale; B, arithmetic scale. (Source A: Center for Disease Control: Poliomyelitis surveillance summary 1977–1978. Issued December 1980, p. 3. B: Center for Disease Control: Immunization Against Disease. JL Conrad [ed]. 1980, p. 3.)

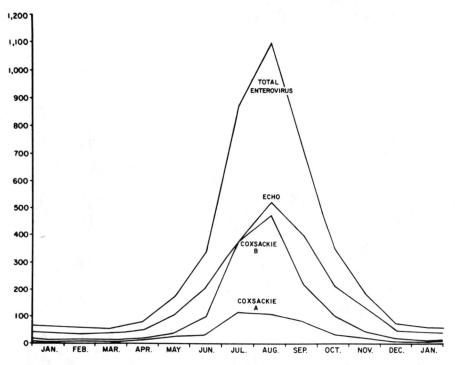

FIG. 4.4. Average numbers of enterovirus isolates; by months, United States, 1972–1975. (Source: Morens DM: Enteroviral disease in early infancy. *J Pediatr* 92:374–377, 1978.)

The shape of the epidemic curve is influenced by the incubation period, the infectious period, the mode of transmission, the number of susceptible persons in the population, and the relative frequency of contact with the source of infection. Curves from common source outbreaks, such as the food-borne illnesses mentioned earlier, often show a rapid increase in the number of affected persons (Fig. 4.5). If the source of the infectious agent is removed, then the frequency of new cases will decrease rapidly unless person-to-person transmission results in perpetuation of the outbreak.

Some viral agents (e.g., influenza, rubella, and measles viruses) that are usually transmitted by respiratory droplets may create large common-source epidemics when conditions are appropriate. The measles outbreak data shown in Fig. 4.6 are an example. In this particular population, low levels of transmission by person-to-person contact occurred for several generations of illness until two children attended a school-wide assembly program during the prodromal phase of their illness. They transmitted measles to over 60 children who became ill about 12 days later.

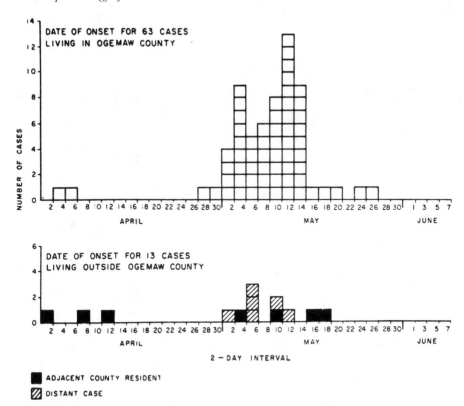

FIG. 4.5. Example of a miniepidemic of hepatitis due to food-borne virus, Ogemaw County (Michigan). (Source: Schoenbaum SC, Baker O, Zdenek J: Common-source epidemic of hepatitis due to glazed iced pastries. *Am J Epidemiol* 104:74–80, 1976.)

Place

The pattern of viral disease transmission differs from one geographic region to another. In some instances, transmission takes place only within limited areas. For example, Colorado tick fever virus is transmitted only at elevations above 5500 feet in a few areas of the Rocky Mountains. However, some persons may travel several thousand miles during the incubation period of their illnesses. Therefore, physicians must obtain careful travel histories from patients with unusual or obscure illnesses.

Conclusions regarding significant geographic clustering should be based on incidence *rates* and not merely on the number of cases reported from an area. Physicians caring for hospitalized patients have erroneously reported increasing incidences of diseases by noting an increase in the number of patients admitted to their institutions. Their observations have resulted from changing referral patterns for patients and not changing incidence rates. Some tertiary care hospitals accept

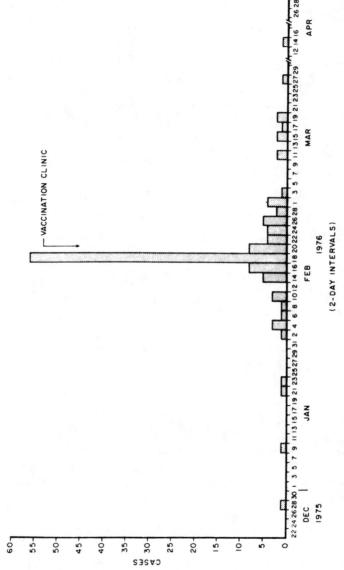

FIG. 4.6. Measles cases by date of onset, Ganado (Arizona) Elementary School, December 27, 1975–April 27, 1976. (Source: Centers for Disease Control: *Measles Surveillance* 1973–1976. Issued July 1977.)

patients from three or more states. Incidence data from these institutions are difficult to interpret. Although disease rates are usually described by place of residence, a more appropriate description might result from determining rates by place of work or school attendance when evidence suggests transmission in those areas.

Person

The analysis of data by factors associated with individuals is often the most extensive portion of epidemiologic reports. When one describes disease patterns, the factors most commonly included are age, sex, race, occupation, socioeconomic status, immune status, dietary habits, sexual habits, and other factors associated with exposure to sources of infection. Persons with immune deficiency disorders are at increased risk of suffering severe complications from viral infections. In general, defects of cellular immunity are more likely to be associated with severe complications and death from cytomegalovirus, herpesvirus, and varicella and measles virus infections.

Persons with agammaglobulinemia tolerate measles infections rather well. However, those with disorders of immunoglobulin synthesis are at high risk for developing paralysis after infection with either wild-type or vaccine strains of poliomyelitis.

Some persons are more efficient transmitters of disease than others. These "disseminators" have been identified for measles, rubella, hepatitis B, and other agents. The host factors associated with increased "shedding" of these viral agents have not been completely defined. In one person, a simultaneous infection with influenza appeared to result in increased transmission of rubella, presumably due to coughing.

ANALYTIC EPIDEMIOLOGY

General

Epidemiologic data are analyzed to test hypotheses regarding the association of various factors with diseases. These hypotheses may have derived from observations of descriptive epidemiologic data, or, more commonly, from anecdotal observations of affected persons.

To determine whether observed associations occur by chance, the investigator must perform appropriate statistical tests. A full discussion of biostatistics is beyond the scope of this chapter. However, every physician should be familiar with a few simple tests such as chi-square analysis, Fisher's exact test, and Student's t-test. To evaluate critically the design used in epidemiologic investigations, physicians should become familiar with the indications for, and limitations of, these tests. If the appropriate calculation shows a low probability that an observed association occurred by chance, then the investigator may be justified in concluding that a significant association exists between the variable and the disease. This significant association is often stated as *rejection of the null hypothesis* (the assumption that there was no association).

Causality

Epidemiologic studies have several objectives: to identify factors associated with disease processes, to try to determine whether a causal relationship exists between the factors and disease, and, eventually, to define measures that may be taken to prevent the disease. The best evidence for determining causality by a suspected agent is a randomized, double-blind, clinical trial in which one group is exposed to the agent and the other group to a placebo. However, when investigating some viral agents, it is often deemed unethical to deliberately expose persons to factors that might result in serious disease. Therefore, studies are often limited to persons naturally exposed to the agent (or factor). Exposed persons are then compared with appropriate control populations to determine the effect of the exposure on the disease in question. The selection of appropriate controls is a key factor in determining the outcome of the study. Ideally, the control group should be as closely identical as possible to the test group in every manner except for the factor or factors in question. In reality, the variability among human subjects usually results in less-than-perfect matching. Statistical tests are useful in evaluating the effect of such matching. In addition to having poorly matched controls, many epidemiologic studies have introduced unanticipated biases. The results of a single epidemiologic study, therefore, are usually insufficient for drawing conclusions regarding causal relationships.

Several authors have reviewed the subject of determining causality from epidemiologic studies. Although many disagree about the interpretation of data, most authorities agree that the following list of criteria are important in considering causal relationships:

1. Human studies. The investigations should be based on data derived from humans and not just from animals.
2. Strength of association. The likelihood that the observed associations occurred by chance must be minimal (i.e., the calculated P values should be extremely low). Also, the associations determined from prospective cohort studies are less likely to be spurious than those noted in retrospective case-control studies.
3. Consistency. The findings from an individual study must be reproducible by other investigators.
4. Dose-response. When appropriate, a gradient of increasing exposure to a factor should be associated with an increasing risk of disease.

TABLE 4.3   Calculation of Risk from Epidemiologic Data

Exposure History	RETROSPECTIVE (CASE-CONTROL) STUDY *Disease*		*Totals*
	$\dfrac{+}{a}$	$\dfrac{-}{b}$	
+	a	b	$a+b$
−	c	d	$\underline{c+d}$
			n

$$\text{Odds ratio} = \frac{ad}{bc}$$

Actual Exposure or Risk Factor	PROSPECTIVE (COHORT) STUDY *Disease*		*Totals*
	$\dfrac{+}{a}$	$\dfrac{-}{b}$	
+	a	b	$a+b$
−	c	d	$\underline{c+d}$
			n

Actual risk of disease

$$\text{in exposed persons} = \frac{a}{a+b}$$

$$\text{in unexposed persons} = \frac{c}{c+d}$$

$$\text{Attributable risk} = \frac{a}{a+b} - \frac{c}{c+d}$$

$$\text{Relative risk} = \frac{\dfrac{a}{a+b}}{\dfrac{c}{c+d}}$$

5. Temporal relationship. The exposure must be at an appropriate time to result in the biologic changes necessary to produce disease.
6. Biologic and epidemiologic sense. The proposed causal relationship must be logical. The agent in question must produce pathophysiologic changes that could result in the disease based on well-recognized biologic principles. Similarly, if the proposed agent is the cause of a disease process, then the epidemiologic description of the disease must be consistent with that hypothesis.
7. Specificity. The agent (or factor) should be associated with the single disease in question and not a broad spectrum of disorders.
8. Analogous relationships. Information on similar agents (or factors) resulting in other disease processes is helpful but not essential.

Investigators may choose to examine the relationship between disease and a factor (or agent) either prospectively (cohort study) or retrospectively (case-control study). Although the data may be presented similarly and analyzed with identical statistical tests, the results of the two approaches must be interpreted differently (Table 4.3). Retrospective case-control studies begin by matching persons who have already developed the disease (cases) with persons known not to have the disease (controls). The relative frequencies of exposures to suspected factors are then calculated. Prospective cohort studies begin with groups (cohorts) known not to have the disease. One group is exposed to the suspected factors and the other cohorts are not. Both (or all) groups are then followed for a specified time to determine whether significant differences in disease rates occur. With data from cohort studies, the *absolute risk* of developing disease can be calculated for persons exposed to the factors $a/(a+b)$ and for persons not exposed to the factor $c/(c+d)$. The *relative risk* for exposed versus unexposed and the *attributable risk* due to exposure to the factor may be determined from cohort studies as shown in Table 4.3.

The data from retrospective case-control studies generally are not utilized for calculation of actual and attributable risk because the risk of disease after exposure may be exaggerated. The investigators have selected for a high proportion of all persons in the study to have disease. Therefore, the numerators in the preceding fractions will be unacceptably large with regard to the denominators.

The chi-square test (or Fisher's exact test for small sample sizes) may be applied to both sets of data to determine the probability of the observed distributions occurring by chance.

Prospective Studies

In testing a hypothesis, prospective or cohort studies have several advantages. The investigator selects a group of persons (cohort) who have been exposed to the factor in question and who have been shown not to have the particular disease. A second cohort, persons who are similar to the first group except that they have not been exposed to the given factor, are chosen simultaneously. Both cohorts are followed for a specified time and evaluated for the presence or absence of disease. Although biases may be introduced in these studies, sampling biases, recall biases, and case identification biases are minimized in appropriately conducted trials. Therefore, evidence from such studies is weighed heavily when attempting to draw conclusions regarding causality.

Unfortunately, cohort studies are not always practical when investigating diseases that occur at low incidence rates or after long latent periods and they are often expensive and time-consuming. When associations between viral infections and cancer are being investigated, the cost may be prohibitive. An exception is the important study carried out by Beasley and co-workers in 1981 of the association between chronic hepatitis B antigenemia and hepatocellular carcinoma (Table 4.4). They followed 19,253 Taiwanese workers who were negative for hepatitis B surface antigen (HBsAg) and 3454 workers known to be HBsAg positive. During the 4½ years of study, the incidence of hepatocellular carcinoma (HCC) in HBsAg-positive workers was 400 times greater than that in antigen-negative workers. This study and other data document-

TABLE 4.4 Incidence of Hepatocellular Carcinoma Among Taiwanese Workers by HBsAg Status on Recruitment

Hepatitis B Surface Antigen	Hepatocellular Carcinoma		Population at Risk	
	+	−		
+	40	3,414	3,454	$P < 0.00001$
−	1	19,252	19,253	
	41	22,666	22,707	

Adapted from Beasley et al.

ing the incorporation of hepatitis B DNA into the DNA of HCC tumor cells have provided convincing evidence of a strong association (and probable causal relationship) between hepatitis B and HCC.

In recent years, extensive epidemiologic studies have been utilized to investigate the relationship between viral infections and chronic diseases. Viral causes have been hypothesized for chronic endocrine disorders (diabetes), neurologic disorders (multiple sclerosis, Alzheimer's disease, and amyotrophic lateral sclerosis), systemic inflammatory processes (rheumatoid arthritis, systemic lupus erythematosus), and cancer (hepatocellular carcinoma, leukemias, lymphomas, cervical carcinoma, and others). Viral agents have been identified as causal agents for some chronic disorders, such as subacute sclerosing panencephalitis (measles), kuru, Creutzfeldt-Jakob disease, progressive multifocal leukoencephalopathy (papovaviruses), Burkitt's lymphoma (Epstein-Barr virus), and hepatocellular carcinoma (hepatitis B virus). Epidemiologic studies have played key roles in establishing these causal relationships. Similar studies will also be important in proving or disproving other postulated relationships.

Retrospective Case-Control Studies

When the incidence rate of a disease is low, associations with factors that might contribute to pathogenesis can be investigated by case-control studies. In this approach, the study is performed *after* the important events have occurred. Persons known to have the disease in question are investigated for exposure to suspect factors. Appropriate control groups are investigated in an identical manner to detect statistically significant differences in their exposures to risk factors before the onset of disease. Ideally, the controls should be identical to the cases in all aspects except for the factor or factors under investigation. In practice this may be difficult. Problems of inappropriate matching have frequently led to incorrect associations in case-control studies. For example, hospitalization rates are higher for smokers than for nonsmokers.

When hospitalized patients are used as controls, higher rates of previous experience with smoking are found than would have been reported by the general population. Therefore, when a disease known to be associated with smoking (e.g., chronic bronchitis or cardiovascular disease) is being investigated, the smoking rates observed among cases might not differ significantly from those of controls. The best controlled studies usually involve matching each case with two or three controls for factors that commonly influence risk or disease (e.g., age, sex, ethnic group).

Case-control studies can be useful for investigating rare complications of viral infections. For example, Reye's syndrome has been known for almost two decades to occur after influenza infections. Serologic studies have indicated that the risk of Reye's syndrome after influenza A infections (H_1N_1) or B viruses was about 2 to 50 per 100,000 infected persons under 20 years of age. Anecdotal observations of children with Reye's syndrome suggested an additional association with exposure to salicylate preparations. Although prospective studies were impractical, several case-control studies documented significantly higher rates of salicylate ingestion among the cases compared with ingestion rates for controls. Although these data do not prove a causal relationship, advisory groups have recommended that salicylates should not be administered to children with varicella or influenzalike infections, and the Food and Drug Administration has required a warning label on all pediatric aspirin preparations. In the next few years, further studies may indicate whether these actions will result in a significant decline in Reye's syndrome.

EPIDEMIOLOGIC HISTORY

Purposes

A careful epidemiologic history may prove invaluable in diagnosing conditions of individual patients or in investigating epidemics of viral diseases. If the infectious agent is known, then the most probable source can be identified. Knowing the time of exposure, and thus the incubation period, may provide a clue toward identifying the infectious agent. Identifying the source of the agent and the type of contact may also lead one to suspect a given agent.

Time

The onset of the first symptoms should be recorded as precisely as possible. If unusual exposures took place, then the time of the exposure should be recorded. If the infectious agent is known, knowledge of the incubation period may allow focusing on the time interval during which exposure was likely to have occurred.

Place

The patient should be queried regarding residence, place of employment, and history of travel within the appropriate time interval. Details of specific activities during travel (e.g., camping) may provide useful clues as to possible animal exposure. Colorado tick fever virus, for example, is known to be transmitted only within a relatively small geographic area, above an altitude of 5500 feet. Affected persons have usually been in close contact with the ground or old wooden cabins. These circumstances lead to possible exposure to the tick vector for Colorado tick fever virus.

Open-ended questions of affected persons during epidemics have often provided clues to common-source epidemics of hepatitis, rubella, and other diseases.

Person

The age, sex, and race of a person is often evident. However, additional factors such as occupation, specific hobbies, recreational activities, and underlying immune status and immunizations should be elicited. Space does not permit the listing of all viral agents and their relative association with each of these factors. Medical and paramedical personnel have an increased likelihood of exposure to numerous viral agents. Laboratory technicians have acquired hepatitis B, influenza, rabies, and Lassa fever viruses, rotavirus and EBV, and probably other agents from clinical specimens.

Animal Exposures

Exposure to animals has been associated with many infections caused by viral agents. A carefully recorded history regarding the timing and intensity of contact is indicated.

Food, Water, and Drug Exposures

Hepatitis A virus and enteroviruses have been found in contaminated foodstuffs. Epidemics of hepatitis A have been traced to contaminated water, salads, doughnut icing, and sandwiches. Waterborne outbreaks of diseases caused by Norwalk agent and rotavirus have also been documented.

Contaminated needles and syringes have long been known to be a source of hepatitis B virus transmission. In recent years, outbreaks of Ebola virus have also been traced to reuse of syringes in Central Africa.

Other

When one traces the source of any infection, the most obvious question to ask affected persons is about known exposure to other persons with similar illnesses. The manifestations of illness are variable for some viral infections (most notably enteroviruses). Therefore, the affected person should be asked about exposure to any illnesses. Questions about contact with hospitals should be included, because these institutions are prime sources of infectious agents. Surgery has been implicated as a source of transmission for hepatitis B virus, cytomegalovirus, rabies virus, and Creutzfeldt-Jakob agent.

LATENT, PERSISTENT, AND RECURRENT INFECTIONS

In humans, most infectious diseases are caused by agents that remain in the host only briefly. For example, most viral agents can be isolated for less than 1 week from the onset of clinical symptoms. Enteroviruses and adenoviruses may persist in the intestinal tract secretions for several weeks or months before being eliminated. A transient local reinfection may develop on reexposure, but the viruses are eliminated more rapidly than during the primary infection. Therefore, an affected person is a source of infectious viral particles only briefly during his or her primary infection. However, some diseases are caused by viruses or viruslike agents that remain latent within the body for many years after the primary infection and are associated with lifelong immunity to reinfection. The epidemiology of these infectious diseases differs altogether from that of conventional viral infections. The viral agents known to persist in humans are listed in Table 4.5. As our knowledge increases, other agents may be discovered that contribute to chronic diseases of humans (see Chapter 9).

The ability of herpesviruses (herpes simplex, varicella-zoster, Epstein-Barr viruses and cytomegalovirus) to persist within our bodies for many years before being reexpressed may provide a unique mechanism to aid in the survival of the virus. After reexpression of these viral agents, transmission to susceptible persons can occur. Therefore, these viruses may be transmitted from infected persons either at the time of the primary infection or years later at times of recurrent infections. The relatively asymptomatic nature of many recurrent infections decreases the likelihood that infected persons will avoid contact with susceptible per-

TABLE 4.5   Viruses and Viruslike Agents Known to Cause Persistent or Latent Infections in Humans

Herpes simplex virus (Types 1 and 2)
Varicella-zoster virus
Cytomegalovirus
Epstein-Barr virus
Rubeola virus
Rubella virus
Papovaviruses (SV40, JC)
Kuru agent
Creutzfeldt-Jakob agent
HTLV

sons. Thus, these viruses are virtually assured of long-term survival because of multiple opportunity for transmission to susceptible persons.

Measles causes an acute and usually self-limited illness. The virus can be recovered during the prodromal period and 1 or 2 days after onset of the rash. The discovery in 1967 by Connolly and his co-workers that measles virus is the causal agent of subacute sclerosing panencephalitis (SSPE) has forced a reconsideration of the concepts regarding viral elimination. The measles virus can persist within the body in a latent form for more than 20 years before the onset of SSPE symptoms. We do not know whether the measles virus persists in all persons after the primary infection or only in those who develop SSPE.

In SSPE, the affected persons are not infectious for other persons by the usual modes of transmission. Therefore, the emergence of the latent virus does not result in secondary cases of measles or SSPE in the contacts of persons who have the latter disease.

The mechanisms of measles virus persistence and subsequent re-emergence several years later are not understood. Epidemiologic studies have yielded some clues that might provide answers. Children with SSPE had measles infections at younger ages than matched controls. The early infections might allow for viral persistence in immature immune systems. Several observations suggest the possibility of additional factors, perhaps a second infectious agent, in the pathogenesis of SSPE. Case-control studies have shown that children with SSPE were significantly more likely to have had close exposure to birds or swine than their matched controls. Unexplained geographic clustering has been noted in several countries (Fig. 4.7), and the risk for children living on farms was six times higher than that for children living in urban, inner-city areas (Fig. 4.8). Further investigations may well unravel the pathogenesis of this intriguing disease before it is eliminated by the widespread use of measles vaccine.

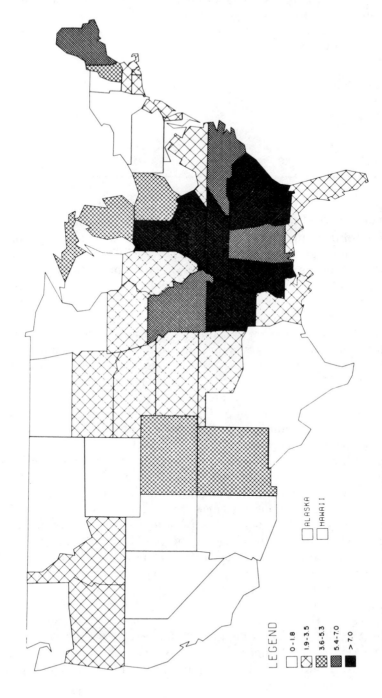

FIG. 4.7. Average annual SSPE incidence (cases per 10 million population <20 years of age), by state, United States 1960–1976. (Source: Modlin JF, Halsey NA, Eddins DL et al. *J Pediatr* 94:231–236, 1979.)

LEGEND

0 - 1.8
1.9 - 3.5
3.6 - 5.3
5.4 - 7.0
> 7.0

ALASKA

HAWAII

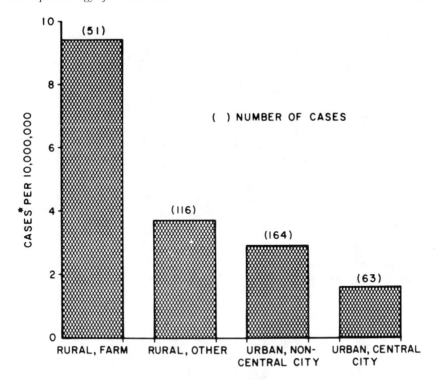

FIG. 4.8. Average annual incidence of subacute sclerosing panencephalitis, by residence at the onset of disease, United States, 1960–1976. (Source: Halsey NA, Modlin JF, Jabbour JT. Subacute sclerosing panencephalitis (SSPE): An epidemiologic review. in JG Stevens, GJ Todaro, CF Fox [eds], *Persistent Viruses,* Academic Press. New York, 1980 pp. 101–114.)

Recommended Readings

Albert RK, Condie F: Hand washing patterns in medical intensive care units. *N Engl J Med* 304:1465–1466, 1981.

American Academy of Pediatrics: *Report of the Committee on Infectious Diseases,* 19th ed. Evanston, IL, American Academy of Pediatrics, 1982.

Beasley RP, Lin C, Hwang L, Chien C: Hepatocellular carcinoma and hepatitis B virus: A prospective study of 22,707 men in Taiwan. *Lancet* 2:1129–1132, 1981.

Benenson AS: *Control of Communicable Diseases in Man,* 13th ed. Washington, DC, American Public Health Association, 1980.

Breman JG, Arita I: The confirmation and maintenance of smallpox eradication. *N Engl J Med* 303:1263–1273, 1980.

Centers for Disease Control: Rubella Surveillance, Jan 1976–Dec 1978. Atlanta, Center for Disease Control, 1980.

Chang T: Recurrent viral infection (reinfection). *N Engl J Med* 284:765–772, 1971.

Child Care Infectious Disease Study Group: Considerations of infectious diseases in day care centers. *Pediatr Infect Dis* 4:124–136, 1985.

Davies PA: Please wash your hands. *Arch Dis Child* 57:647–648, 1982.

De-The G: The epidemiology of Burkitt's lymphoma: Evidence for a causal association with Epstein-Barr virus. *Epidemiol Rev* 1:32–54, 1979.

Evans AS (ed): *Viral Infections of Humans: Epidemiology and Control,* 2nd ed. New York, Plenum Medical Book Company, 1982.

Feinstein AR: *Clinical Epidemiology: The Architecture of Clinical Research.* Philadelphia, Saunders, 1985.

Glezen WP, Denny FW: Epidemiology of acute lower respiratory disease in children. *N Engl J Med* 288:498–504, 1973.

Goscienski PJ: Zoonoses. *Pediatric Infect Dis* 2:69–81, 1983.

Koster FT, Curlin GC, Aziz KMA, Haque A: Synergistic impact of measles and diarrhea on nutrition and mortality in Bangladesh. *Bull WHO* 59:901–908, 1982.

Krugman S, Katz S: *Infectious Diseases of Children.* Philadelphia, Saunders, 1982.

Marks JS, Serdula MK, Halsey NA, et al: Saturday night fever: A common-source outbreak of rubella among adults in Hawaii. *Am J Epidemiol* 114:574–583, 1981.

Mausner JS, Bahn AK: *Epidemiology: An Introductory Text.* 2nd ed. Philadelphia, Saunders, 1984.

Meiklejohn G: The end of smallpox. *Adv Int Med* 23:385–404, 1978.

Modlin JF: Fatal echovirus 11 disease in premature neonates. *Pediatrics* 66:775–780, 1980.

Moore M: Enteroviral disease in the United States, 1970–1979. *J Infect Dis* 146:103–108, 1982.

Nathanson N: Slow viruses and chronic disease: The contribution of epidemiology. *Public Health Rep* 95:436–443, 1980.

Ramia S: Transmission of viral infections by water route: Implications for developing countries. *Rev Infect Dis* 7:180–188, 1985.

Salk D: Eradication of poliomyelitis in the United States: I. Live virus vaccine-associated and wild poliovirus disease. *Rev Infect Dis* 2:228–242, 1980.

Starko KM, Ray CG, Dominquez LB, Stromberg WL, Woodall DF: Reye's syndrome and salicylate use. *Pediatrics* 66:859–886, 1980.

Strangort K: Respiratory illness in preschool children with different forms of day care. *Pediatrics* 57:191–196, 1976.

Yorke JA, Nathanson N, Pianigiahi G, Martin J: Seasonality and the requirements for perpetuation and eradication of viruses in populations. *Am J Epidemiol* 109:103–123, 1979.

5

Respiratory Tract Infections

THOMAS R. CATE

Viral infections of the respiratory tract can be asymptomatic, cause mild or moderate discomfort, or cause severe, life-threatening diseases. Host factors can have an important influence on the clinical manifestations. For example, the small caliber of an infant's bronchioles makes them easily obstructed by the inflammatory response to respiratory syncytial virus (RSV) infection, producing the syndrome of bronchiolitis. Adults who have small-airways disease secondary to prolonged smoking also tend to have more frequent and severe lower-tract symptoms (cough, sputum, dyspnea) with respiratory viral infections than normal persons. Having had a previous infection is also an important factor. Secretory immunoglobulins induced by prior infection help prevent reinfection, but the secretory antibody titer decreases more rapidly than the serum antibody level; thus, reinfections occur with respiratory viruses such as parainfluenza or RSV, although residual, partial immunity can reduce disease severity. Impaired cell-mediated immunity has been associated with prolonged shedding of respiratory viruses and with severe virus pneumonia, most frequently with cytomegalovirus, but also with adenovirus.

Despite the importance of host factors in modifying clinical responses to respiratory viral infections, particular viruses or virus groups are typically associated with selected respiratory disease syndromes. The type of illness and epidemiologic considerations often permit narrowing the list of viruses to be considered in a particular case. In the following sections, the major respiratory viruses are reviewed according to their most distinctive syndromes and epidemiologic patterns.

TABLE 5.1   Illnesses Associated with Respiratory Virus Infections

Syndrome	Signs and Symptoms	Typical Viruses
Rhinitis (coryza, common cold)	Nasal obstruction and discharge, sneezing	Rhinoviruses, coronaviruses
Pharyngitis-tonsil-litis	Sore throat, pharyngeal erythema, ± tonsillar enlargement, ± exudate, cervical adenitis	Adenoviruses, Coxsackievirus A, Herpes simplex virus, Epstein-Barr virus
Croup	Coryza, hoarseness, inspiratory stridor, expiratory cough ("brassy"), and minimal fever in children 3 months to 3 years old	Parainfluenza viruses, influenza viruses, respiratory syncytial virus
Tracheobronchitis	Paroxysmal, nonproductive cough, substernal discomfort	Children: respiratory syncytial virus; parainfluenza viruses Children and adults: influenza viruses
Bronchiolitis	Coryza, tachypnea, wheezing, weak cough, and hyperaeration on chest x-ray in an infant less than 1 year old	Respiratory syncytial virus, parainfluenza virus type 3, adenoviruses
Pneumonia	Cough, fever, headache, malaise lethargy, dyspnea, and pulmonary infiltrate on chest x-ray	Children: respiratory syncytial virus, parainfluenza viruses Children and adults: influenza viruses, adenoviruses

COMMON COLD

The Syndromes

The common cold (coryza, nasal catarrh), the most frequent respiratory viral illness, is due to infection of the nasal epithelium. That infection interferes with ciliary activity, causes sloughing of epithelial cells, and induces mucosal inflammation with resultant hyperemia, edema, and increased vascular permeability. These events cause the characteristic symptoms (Table 5.1). The illness lasts from a few days to a week, though residual cough may persist for an additional week or two.

Mild throat irritation often occurs during a cold, but a prominent sore throat, pharyngeal exudate, or cervical lymphadenitis suggest other etiologies. Certain viruses can cause significant pharyngitis, including adenoviruses, herpes simplex virus (associated gingivostomatitis with ulcers), Epstein-Barr virus (pharyngitis is often a prominent component of infectious mononucleosis), and selected group A Coxsackie viruses (herpangina with pharyngeal vesicles that ulcerate). However, differentiation of the pharyngitis caused by viruses from that caused by group A beta-hemolytic streptococci is both important and difficult, often requiring a throat swab for bacteriologic examination. *Chlamydiae* and mycoplasma also account for a significant proportion of cases of sore throat, which, along with those of streptococcal origin, can be effectively treated with antibiotics.

Agents responsible for many common colds remain unidentified. Almost all viruses that can infect the respiratory tract can occasionally cause a cold, but the two virus groups responsible for the largest number are rhinoviruses and coronaviruses.

Rhinoviruses

CLASSIFICATION AND BIOLOGIC PROPERTIES

Rhinoviruses are picornaviruses and share many properties with the other genus of this family, the enteroviruses. The virions are nonenveloped (ether-resistant) icosahedrons, 20 to 30 nm in diameter, and contain a single-stranded RNA genome. Rhinoviruses differ from enteroviruses by exhibiting acid lability (unstable at pH 3) and greater density in cesium chloride gradient centrifugation analysis.

Rhinoviruses are relatively stable at room temperature and can survive for hours on various surfaces. Human rhinoviruses infect only humans and higher primates, and they can be grown only in cell or organ (embryonic nasal or tracheal ciliated epithelium) cultures from these species. They grow in the cell cytoplasm, some only in cells of human origin (H strains), some also in monkey cells (M strains), and some only in organ cultures. In the human host, rhinoviruses preferentially infect nasal ciliated columnar cells, and optimal conditions for their growth in vitro are a temperature of 33 or 34°C, pH 6.8 to 7.3, and incubation of cell culture tubes on a roller drum to increase oxygenation. These conditions parallel the microenvironment of the nasal passages. Rhinoviruses are almost never recovered from feces or other nonrespiratory sources. A semicontinuous human embryonic diploid fibroblast cell strain (WI-38) is commonly used for primary isolation of rhinoviruses, virus growth being recognized by granular cytopathic changes followed by cell death.

ANTIGENIC PROPERTIES

More than 100 rhinovirus serotypes have been identified. They have no common antigen. Hemagglutination by certain serotypes under special conditions has been described, but the test has not been used extensively for measuring antibody. Rhinovirus serotyping and testing for antibody responses is done using neutralization tests. Cross-reactions exist between some serotypes, and evidence for antigenic drift (i.e., change in antigenic properties) has been detected. Antibody responses to infection can usually be found in serum or nasal secretion by means of neutralization tests made sensitive by the use of small doses of virus, though reinfections likely occur as secretory antibody titers fall with time.

EPIDEMIOLOGY

Rhinovirus infections occur worldwide. Several serotypes may circulate in a given population and be gradually replaced by others with time. Persons average about one rhinovirus infection per year, but rates are highest in young children and decrease with age. Although infections occur year-round, they peak during the fall and spring in temperate climates.

Infection spreads well during close or prolonged contact, as in daycare centers and within the family, but less well among adults at work. Explosive outbreaks of rhinovirus infection have occurred among military recruits, but the infections more commonly spread gradually with 2- to 5-day intervals between onsets. Rhinoviruses can survive for hours on hands and environmental surfaces, and volunteers have readily infected themselves by touching contaminated fingers to their noses or eyes; this mode of transmission seems the most likely. Only rarely have rhinovirus colds been transmitted experimentally by means of aerosols generated by sneezing, coughing, or singing. Virus titers in nasal secretions peak during the period of maximal illness severity, and transmission appears to occur most frequently at that time. The incubation period is 1 to 4 days.

DISEASE

The common cold is the most frequent manifestation of rhinovirus infection. Symptomatic lower respiratory tract diseases such as tracheobronchitis and pneumonia occur only occasionally, mostly in young children. Adults who smoke or have chronic obstructive lung disease are also more likely to develop lower respiratory tract symptoms with rhinovirus infection than persons without these conditions.

Coronaviruses

CLASSIFICATION AND BIOLOGIC PROPERTIES

Coronaviruses are a family of roughly spherical, enveloped, 80- to 160-nm, single-stranded RNA viruses. Their name derives from the distinctive, club-shaped projections from the virion that give it a crownlike appearance. Replication occurs in the cytoplasm of cells, with budding into cytoplasmic vacuoles. Members of the family infect humans and many lower animals, but individual strains show species specificity (e.g., human strains infect only humans or cultured human cells).

Human coronaviruses are difficult to cultivate. Many have been detected only by inhibition of ciliary activity in organ cultures of tracheal or nasal epithelium. Some will grow and produce a late, nondescript cytopathic effect in diploid fibroblasts or in a heteroploid embryonic lung cell line (L-132). The viruses grow best at 33°C. They are unstable in acid, ether, and heat.

ANTIGENIC PROPERTIES

At least three antigenic types of human coronaviruses have been identified in neutralization tests, and additional types are known to exist. Human strains seem unrelated to animal strains as determined by neutralization tests, but some cross-reactions exist in complement-fixation assays. Two human strains adapted to growth in suckling mouse brain agglutinate red blood cells; other strains do not. Antibody responses, measured by complement-fixation, hemagglutination-inhibition, or neutralization, occur after infection.

EPIDEMIOLOGY

Knowledge of the epidemiology of coronaviruses is incomplete. Infections that have been identified occurred predominantly during the winter and spring. Familial contact facilitates transmission, and all age groups are affected.

DISEASES

Epidemiologic and volunteer studies indicate that human coronaviruses cause the common cold syndrome. Apparently, the viruses are not significant causes of lower respiratory tract disease.

LARYNGOTRACHEOBRONCHITIS, BRONCHIOLITIS, AND PNEUMONIA IN CHILDREN

The Syndromes

LARYNGOTRACHEOBRONCHITIS (CROUP)

Laryngitis is a frequent and inconvenient component of common colds in older children and adults. In young children (3 months to 3 years), it may presage the more severe syndrome known as laryngotracheobronchitis, or croup. The subglottic airway with its semirigid cartilagenous encasement is small at this age and is subject to partial obstruction when mucosal inflammation occurs. This subglottic narrowing produces the croup syndrome (Table 5.1). The clinical course of croup often fluctuates, sometimes from hour to hour, and usually lasts for 3 or 4 days. Respiratory failure necessitating tracheostomy is infrequent. Parainfluenza viruses are most frequently implicated in croup, though RSV, influenza virus, and occasionally other viruses can cause the syndrome.

Distinguishing croup from *Hemophilus influenzae* epiglottitis is particularly important. The latter disease can progress rapidly to complete airway obstruction and usually requires an endotracheal tube or tracheostomy in addition to antibiotic therapy. Typically, the child with *H.*

influenzae is more than 3 years old. The illness is of rather sudden onset
and is manifested by fever, toxic appearance, leaning forward in bed
with chin protruding, breathing tentatively and with inspiratory stri-
dor, and drooling because of sore throat and difficulty in swallowing.
Hoarseness is usually present, but not a barking cough or a history of
coryza. Examination will reveal the swollen, erythematous epiglottis, but
such examination should only be done with provisions at hand for rapid
insertion of an artificial airway, since it may precipitate complete airway
obstruction. X-rays of the neck show the swollen epiglottis with normal
subglottic airway in persons with *H. influenzae* epiglottitis as opposed to
a narrowed subglottic airway with normal epiglottis in patients with
croup.

BRONCHIOLITIS

Bronchiolitis, like croup, is an age-specific syndrome; it usually occurs in infants 1 to 6 months of age and is uncommon in children more
than 1 year old. The bronchiolitis syndrome (Table 5.1) occurs because
of partial obstruction of small bronchioles by mucosal inflammation,
sloughed epithelial cells, and mucus. About 1 or 2% of infants with this
illness die of it, especially those having underlying cardiopulmonary
disorders. Most infants recover from the acute phase of bronchiolitis
within a week, though complete recovery may require a few additional
weeks. Some children have recurrent episodes of wheezing after having
had bronchiolitis (allergic predisposition?), and some authorities believe
small-airways disease may persist indefinitely.

PNEUMONIA

In infants with viral pneumonia (Table 5.1), symptoms of tracheobronchitis and bronchiolitis are often manifested in various degrees of
severity, whereas older children will likely have symptoms of tracheobronchitis in addition to fever, lethargy, and tachypnea. Bacterial su-
perinfection is uncommon in viral pneumonia of children.

Although influenza viruses and adenoviruses participate in causing
laryngotracheobronchitis, bronchiolitis, and pneumonia in children, they
are the primary causes of lower respiratory tract virus disease in adults,
and will be described later in association with adult diseases. Because
parainfluenza viruses and respiratory syncytial virus are the major
pathogens in children and do not commonly cause severe disease in
adults, they are described next.

Parainfluenza Viruses

CLASSIFICATION AND BIOLOGIC PROPERTIES

Parainfluenza viruses are members of the Paramyxoviridae family and
share the Paramyxovirus genus with mumps virus and other viruses of

animals. Other Paramyxoviridae genera are Morbillivirus, containing measles virus, and Pneumovirus, containing RSV.

All Paramyxoviridae are roughly spherical, enveloped (ether-labile) viruses that are typically 125 to 250 nm in size (range 100–800 nm) and have short spikes projecting from the outer lipid envelope. All are single-stranded RNA viruses that multiply and produce inclusions in the cell cytoplasm, bud from the cell surface, and cause fusion of cells. Paramyxoviridae are unstable at 4°C in medium without protein; even with protein-containing carrier medium, recovery of virus from respiratory secretions may improve if tissue culture tubes are inoculated at the patient's bedside without delay. In cell cultures, Paramyxoviridae can establish a persistent infection that cannot be eradicated by addition of antibody. The virus multiplies in the cytoplasm and is released at the cell surface, but it does not prevent cell division; abundant virus components in the cytoplasm randomly segregate into daughter cells at the time of division, leading to persistent infection.

Parainfluenza viruses have two types of surface spikes: one (HN) exhibits hemagglutinin and neuraminidase activities, whereas the other (F) exhibits fusion and hemolysin activities in vitro. Primary human embryonic kidney and primary rhesus monkey kidney cell cultures have been used for recovery of parainfluenza viruses, but a continuous cell line, the LLC-MK2 line of rhesus monkey kidney cells, has more recently been found to serve as well for this purpose with addition of trypsin (2 μg/ml) to the medium. Trypsin helps to cleave the HN and F glycoproteins, a necessary step for virus infectivity. Use of LLC cells has the advantages of ready availability and avoidance of confusion with simian parainfluenza virus SV5, which can be latent in primary rhesus monkey kidney cells. Virus-induced cytopathic effect and cell fusion may occur slowly, if at all, but growth of virus can be recognized by hemadsorption of guinea pig red blood cells to the surface of infected cells, or by immunofluorescence. Parainfluenza viruses can be adapted to grow in a variety of cell lines and in embryonated eggs, but such techniques are not useful for primary isolation. Related parainfluenza viruses infect a variety of animal species.

ANTIGENIC PROPERTIES

The HN and F surface antigens and the internal nucleoprotein (NP) antigen all differ among the parainfluenza viruses. Though cross-reactions exist, there is no common antigen; the viruses are antigenically stable. Four different types of parainfluenza virus infect humans. For economy and convenience, tests usually used for measuring antibody to parainfluenza viruses are hemagglutination inhibition and complement fixation, although neutralization and other tests are available. Type-specific antibody responses are demonstrable after a person's initial infection with a parainfluenza virus, but heterotypic responses occur after subsequent infections with different types. Results in adult-volunteer studies, and occurrence of infections in infants despite the presence of maternally acquired serum antibody, suggest that secretory antibody is important in preventing infection. Nonetheless, prolonged persistence of virus shedding has been noted despite the presence of both serum and secretory antibody.

EPIDEMIOLOGY

Parainfluenza viruses are distributed worldwide. Types 1 and 2 characteristically produce epidemics in the fall of every other year in temperate climates, sometimes in alternating years and sometimes both in the same year. Parinfluenza virus type 3 causes endemic disease the year-round, sometimes with increased frequency in the spring. About 50% of infants will have been infected with type 3 by 1 year of age and the remaining 50% during the next 5 years. About 80% of children will have been infected with types 1 and 2 by age 10 years. Type 4, which has been associated with upper respiratory tract illness, is difficult to recover, and little is known about its epidemiology. The viruses spread by close contact, but whether or not small-particle aerosol plays a role is unknown. In volunteer studies, the incubation period has been 3 to 8 days.

DISEASE

Many parainfluenza virus infections are asymptomatic. The most common disease caused by these viruses is coryza, particularly during reinfections, but also often during primary infections. Sometimes the coryza will be associated with high fever, necessitating a workup for sepsis. Parainfluenza viruses are also responsible for one third to one half of the cases of croup, and type 3 is important as a cause of bronchiolitis and pneumonia of children.

Respiratory Syncytial Virus (RSV)

CLASSIFICATION AND BIOLOGIC PROPERTIES

RSV is classified in the Pneumovirus genus of the Paramyxoviridae and shares the general properties of that family. One important distinction from parainfluenza viruses is that, despite the presence of surface spikes, RSV has no detectable hemagglutinating, hemadsorbing, hemolytic, or neuraminidase activity. However, cell fusion caused by RSV produces the characteristic syncytia and giant cells. RSV is also distinct from parainfluenza viruses in having slightly smaller, more spherical virions and occasional long, filamentous forms. RSV causes illness only in humans and chimpanzees, although inapparent infection can be produced in other mammals. The virus grows and causes syncytia in a variety of human cells and in monkey and bovine kidney cells. For primary isolation, a human heteroploid cell line such as HEp-2 is generally used, but sensitivity of the cells for RSV can vary and must be continually monitored. RSV is very labile and a prime example of a virus whose recovery one can increase by inoculating tissue culture tubes without delay.

ANTIGENIC PROPERTIES

Antigenic variants of RSV can be detected in neutralization assays, but cross-reactions are too great for division into distinct serotypes. The internal NP antigen detectable by complement fixation is common to all RSV viruses. RSV antigens are distinct from those of other Paramyxoviridae.

The complement-fixation text is useful for detecting serum antibody responses to infection. Neutralizing antibody responses also occur in serum and in respiratory secretions. Reinfections are common, presumably as secretory IgA antibody levels fall, but disease due to reinfection is milder than that due to primary infection.

EPIDEMIOLOGY

RSV, distributed worldwide, causes yearly outbreaks of infection, usually during the winter months in temperate climates. Although influenza virus and RSV epidemics tend to occur at similar times of the year, one will usually displace the other so that the two are not concurrent. Infection with RSV occurs in almost 100% of children during the first 5 years of life, about one third of the infections occurring in the first year. Because nosocomial infections with RSV are well documented, considerable effort has been given to defining the means of transmission. The major mode of transmission seems to require close contact, suggesting that it may be by large droplets or by self-inoculation of eyes or nose after touching contaminated surfaces or skin. The period of incubation to the onset of disease is from 2 to 8 days.

DISEASE

RSV is rarely recovered from asymptomatic children. The most distinctive diseases associated with primary infection are bronchiolitis, pneumonia, tracheobronchitis, and croup, in that order; coryza accompanied by high fever is also common. Reinfection with RSV occurs frequently and, though the associated disease tends to be milder, lower respiratory tract manifestations are often present in young children and adults who have chronic obstructive lung disease. Similarly, maternally acquired serum antibody does not prevent RSV infection of infants, though recent data suggest that it may reduce the severity of disease.

Some investigators suggest that immunopathologic processes might contribute in a major way to RSV disease. One reason for this suggestion is the observation that young children who received an experimental, alum-precipitated, killed-virus vaccine developed more severe disease during subsequent natural infection than those who were unvaccinated; the phenomenon did not occur in older children, perhaps because they had had natural RSV infection before vaccination. Results of studies of cell-mediated immunologic responses to live and killed virus in mouse-influenza models suggest a possible explanation. Cytotoxic T cells from mice exposed to infectious virus were capable of conferring protection against challenge in adoptively immunized syngeneic hosts; in contrast, T cells from mice vaccinated with killed virus mediated delayed hypersensitivity but not cytotoxicity, and they increased disease severity in adoptive hosts after challenge. However, even if a similar phenomenon occurred in young children whose first exposure to RSV was in the form of inactivated virus vaccine, it might have little to do with the pathogenesis of naturally occurring RSV disease in the absence of vaccine. Two additional immunopathologic processes have been suggested: an Arthus-type reaction between virus and maternally acquired serum antibody and IgE-mediated bronchospasm on reinfection. Evidence, however, has also been adduced against the latter processes.

INFLUENZA SYNDROME AND ATYPICAL PNEUMONIA IN ADULTS

The Syndromes

INFLUENZA

Influenza virus infection may cause many different signs and symptoms, but characteristic influenza is tracheobronchitis accompanied by malaise, myalgias, headache, and fever. The respiratory disease is due to replication of virus in columnar epithelial cells of the respiratory tract, producing cell death and inflammation. The systemic symptoms may be due to products released by mononuclear cells during the inflammatory response (e.g., endogenous pyrogen); influenza viruses are rarely detectable in blood. The acute illness lasts from 3 to 5 days, but cough and lassitude usually persist for 1 or more additional weeks.

The distinguishing feature of the disease, influenza, rests not so much on the individual case but on the occurrence of epidemics involving all age groups and causing recognizable increases in mortality. Severe or complicated influenza virus disease leading to death occurs most frequently in very young and elderly persons, and those of any age weakened by underlying disease, particularly of the heart or lungs.

Severe influenza virus diseases in the very young include pneumonia, bronchiolitis, and croup, and pulmonary complications are also largely responsible for mortality in other high-risk persons. Influenza viruses, themselves, can cause severe pneumonia in adults, although, fortunately, this happens rarely. Secondary pneumonia due to a pneumo-coccus, *Staphlyococcus aureus, Hemophilus influenzae,* or other bacteria is relatively common during influenza virus infection of the lower respi-ratory tract in adults.

ATYPICAL PNEUMONIA

The "atypical pneumonia" syndrome (that is, atypical of bacterial pneumonia) is rarely fatal and is frequently a "walking pneumonia" not necessitating hospitalization. This illness is most common from late childhood to middle age. The patient often presents with symptoms of tracheobronchitis (Table 5.1) and a patchy pulmonary infiltrate with normal white blood cell count or slight neutrophilia. The illness can last for a few weeks with fluctuating symptoms; complete recovery may re-quire from 1 to 3 months.

Atypical or interstitial pneumonia may be caused by *Mycoplasma pneu-moniae* or an adenovirus, and less frequently by influenza, other respi-ratory viruses, or other agents. Clinical differentiation between the causative agents is not possible with any certainty. Pharyngitis and cer-vical adenitis can be prominent with both *Mycoplasma pneumoniae* and adenovirus infections. Myringitis, especially bullous myringitis, is suggestive of *Mycoplasma pneumoniae,* and a variety of skin rashes in-cluding erythema multiforme major (Stevens-Johnson syndrome) have been associated with this organism. Granular bulbar and palpebral con-junctivitis is strongly suggestive of adenovirus infection. Cold aggluti-nins, serum IgM antibodies that agglutinate human group O red blood cells in the cold, are elevated to titers of 1:32 or greater and increase in titer fourfold or more in about 75% of patients with *Mycoplasma pneumoniae* atypical pneumonia but also in about 20% of patients with adenovirus atypical pneumonia. Influenza viruses and adenoviruses, which cause the most serious respiratory virus diseases of adults, are described next.

Influenza Viruses

CLASSIFICATION AND BIOLOGIC PROPERTIES

The Orthomyxoviridae family contains only the influenza viruses. There are three types (genera), A, B, and C, and three major subtypes (species) of human type A viruses. Other influenza A virus subtypes cause natural infections in nonhuman hosts, but not types B or C. Numerous antigenic variants of type A and type B viruses exist and continue to appear, but antigenic differences between type C strains are only mi-nor. Nomenclature of individual strains is covered under antigenic properties in the next subsection.

Influenza virions are somewhat heterogeneous, having roughly spherical forms 80 to 120 nm in diameter and long filamentous forms of similar diameter. They are enveloped (ether-senstive) with spikelike glycoprotein projections from their surfaces. The genome consists of single-stranded RNA in eight segments for types A and B, and seven segments for type C. Different steps in viral replication occur in the host-cell nucleus and in the cytoplasm, and an early replicative step utilizes functions of host-cell DNA. Viral components are assembled at the host-cell cytoplasmic membrane, and virions bud from this surface. When a cell is infected simultaneously with two influenza viruses of the same type, gene reassortment occurs, yielding daughter viruses with variable numbers of gene segments from each parent. By using parent viruses with selected phenotypic properties, and by defining the origin of gene segments (e.g., by RNA gel electrophoresis) and the phenotype of daughter viruses, investigators have determined the virus proteins coded for by each gene segment.

The surface spikes of influenza A and B viruses are of two different types, namely, hemagglutinin (H) and neuraminidase (N). The viral hemagglutinin attaches to specific glycoprotein cell receptors containing terminal N-acetyl neuraminic acid (NANA) residues; this interaction underlies viral agglutination of erythrocytes (hemagglutination), as well as attachment of virus to susceptible cells, as the first step in viral replication. Antibody response to the hemagglutinin can both inhibit hemagglutination and neutralize virus infectivity. The neuraminidase removes NANA from erythrocyte receptors after viral attachment, thus releasing virus and rending the cells no longer agglutinable. The functions of neuraminidase during infection seem to be the release of virus from the infected cell and the prevention of aggregation of progeny virions. Glycoprotein in respiratory secretions can bind to the hemagglutinin and thus block viral attachment to susceptible cells. Removal of NANA from these sialoproteins may be another important activity of the viral neuraminidase in vivo. Antibody response to neuraminidase does not neutralize virus infectivity but does reduce virus plaque size in vitro and the severity of disease in experimental animals. Influenza C virus surface spikes are of only one type and mediate both hemagglutin and receptor-destroying activity; the cell receptor for influenza C virus differs from that for types A and B.

Influenza viruses are relatively unstable at temperatures greater than 37°C. They survive for a few hours to days in protein-containing (transport) medium at 4°C and for longer periods at −70°C. The virus has traditionally been grown in embryonated chicken eggs or in primary rhesus monkey kidney cell cultures. The continuous Madin Darby canine kidney (MDCK) cell line, with addition of trypsin to the culture medium, also serves well for growth and primary isolation of influenza viruses. Trypsin cleaves the H glycoprotein, a portion of which must be exposed to mediate viral entry into the cell. MDCK cells have the advantages of ready availability and absence of latent hemadsorbing viruses, such as SV5. Influenza viruses do not cause development of inclusions or any characteristic cytopathic effect, although they can cause cell degeneration and death. Virus growth is usually detected by means of hemadsorption of red blood cells to infected cells or by the hemagglutinating activity of fluid harvested from eggs.

ANTIGENIC PROPERTIES

Influenza viruses have a type-specific internal nucleoprotein (NP) antigen that is relatively stable antigenically. The corresponding antibody can be used to identify newly isolated viruses as belonging to type A, B, or C, usually by means of the complement-fixation (CF) test. Antibody response to the NP antigen occurs after infection and can also be used for retrospective serodiagnosis. Antibody response to NP antigen does not persist after infection for longer than a few months, at most, and does not confer protection against reinfection. Another internal antigen, the matrix (M) protein, is also type-specific. Immunologic assays, using M protein, or antibodies to it, provide information comparable to that with NP antigen.

A unique characteristic of influenza viruses is the antigenic variation exhibited by the H and N surface antigens. Two different degrees of antigenic variation may occur. Minor changes, termed antigenic drift, are seen every year or so in the H or N antigen, or both. Antigenic drift probably is due to point mutations in viruses as they circulate in a partially immune population and occurs within all three types of influenza virus, although to the least extent within type C. Major changes, or antigenic shifts, occur at longer, irregular intervals in either H antigen alone, or in both H and N antigens. A shift is defined by the appearance of a strain whose surface antigens show almost no immunologic cross-reactivity with those of earlier strains and has occurred only among type A viruses. A current theory to explain antigen shift is that genetic reassortment occurs when an animal and a human strain both infect a single host. A new type A virus emerges, which contains genes from the animal strain that code for one or both surface antigens, and genes from the human strain that govern infectivity and transmissibility in humans. Populations with little or no previous exposure to the animal virus antigens are fully susceptible.

Antibody responses to influenza viruses are commonly measured by a complement-fixation, hemagglutination-inhibition, or neutralization test. Complement-fixation can be used to measure antibody responses against the group-specific NP antigen or against an individual strain if isolated H subunits are used as the antigen. The hemagglutination-inhibition and neutralization assays must use the same virus variant that caused the infection, or a closely related one, to detect antibody responses; anti-H antibody is measured by the latter tests, although anti-N antibody can influence results unless a reassortant test virus with an unrelated N antigen is used. Sera for hemagglutination-inhibition assays must be pretreated to remove nonspecific sialoprotein and protein inhibitors; receptor-destroying enzyme (a neuraminidase from *Vi-*

brio cholerae) and heat (56°C for 30 minutes) are commonly used for this purpose. Anti-N antibody can be measured by enzyme-inhibition assays. Radioimmunoassays (RIA) and enzyme-linked immunosorbent assays (ELISA) have also been used to measure antibody, especially in respiratory secretions.

The single best correlate of immunity to influenza-virus illness is the serum antibody titer against the virus measured in hemagglutination-inhibition or neutralization tests, although data from volunteer studies suggest that secretory antibody can contribute to resistance. Cell-mediated immunologic responses with both type and subtype specificity occur after influenza-virus infection. However, immunity to reinfection with influenza viruses is rather specific for closely related viruses of the same subtype. Cell-mediated immunity therefore probably does not play a significant role in resistance to reinfection, though it may well decrease the severity of disease by initiating recovery mechanisms more rapidly, that is, anamnestically.

EPIDEMIOLOGY

An important characteristic of influenza is its occurrence in epidemics, which are now explicable by the molecular biology of the virus. Only group A viruses have caused worldwide pandemics of influenza, because of successive antigenic shifts that each time allow the emerging virus strain to evade the herd immunity of the population. The result has been extremely high attack rates, with high morbidity and mortality. In the 50 years since 1933, when human influenza virus was first isolated, two antigenic shifts have occurred. In strains isolated during the pandemic of 1957, both the H and N antigens showed major antigenic divergence from those of earlier strains; and in strains isolated during the pandemic of 1968, H antigen had changed markedly, but the N antigen remained as before. Viruses recovered in the 1933–1957 period have been designated H_1N_1 subtypes, including viruses formerly designated H_0N_1. The change that occurred in 1947 from H_0N_1 to H_1N_1 has been downgraded to a "drift" because of the relatively close antigenic relationship of H_0 to H_1. Viruses recovered in the 1957–1968 period are designated H_2N_2, and those since 1968 as H_3N_2. Reappearance of H_1N_1 viruses in 1977 (Russian influenza) led to a previously unknown or unrecognized situation in which two influenza group A subtypes (H_3N_2 and H_1N_1) cocirculated in the population for long periods.

Serum antibody titers, regardless of the age of the subject, tend to be highest against the influenza A virus subtype that caused the first childhood infection, a phenomenon called "original antigenic sin." Subsequent infections with different influenza A viruses broaden the antibody response, but also boost titers against the first virus. From studies correlating chronological age with antibody titers to influenza A subtypes, it has been deduced that the strain causing the 1918–1920 world pandemic of influenza was similar to the first influenza virus ever recovered in the laboratory, namely, a strain isolated from swine in 1931. This virus is antigenically related to human H_1N_1 strains, with which it is now grouped.

Less severe and more localized influenza epidemics have occurred because of antigenic drift in prevalent strains of viruses to which a substantial portion of the population already had partial immunity. Attack rates during such epidemics have often been 20 to 30%. Epidemics tend to occur at from 1- to 3-year intervals with type A viruses, and 3- to 6-year intervals with type B viruses. These epidemics generally occur during winter months and last 5 or 6 weeks. Even when epidemics do not occur, some influenza A or B virus infections or both are evident each winter in large populations, sometimes with focal outbreaks. Influenza C viruses cause only limited or focal outbreaks of relatively mild respiratory disease.

Influenza A virus epidemics tend to have a fairly typical pattern within a community. The first sign will be increased occurrence of febrile respiratory disease in school-aged children, as reflected in school absenteeism. These children apparently bring the virus home and infect preschool children and adults. Pediatric hospital admissions for pneumonia also rise early and remain elevated for the first half of the epidemic (2 or 3 weeks). Adult disease manifested by increases in industrial absenteeism for respiratory illness, pneumonia admissions to adult hospital services, and deaths from pneumonia and influenza predominate during the second half of the epidemic. Highest overall attack rates for infection will be in children, but most complications and deaths occur in the elderly and other high-risk groups. Type B influenza virus epidemics differ because their predominant effects are seen in children.

Influenza viruses can undoubtedly be transmitted by close contact, but documented outbreaks exist in which transmission was almost certainly by means of small-particle aerosols. The explosive nature of influenza virus epidemics suggests that aerosol is a major means of transmission. The incubation period is usually brief, from 1 to 4 days.

DISEASE

Respiratory disease syndromes caused by influenza viruses range from the common cold to overwhelming pneumonia, and they occur in epidemics that affect individuals in any age group. Although influenza viruses cause severe diseases and death in young infants and children, excess morbidity and mortality are often not apparent because other viruses of similar virulence for the very young are displaced. In elderly persons, and in those with chronic debilitating disease, mortality due to influenza and its complications in excess of expected levels for a given season serves as one of the standard indicators of the extent of epidemic influenza. Secondary bacterial pneumonia is responsible for most mortality during periods of epidemic influenza. Even when findings of pneumonia are absent, evidence of disease of the small airways is commonly encountered during the course of influenza and may be critical in patients who already suffer from chronic cardiopulmonary disease. Rare complications of influenza-virus infection include myositis, myocarditis, and encephalopathy.

Reye's syndrome is a relatively unusual complication in children (1 case per 100,000 population per year in persons less than 18 years of age) a few days after the onset of influenza, chickenpox, or nonspecific upper respiratory or gastrointestinal illnesses. The syndrome consists of nausea, vomiting, and lethargy progressing to coma, and it is fatal in up to 40% of patients. Autopsy findings include hepatomegaly with fatty degeneration of hepatocytes, cerebral edema, and little inflammation. Although the pathogenesis is unclear, recent evidence suggests an association of Reye's syndrome with use of aspirin for treatment of the preceding acute febrile illness.

Adenoviruses

CLASSIFICATION AND BIOLOGIC PROPERTIES

Adenoviruses of humans share the Mastadenovirus genus of the Adenoviridae family with other adenovirus species infecting lower mammals. A common group-specific antigenic determinant is shared by members of the Mastadenovirus genus, but this antigen is not present in avian adenoviruses, which are placed in a separate genus. Adenoviruses are relatively species-specific. However, human adenoviruses can cause asymptomatic infections in laboratory animals, and large amounts of certain adenovirus serotypes can produce tumors when injected into rodents. No evidence for oncogenicity of advenoviruses has been found in humans.

Adenoviruses are nonenveloped (ether-resistant) icosahedrons, 60 to 90 nm in diameter, that contain a double-stranded DNA genome. The capsid of each virion is composed of 240 hexons with both type-specific and group-specific antigens, and 12 pentons at the vertices of each angle with group-specific antigen and toxic activity for cells (cell-detaching factor); extending from each penton are fibers that have type-specific antigen and hemagglutinin and are capable of blocking both host- and virus-directed synthesis of DNA, RNA, and protein. Human adenoviruses may be divided into three groups based on their agglutination reactions with rhesus and rat red blood cells, these groups corresponding with other properties of the viruses. The virions are stable in protein-containing medium at 4°C but become unstable when purified.

Adenoviruses replicate and are assembled in the host-cell nucleus, where they produce large basophilic inclusions (in contrast to the acidophilic nuclear inclusions of herpesviruses). Virus is released at cell lysis, which is delayed because the host cells remain metabolically active even though they cannot replicate DNA or RNA. Latent infections of lymphoid tissue frequently occur in vivo (the name adenovirus derives from the fortuitous recovery of these agents in explanted adenoid tissue, which was being studied as a potential source of cells for propagation of viruses). The mechanism of this latency is uncertain, but it involves release of only small amounts of infectious virus and limitation of infection to only a few cells.

A variety of human cells in culture support the growth of human adenoviruses. Primary embryonic kidney and several continuous cell lines, including HEp-2, are satisfactory. Replication of adenovirus produces a characteristic cytopathic effect (grapelike clusters of rounded, refractile cells), which develops slowly over several days to a week. In persons with adenoviral respiratory illness, virus is present in respiratory secretions and can be recovered from throat or conjunctival swabs and from stool or rectal swab specimens. Virus may persist in the tissues and be shed for some time after clinical recovery. Cell cultures inoculated with clinical specimens should be maintained and observed for 3 to 4 weeks with passage to fresh cells as necessary ("blind" passage) before being pronounced negative.

ANTIGENIC PROPERTIES

Thirty-nine human adenovirus serotypes have been identified, many of them associated with specific clinical syndromes. More than half the serotypes, however, are not known to cause disease. Serum antibody responses develop after acute infection. Since adenoviruses can be recovered from healthy persons, detection of an antibody response is important for substantiating acute adenoviral infection. In complement-fixation tests, any adenoviral serotype may be used because all share the same group-specific antigen, but response to a new infection may be obscured by persistent antibody resulting from past adenoviral infections. Neutralization and hemagglutination inhibition tests are type-specific, and therefore require use of virus of the same serotype as that isolated from the patient, or a panel of commonly encountered serotypes that might include the one responsible for the infection. Nonspecific inhibitors must be removed from sera before they can be used in HI tests. Resistance to reinfection is strictly type-specific.

EPIDEMIOLOGY

Human adenoviruses infect only humans and are endemic, with a tendency to increase in activity toward late fall and winter. Most persons are infected with one or more serotypes by age 15, and 50% or more of surgically removed adenoids will yield an adenovirus when explants are cultured. The virus can spread by means of infectious secretions transmitted by close contact, including use of contaminated ocular instruments. Swimming pools can serve as a means for transmitting conjunctivitis (feces-to-eye). The role of virus that is ingested or deposited in the nasopharynx in causing disease other than mild coryza or pharyngitis is unclear. Nasal instillation of adenoviruses into volunteers produces asymptomatic infection or mild upper respiratory tract illnesses, and inoculation with enteric-coated capsules produces asymptomatic intestinal infection. Inoculation of the lower respiratory tract through small-particle aerosol can produce the full range of adenovirus respiratory disease in volunteers, and winter epidemics among military recruits are compatible with this mode of spread. The incubation period for illness is 1 or 2 weeks.

DISEASE

Most infections with adenoviruses are asymptomatic; only about 10% of respiratory illness seems to be caused by these viruses. Serotypes 1, 2, 5, and 6 predominate in young children and may cause coryza, pharyngitis, or occasional severe lower respiratory tract disease. Serotypes 3, 4, and 7 are associated with pharyngoconjunctival fever, influenza syndrome, and atypical (interstitial) pneumonia in older children and young adults. The whooping-cough syndrome has been associated with adenoviral infection in the absence of *Bordetella pertussis*. Epidemic keratoconjunctivitis caused by serotype 8 or 19 has occurred among persons who work in dusty environments that produce conjunctival irritation. Gastrointestinal symptoms are occasionally prominent in persons having adenoviral respiratory illnesses, and lymphoid hyperplasia caused by infection in the gastrointestinal tract can be a focus for precipitating intussusception. Hemorrhagic cystitis caused by serotypes 11 or 21 and overwhelming, disseminated adenovirus disease in immunocompromised patients have been described.

DIAGNOSIS

The time required for laboratory diagnosis of the specific cause of acute respiratory illness today almost invariably exceeds the duration of the illness. Techniques for rapid diagnosis, such as immunofluorescent staining of exfoliated cells, will not likely become generally available until antiviral drugs are available that require specific diagnoses for proper use. Reasons for establishing a diagnosis in the absence of specific antiviral drugs include awareness of the agents currently causing disease in the community, so that later patients with similar illness may be properly treated, and documenting the cause of unusual or severe disease.

Specimens for virus isolation should be collected as early as possible during acute respiratory illness when virus titers are usually maximal. Included may be nose, throat, and conjunctival swabs; nasal wash and gargle specimens; sputum; stool or rectal swab; and pleural fluid, if present. Consultation with laboratory is essential in determining which specimens to collect and which viruses are most likely to be involved.

Swab or liquid specimens should be placed in or diluted with protein-containing transport medium such as veal infusion broth with 0.5% bovine serum albumin. Antibiotics such as penicillin 100 U/ml, streptomycin 100 μg/ml, and amphotericin 2.5 μg/ml are added to the medium to prevent bacterial or fungal overgrowth. Streptomycin and amphotericin should be omitted if the presence of *Mycoplasma pneumoniae* is suspected. Specimens should be kept refrigerated and transported to the laboratory as soon as possible in wet ice.

A 10- to 20-ml clot tube of blood should be collected at the time of acute illness and sent to the laboratory for separation and storage as an acute phase serum. A convalescent blood sample should be collected 2 or 3 weeks later. Even if a virus is recovered, the possibility exists that it represents shedding from some prior infection unless an antibody response also develops in relationship to the acute illness. If no virus is recovered, the paired sera are the only means for establishing a diagnosis by detection of an antibody response to viruses that seem likely on clinical or epidemiologic grounds.

TREATMENT

With the exception of amantadine use, treatment of viral respiratory tract disease is nonspecific and aimed primarily at alleviating symptoms and reducing the probability of complications. Some of the malaise and lethargy may respond simply to rest. Such confinement may also help reduce transmission to persons outside the home, because the largest amount of virus in respiratory secretions (and, hence, the greatest contagiousness) is usually associated with maximal symptoms. Until the relationship of Reye's syndrome to aspirin ingestion is clarified, myalgias and headache are best treated with acetaminophen. Fever does not usually require therapy per se, except in infants and young children in whom febrile convulsions are a potential danger.

Measures to promote clearance of secretions are the mainstays for preventing secondary complications. Increased hydration and humidified air will help loosen secretions. Vigorous blowing of the nose should be avoided, because it may force infected secretions into eustachian tubes and sinuses. A nasal-suction bulb facilitates removal of secretions from infants and young children. Phenylephrine nose drops or sprays help relieve nasal obstruction, as will oral medications such as pseudoephedrine. Use of antihistamines may make secretions more viscid and difficult to clear, though they may be of help in some persons with excessive, thin secretions. The mild pharyngitis frequently associated with viral respiratory disease can be relieved by gargling with warm saline (½ teaspoon of salt to a glass of warm water). Whether tracheobronchial secretions are cleared better by expectorants or by hydration and humidified air remains a question. Cough suppressants should be reserved for treatment of excessive coughing that is painful or tiring, or that interferes with sleep.

Persons with croup, bronchiolitis, and viral pneumonia will usually require hospitalization for appropriate respiratory therapy. Antibiotics are generally contraindicated in the treatment of viral respiratory disease. Though it is necessary to be alert to secondary bacterial infections, use of antibiotics before the onset of those infections subjects the patients to unnecessary risks of side effects and infection with a resistant organism. Tetracycline or erythromycin, however, may be used for treating patients with the atypical pneumonia syndrome in circumstances that are not associated with epidemic adenovirus infections because of the predominance of *Mycoplasma pneumoniae* in causing such disease.

Amantadine (Symmetrel) is an oral medication effective for both treating and preventing illness due to type A influenza virus. It is not effective against type B or other respiratory viruses. During an influenza A virus epidemic, all high-risk persons who develop influenza should receive amantadine treatment. To be effective, amantadine should be begun within 48 hours of clinical onset in a dose of 100 mg every 12 hours and continued for 5 to 7 days. When used in this way, the drug will hasten clearing of fever, symptoms, and dysfunction of small airways. The mild side effects of the medication are not recognizable because of the symptoms of influenza. Amantadine is excreted through the kidneys, and one must therefore be cautious in its use in persons with impaired renal function.

PREVENTION

Several nonspecific measures may help to reduce the transmission of viral respiratory disease. Resting and reducing contact with persons outside the home during acute illness can be beneficial, not only therapeutically but also in reducing the probability of transmission. Close contact can be an important means of spread of all the respiratory viruses. Use and careful disposal of paper tissues and then washing one's hands help reduce the burden of virus in the local environment. For viruses that spread by small-particle aerosol, fresh-air ventilation will reduce virus concentrations in the air.

Nosocomial respiratory viral infections are a common problem on pediatric wards and on all wards during an influenza virus epidemic. When there is an influenza or RSV epidemic, elective admissions should be deferred, particularly for high-risk patients. Cohorting of patients with respiratory illness and of personnel caring for them can help reduce probabilities of transmission. Attention should be paid to ventilation of the hospital. Air from rooms on wards housing patients with acute respiratory illness should not circulate to other patient areas. Care must be given to separate isolation of patients who have complicating bacterial superinfections, particularly those involving staphylococci, group A beta-hemolytic streptococci, or meningococci. Careful handwashing should be routine in caring for persons with acute respiratory disease. Use of a gown is indicated if soiling is likely. Use of a mask is indicated in caring for patients with influenza if the persons providing the care have not already had the illness, and always in caring for patients with secondary pneumonia due to particularly virulent bacteria.

Megadoses of vitamin C have been recommended as a means for increasing resistance to acute respiratory illnesses, but no data exist to substantiate the value of such therapy.

Specific measures for preventing acute respiratory tract disease are limited to inactivated influenza-virus vaccine (Chapter 13) and amantadine. Live adenovirus 4 and 7 in an enteric capsule will produce an asymptomatic gastrointestinal infection and immunity to respiratory tract disease due to these viruses in military recruits, but objections have been raised to the general use of such vaccines because of the rodent oncogenicity of certain serotypes and the relatively minor overall contribution of adenoviruses to human disease.

Amantadine, mentioned previously for use in treatment, is also effective for preventing infection with influenza A virus. Major candidates for amantadine prophylaxis during an influenza A virus epidemic are high-risk persons who either have not received vaccine or have received vaccine that differs antigenically from the epidemic virus. The medication is given orally in a dose of 200 mg/day in adults (about 6 mg/kg, not to exceed 150 mg/day, in children from 1 to 9 years of age), either as a single dose or in split doses. Doses should be reduced in patients who have reduced renal function due to disease or increasing age. Amantadine is effective prophylactically only while it is being taken, and it must be continued for the duration of the epidemic, usually about 6 weeks. It does not interfere with antibody responses to vaccine, and it may be used to provide protection for the 2 or 3 weeks that it takes for vaccine to become effective if vaccination has been delayed until influenza is already present in the community. Side effects such as insomnia, light-headedness, nervousness, difficulty in concentrating, and drowsiness occur in 5 to 10% of persons; these side effects usually develop in the first 2 or 3 days of treatment, often recede while treatment is continued, and disappear when the medication is stopped. More serious nervous system side effects, such as seizures and hallucinations, may occur in persons with high blood levels. However, experience with prolonged amantadine treatment of patients with Parkinson's disease attests to the general safety of the medication.

Recommended Readings

Belshe RB (ed): *Textbook of Human Virology.* Littleton, MA, PSC Publishing Co., Inc., 1984.

Davis BD, Dulbecco R, Eisen HN, Ginsberg HS: *Microbiology,* 3rd ed. Hagerstown, MD, Harper & Row, 1980.

Evans AS (ed): *Viral Infections of Humans, Epidemiology and Control,* 2nd ed. New York, Plenum Medical Book Co, 1982.

Hoeprich PD: *Infectious Diseases.* 3rd ed. Philadelphia, Harper & Row, 1983.

Kendig EL Jr, Chennick V: *Disorders of the Respiratory Tract in Children,* 3rd ed. Philadelphia, Saunders, 1977.

Kilbourne ED (ed): *The Influenza Viruses and Influenza.* New York, Academic Press, 1975.

Knight V (ed): *Viral and Mycoplasmal Infections of the Respiratory Tract.* Philadelphia, Lea & Febiger, 1973.

Lennette EH, Schmidt NJ (eds): *Diagnostic Procedures for Viral, Rickettsial and Chlamydial Infections,* 5th ed. Washington, DC, American Public Health Association, 1979.

Mandell GL, Douglas RG Jr, Bennett JE: *Principles and Practice of Infectious Diseases,* 2nd ed. New York, Wiley, 1985.

Stuart-Harris CH, Potter CW (eds): *The Molecular Virology and Epidemiology of Influenza.* New York, Academic Press, 1984.

6

Viral Diseases of the Central Nervous System

ROBERT E. SHOPE

ACUTE CLINICAL SYNDROMES

Viral infections of the central nervous system (CNS) present as aseptic meningitis, poliomyelitis, or encephalitis. However, inapparent (subclinical) infection is by far the most common outcome of exposure to a virus, especially viruses having the potential to invade the CNS, because they may encounter difficulty in crossing the blood-brain or the blood-cerebrospinal-fluid barriers to reach the target organ. The inapparent infection is detected in random serosurveys or in testing in contacts of cases or normal populations during outbreaks of CNS disease. Thus, during epidemics of poliomyelitis only 1 of 1000 or more infected persons develops signs, and similarly during epidemics of St. Louis encephalitis only 1 clinical case exists for every 80 to 800 inapparent infections. This circumstance creates an appearance of clustered but sporadic disease when actually a relatively large proportion of the population may be infected during a short period.

Aseptic Meningitis

Some viruses replicate only in the leptomeninges. Inflammation of the meninges leads to fever, headache, stiff neck, and at times an elevation of the numbers of white blood cells in the cerebrospinal fluid (CSF) and an abnormal protein level. Aseptic meningitis is characteristic of Coxsackie virus and echo virus infections and may also be found with mumps and other childhood infections. Why these viruses are limited in their spread in the CNS whereas others, such as arboviruses, rabies and herpes viruses, invade the brain parenchyma is not known. Ability to invade may relate to the more stringent neurotropism of these latter viruses. Usually patients with aseptic meningitis have a good prognosis

TABLE 6.1   Reported Cases of Viral Encephalitis in the United States, 1980–1981

Virus Infections	1980		1981	
	Subtotal	Total	Subtotal	Total
Arthropod-borne		182		125
Western encephalitis	0		19	
Eastern encephalitis	8		0	
St. Louis encephalitis	125		15	
California group encephalitis	49		91	
Childhood infections		40		38
Measles	3		1	
Mumps	19		7	
Chickenpox	18		30	
Rubella	0		0	
Enterovirus		43		82
Herpes simplex virus		40		97
Other known causes		43		36
Indeterminate cause		1039		1156

Source: Centers for Disease Control.

and do not demonstrate severe changes in mental alertness, convulsions, or other focal neurologic signs. The syndrome, however, is part of a spectrum of disease; cases diagnosed as aseptic meningitis may include unrecognized encephalitis, in which virus has spread from the meninges to the brain tissue.

Encephalitis

Viral encephalitis, an inflammation of the brain parenchyma, is secondary to viral invasion of neurons, glial elements, or other brain cells. This syndrome is characteristic of arboviral encephalitis, herpes simplex encephalitis, rabies, and some of the CNS infections with childhood disease viruses. Table 6.1 shows the relative frequency of occurrence of these agents in the United States. A patient with encephalitis often presents with convulsions, focal neurologic signs, or altered consciousness or personality, and almost always has elevated protein levels and cell counts in the CSF at some time during the infection. The meninges may also be inflamed, producing the syndrome of meningoencephalitis. Encephalitis is a reportable disease in the United States.

Poliomyelitis

Some viruses, especially poliovirus and other enteroviruses, have as their primary target the gray matter of the spinal cord or brainstem. This localization produces flaccid paralysis, fever, and autonomic disturbances in a patient otherwise alert and without sensory deficit. The paralysis may be ascending (simulating Guillain-Barré syndrome) and, in the most serious complication, paralysis may extend to the muscles of respiration. Although this syndrome is rare in the developed countries because of control through immunization with poliovirus vaccine, it remains common in parts of the Third World.

Because the differential diagnosis, causative viral agent (treatable in the case of herpes simplex encephalitis), prognosis, public health implications, and management differ for each syndrome, the clinician must distinguish between aseptic meningitis, encephalitis, and poliomyelitis.

PATHOGENESIS OF VIRAL INFECTIONS OF THE CNS

The CNS is a privileged organ that is uniquely protected from the outside world and even from the rest of the body. It is infrequently invaded by infectious agents because few entryways are available. The blood-brain barrier is an anatomic feature that includes tight cerebral capillaries that do not permit passage of particulate matter and impermeable basement membranes that impinge directly against astrocyte processes. The blood-cerebrospinal-fluid barrier consists of tightly junctured choroid plexus epithelium that buffers the CSF from blood vessels. Only in the olfactory mucosa are nerve cells (receptor cells with olfactory rods) exposed to the exterior through the cribriform plate.

Human beings regularly harbor viruses such as poliovirus and herpes simplex, measles, rubella, and Epstein-Barr viruses, which have the potential to cause CNS infection. Most primary systemic infections with these agents, however, are self-limited because resulting antibody formation blocks viral access to the CNS. Some investigators believe that the blood-brain and blood-cerebrospinal-fluid barriers usually prevent entry of viruses.

We do not know why only a small proportion of infected persons develop viral CNS infections. These persons may have defective barriers, or they may have some other genetic or environmentally controlled factor that permits the virus to invade the CNS.

How, then, do viruses enter the CNS? The experimental evidence is incontrovertible that rabies virus enters through the peripheral nerves. In animals, for example, the infection is contained locally at the inoculation site if the nerve trunk is severed. Rabies virus may replicate first in the muscle or go directly to specific receptor sites at the neuromuscular junction and possibly in the sensory nerve endings. The virus proceeds by centripetal axonal spread to the motor neuron or dorsal root ganglion. In a somewhat similar manner, activation of herpes simplex virus, which is latent in the trigeminal ganglion, may permit the virus to proceed by axonal spread through a recurrent branch of the ophthalmic division of the trigeminal nerve, which supplies the meninges over the base of the middle and anterior fossae. This process would explain the temporal-lobe localization of the initial brain infection in many patients with herpes simplex encephalitis.

One can also demonstrate in animals that intranasal infection with various viruses leads initially to infection of the olfactory bulbs. This route may explain a few cases of rabies with aerosol exposure in bat caves or in laboratories. Furthermore, it is a possible route of CNS infection with arboviruses and others in which viremia could seed the nasal mucosa, and thus lead to CNS invasion by the virus. This route, however, seems to be a minor pathway for most viruses causing CNS infections.

Two other possible mechanisms of virus entry exist, and both require viremia, the usual antecedent of viral CNS infections. In the first mechanism the viremia seeds the vascular endothelium of the small vessels where virus is transported across infected vascular endothelium or in infected leukocytes that migrate into the brain. Alternatively, viremia seeds the choroid plexus epithelium, and the infection proceeds into the CSF. In the second mechanism, viremia seeds muscle and, from muscle, virus particles attach to the neuromuscular junctions and are carried by the axoplasm into the motor areas of the CNS. For each of these mechanisms to be effective, virus must evade multiple defense mechanisms such as circulating antibody, interferon, and virus clearance by the reticuloendothelium system. One should not be surprised, therefore, that viruses with the potential to invade the CNS so often cause only abortive febrile systemic disease without CNS infection.

Once CNS infection is started, the virus may spread rapidly. Spread is by the CSF or by direct cell-to-cell budding after carriage through axonal or glial processes. The specific pattern of acute CNS disease appears to depend on virus tropisms. Those viruses that infect neurons (such as rabies) cause encephalitis, whereas viruses with predilection for meningeal tissue (such as echo virus and Coxsackie virus) may preponderantly cause aseptic meningitis. Some CNS viral pathogens are encephalitogenic mostly in children, others primarily in persons over 55 (i.e., St. Louis encephalitis virus). These variations of pathogenicity that depend on the age of the patient probably relate to cell tropisms or immunocompetence.

The same blood-brain barrier that impedes entry of viral particles also impedes entry of circulating antibody. After CNS infection begins, however, the barrier breaks down, and not only does circulating antibody enter, but also limited numbers of B-cells enter the brain, differentiate to plasma cells, and lead to oligoclonal antibody formation, sometimes to extremely high titers in the brain and CSF. This antibody may be a factor in limiting spread of infection, but at the same time may enhance cell damage through cell-mediated mechanisms, such as antibody-dependent cellular cytotoxicity (ADCC).

SPECIFIC CAUSES OF VIRAL CNS DISEASE

Enteroviruses

ECHO VIRUSES, COXSACKIE VIRUSES

About three fourths of the cases of reported aseptic meningitis are caused by echo viruses, and others by the closely related Coxsackie viruses. These are enteroviruses, spherical single-stranded RNA viruses ranging from 20 to 30 nm in diameter. They do not have a lipid envelope, resist the action of bile, and thus are stable in the gut. After a person is infected through the fecal-oral route, initial viral replication occurs in the pharynx and intestine, during which time no evidence of illness is seen or the patient may report a prodromal low-grade fever with mild respiratory or gastrointestinal symptoms. The pharyngeal and enteric lymph nodes are infected next, and the virus presumably then spreads by viremia to the CNS. The target organs are the meninges and ependyma, and because these agents are not highly neurotropic, they rarely cause frank encephalitis.

The disease is common in children, and in the United States it occurs usually in the summertime. The patient characteristically presents with fever, headache, nausea, and vomiting. The deep tendon reflexes are exaggerated and stiff neck and positive Brudzinski's and Kernig's signs are noted, especially in older children. The CSF may show slightly elevated protein and increased number of leukocytes, initially polymorphonuclear leukocytes and later lymphocytes. Two variations are seen and these may have diagnostic significance. Echo virus infections, especially echo virus 9, a classic epidemic of which occurred in Milwaukee in 1957, sometimes present with a striking maculopapular rash on the trunk and face. Coxsackie B virus infections, on the other hand, may be accompanied by severe myalgia, pleurodynia, or orchitis. During outbreaks of meningitis in children, Coxsackie epidemic pleurodynia is sometimes seen in adults.

Newborn infants are especially vulnerable to Coxsackie B virus infections. In addition to aseptic meningitis, these patients tend to have generalized infections characterized by hepatitis, pancreatitis, and myocarditis, often ending in death. Infection may be transplacental or from infant to infant in the newborn nursery. Although some investigators report that children with echo virus or Coxsackie virus meningitis have lowered intelligence later in life, these findings are not consistent in all studies.

ENTEROVIRUS 70

Hemorrhagic conjunctivitis caused by enterovirus 70 has been recognized during the past 15 years in Africa, Europe, Asia, and Latin

America. It has been called "Apollo disease" because the first massive epidemic in 1967 in Ghana coincided with the U.S. space landing on the moon. The virus infects the conjunctiva and rarely the gut. In a small proportion of affected patients, motor paralysis similar to polio-myelitis appears from 2 weeks to a month after the initial conjunctivitis. The CNS disease is preceded for 1 to 3 days by fever, malaise, head-ache, vertigo, and stiff neck. An asymmetrical flaccid paralysis usually starts in the upper portions of the legs, but the arms and lower por-tions of the legs may also be affected. Deep tendon reflexes are dimin-ished. Cranial nerves may also be involved. The disease is most fre-quent in 20- to 40-year-old persons, males more commonly than females. The CSF contains increased protein and white blood cells, usually lym-phocytes. Death rarely results, but permanent sequelae, consisting of paralysis and muscle atrophy, occur in about 25% of patients with CNS disease. The virus is transmitted by contact with eye secretions, usually by contaminated hands or ophthalmologic instruments.

POLIOMYELITIS

Paralytic poliomyelitis has been rare in the United States since 1961. Wild-type polioviruses have been almost completely replaced by the highly successful oral polio vaccine strains that have colonized the guts of vaccinated children and in many instances also are transmitted to nonimmunized contacts. Although the vaccines comprising attenuated poliovirus types 1, 2, and 3 are remarkably safe and do not revert to virulence, they may, rarely, be associated with illness. Of the cases of paralytic poliomyelitis reported in 1981 in the United States, six of seven were in vaccinees or their contacts and must be considered endemic vaccine-associated cases. The paralytic syndrome is also still seen occa-sionally after infection with enteroviruses other than poliovirus, and poliomyelitis is still found in Third World countries where vaccination is not practiced or has not been achieved completely.

Poliomyelitis follows infection of neurons by enteroviruses, causing direct damage to the cells, and a resulting immune-cell response, pri-marily in the medulla and spinal cord. The patient may have a tran-sient prodromal minor illness consisting of fever, malaise, headache, sore throat, vomiting, and abdominal pain. This systemic enterovirus infec-tion precedes invasion of the nervous system. The illness then returns with greater severity and includes stiff neck and pain in the neck, back, and legs. Paresthesias may also result, indicating limited viral invasion of the sensory components of the nervous system. Kernig's and Brud-zinski's signs are often positive. The CSF may contain elevated protein levels and increased numbers of mononuclear cells. At this stage, if crit-ical motor neurons have escaped infection, many patients never be-come paralyzed.

As the disease progresses, acute pain presages or accompanies paral-ysis in one or more extremities. Muscle fasciculation, weakness, increased deep tendon reflexes, and loss of abdominal and cremasteric reflexes indicate onset of paralysis. These signs usually develop fully within 2 days, but the onset of paralysis may be gradual, over a few days. The full flaccid paralysis does not ordinarily progress further after the fever subsides.

In spinal poliomyelitis, the principal paralyzed muscle groups are those of the extremities, bladder, abdomen, back, and diaphragm. Paralysis of the intercostal and diaphragmatic muscles requires immediate life-support efforts to maintain respiration. In bulbar poliomyelitis crani-ally innervated muscles are paralyzed, and loss of medullary function leads to circulatory and respiratory failure. This failure may be mani-fested by Cheyne-Stokes respiration, acid-base imbalance, tachycardia, and loss of blood pressure. Severe poliomyelitis of the bulbar type re-sults in a 5 to 10% mortality. The condition of survivors stabilizes after the end of the febrile stage, and some recovery of function may occur over a 1- or 2-year period. Sequelae include permanent muscle atrophy and deformity.

Poliomyelitis may occur in any age group. The apparent to inappar-ent infection ratio is higher in adults than in children. Where natural infection with wild-type virus is still common, epidemics occur prepon-derantly in children less than 3 years of age.

Childhood Disease Viruses

MEASLES (RUBEOLA)

Measles is associated with encephalitis in about 1 out of 2000 infec-tions. Encephalitis occurs more frequently in children more than 10 years old than in younger age groups. With the widespread use and effec-tiveness of attenuated (live) measles virus vaccine in the United States and Europe, measles has decreased to the point at which it is now rare. Measles and its encephalitic complication are still prevalent in less de-veloped countries. Measles encephalitis is characterized by sudden on-set during the eruptive phase, often about 6 days after appearance of the rash. The patient has headache, confusion, and, in severe cases, convulsions and coma. The CSF protein level and white blood cell count are elevated, with lymphocytes usually predominating. Mortality is about 15%, and about 25% of cases have sequelae. Measles of early childhood may rarely be followed years later by a chronic, progressive neurologic disorder, subacute sclerosing panencephalitis (SSPE). SSPE is described in Chapter 9.

MUMPS

Mumps virus was the most common cause of reported encephalitis in the United States until the introduction of attenuated mumps virus vac-cine in 1967. Since then the decline in mumps and mumps meningoencephalitis has been steady. Nevertheless, meningoencephalitis is diagnosed in about 3 out of 1000 cases of mumps, and only about 50% of children in the United States receive the mumps vaccine. The disease is a manifestation of direct infection of the CNS by mumps virus, which is presumably seeded during the viremic phase. CNS signs are evident before, at the same time as, or after parotitis, and a minority of cases do not manifest parotitis, orchitis, or other signs of mumps. The clinical course can be that of meningitis, meningoencephalitis, or encephalitis. The prognosis for full recovery is excellent.

RUBELLA (GERMAN MEASLES)
   Rubella virus infection may also be associated, although rarely, with encephalitis after acquired infection as well as congenital infection. The use of rubella virus vaccine has rendered these forms of encephalitis even rarer.

VARICELLA
   Meningoencephalitis, an uncommon complication of varicella (chickenpox), usually manifests as acute cerebellar ataxia, a benign syndrome that starts a few days after onset of the rash. Although direct varicella-zoster virus infection of the CNS may occur, some investigators postulate that most cases are associated with an allergic phenomenon of unknown pathogenesis.

Arboviruses

Arboviruses are a common cause of reported encephalitis in the United States. Table 6.2 shows the numbers of cases by diagnosis between 1955 and 1982. Individual cases cannot be distinguished clinically from other causes of viral encephalitis, such as herpes simplex, childhood disease viruses, and enteroviruses. Hallmarks of arboviral encephalitis are summer/fall prevalence, a history of exposure to arthropods, and distribution limited to the area of occurrence of specific mosquito or tick vectors. Arboviral encephalitides are zoonotic diseases transmitted to humans by blood-feeding arthropods, usually mosquitoes, but for some viruses the vector is a tick. Human beings are dead-end hosts, that is, they do not maintain the transmission cycle between arthropods and vertebrate animals. Most cases of arboviral encephalitis are sporadic, focal, and found where changes in the ecology, such as irrigation, flooding, or discarded tires, create breeding places for mosquitoes. Alternatively, infections occur when people go to the woods or water for recreation or build their homes in forested suburban areas. This principle holds for viral encephalitides: LaCross, eastern, Powassan, and the rural forms of St. Louis, western, and Venezuelan. Large epidemics of St. Louis encephalitis and western encephalitis occur under conditions of heavy snow runoff or river flooding, or, in the case of St. Louis encephalitis, when

TABLE 6.2 Reported Cases of Arboviral Encephalitis in the United States

Year	Western Encephalitis	Eastern Encephalitis	St. Louis Encephalitis	LaCrosse Encephalitis
1955	37	15	107	
1956	47	15	563	
1957	35	5	147	
1958	141	2	94	
1959	14	36	118	
1960	21	3	21	
1961	27	1	42	
1962	17	0	253	
1963	56	0	19	1
1964	64	5	470	42
1965	172	8	58	59
1966	47	4	323	64
1967	18	1	11	53
1968	17	12	35	66
1969	21	3	16	67
1970	4	2	15	89
1971	11	4	57	58
1972	8	0	13	46
1973	4	7	5	75
1974	2	4	74	30
1975	133	3	1815	160
1976	1	0	379	47
1977	41	1	132	65
1978	3	5	26	109
1979	3	3	32	139
1980	0	8	125	49
1981	19	0	15	91
1982[a]	9	12	34	130

[a]Data preliminary. Between 1963 and 1975, a total of 26 cases of encephalitis caused by Venezuelan encephalitis, Tensaw, and Powassan viruses were reported.
Source: Center for Disease Control.

hot, dry weather creates urban conditions favorable for the sewage-breeding mosquito, *Culex pipiens.*

LACROSSE VIRUS

LaCrosse virus is the most common cause of arboviral encephalitis in the United States during most summers. California encephalitis, snowshoe hare, and Jamestown Canyon are LaCrosse-related viruses and sometimes cause a similar syndrome. Collectively, these viruses cause disease called "California group encephalitis." They belong to the family Bunyaviridae and have 3-segmented single-stranded RNA genomes contained in spherical, enveloped particles of about 90-nm diameter. The viruses form in association with the Golgi apparatus in the cytoplasm.

LaCrosse virus infects all age groups; however, the disease is limited primarily to children with the median age of 5 years but range from neonate to 17 years. About two thirds of cases are males, probably because of the greater exposure of males to mosquitoes. The disease causes a sudden onset of fever, malaise, headache, nausea and vomiting, and stiff neck. About 50% of hospitalized patients present with seizures, or have convulsions at some time during the course of illness. Other focal neurologic findings are relatively uncommon. Mild elevation of CSF protein level and mononuclear cell count is the rule.

The convulsions are often refractory to anticonvulsant medications and may necessitate intubation and general anesthesia for control. Cerebral edema is a feature of the rare fatal cases, leading to the recommendation that these patients not be overhydrated and even that dehydration measures be used. No specific treatment or vaccine is available.

The presumptive diagnosis is made by demonstration of LaCrosse virus-specific IgM in serum or spinal fluid on admission or during the first week of hospitalization, or by a rise in antibody titer to LaCrosse virus by neutralization, hemagglutination-inhibition, complement-fixation, immunoelectrophoresis, or ELISA tests. Although encephalitis caused by LaCrosse virus clinically resembles that caused by herpes simplex virus, the physician should resist the temptation to biopsy the brain unless the LaCrosse IgM titer is negative, because cerebral edema may be a prominent feature of LaCrosse disease. The age, geographic distribution, and season (i.e., sporadic cases occurring in children during July, August, and September in areas adjacent to deciduous hardwood forests in the Middle West and Northeast United States) should engender a high index of suspicion.

The epidemiology of LaCrosse encephalitis explains its distribution and seasonality. The virus is maintained in *Aedes triseriatus* mosquitoes, in which it is transmitted from the female through the egg to the progeny, and from the male by sexual transmission to the female. The mosquito lays its eggs in stagnant water found in tree-holes of oak and hickory forests and in discarded tires. In the hyperendemic areas of Minnesota, Iowa, Wisconsin, Ohio, and New York, the adult female emerges early in June. The mosquito can transmit immediately when it takes its first blood meal, and through the summer until frost. The insects also infect chipmunks and squirrels, which serve to amplify the transmission cycle. Cases cluster in communities where houses abut hardwood forests or where yards contain discarded tires or recreational tire swings.

Encephalitis that resembles LaCrosse encephalitis is diagnosed serologically as caused by snowshoe hare virus in Canada and by California encephalitis virus in California. Relatively milder aseptic meningitis and encephalitis, primarily in adult patients, has been attributed to Jamestown Canyon virus infection in Michigan, Ontario, and New York state.

ST. LOUIS ENCEPHALITIS VIRUS

St. Louis encephalitis virus is in the family Flaviviridae, which contains enveloped viruses having single-stranded RNA in spherical particles of 30 to 40 nm in diameter. The particles form in intracytoplasmic vesicles.

St. Louis encephalitis virus is the major cause of epidemic arboviral encephalitis in the United States. Reported sporadic cases in rural areas number 150 or fewer in most years; however, about every 10 years, major urban epidemics occur. In 1975, confirmed cases numbered 1815, with about 9% mortality, as diagnosed in an epidemic encompassing the urban areas of much of east-central United States and Ontario, Canada.

St. Louis encephalitis virus infections occur at all ages; however, age-specific rates of morbidity and mortality are greater in persons more than 55 years old. The reason for the occurrence of disease and its greater severity in older persons is not known. The disease has a sudden onset with fever, headache, and malaise. These symptoms may precede neurologic signs such as an obtunded state, stiff neck, and sometimes focal neurologic deficits. Convulsions are reported in about 5% of patients. The CSF may show normal findings, but protein is usually elevated in the range of 45 to 100 mg/100 ml and a mononuclear cell count between 5 and 500 is found. The presumptive diagnosis is made by demonstrating specific IgM in serum or CSF on the patient's admission to the hospital or during the first week of illness. Alternatively, a rise in serum antibody level during convalescence is diagnostic.

A few patients have had the syndrome of abnormally increased secretion of antidiuretic hormone (ISADH), characterized by hyponatremia. These patients showed good response when treated with water restriction.

Patients who develop elevated temperature and coma for more than 48 hours have a poor prognosis, especially associated with convulsions. Mortality ranges between 3 and 25%, with the higher rates occurring among persons more than 55 years of age. No specific treatment has proved effective and no vaccine is available. Sequelae, including neurasthenia and motor and personality changes, are found in 30 to 50% of patients, but most patients return to normal function within 3 years after infection.

Between July and September, the index of suspicion of St. Louis encephalitis should be high in elderly patients exposed to mosquitoes in the western United States and in urban and suburban areas of major cities such as Tampa, Memphis, St. Louis, Chicago, Dallas, and Houston. The virus is maintained in rural areas of western United States by transmission between *Culex tarsalis* mosquitoes and birds. Some evidence exists that the virus winters-over in the mosquito egg and in hibernating adult female mosquitoes. Neither mechanism, however, has been shown with reliable frequency. *Culex tarsalis* breeds in ground pools.

Large rural mosquito concentrations emerge in irrigation water and in river floodwaters. *Culex nigripalpus* plays a similar role in transmission of St. Louis encephalitis virus in Florida.

The urban epidemics are maintained by a different mosquito complex, *Culex pipiens pipiens* in the northern cities of North America and *Culex pipiens quinquefasciatus* in the more southern cities. This mosquito complex requires high organic content of its breeding sites, such as that found in sewage, hence its urban distribution.

WESTERN ENCEPHALITIS VIRUS

Western encephalitis, limited in the United States and Canada to the middle and western regions, is a mosquito-borne disease that occurs sporadically in rural areas and in rural epidemic form during flooding. Human disease is often preceded or accompanied by encephalitis in horses. Cases occur in the summer, and epidemics peak somewhat earlier than those of St. Louis encephalitis. Western encephalitis is transmitted by the same mosquito, *Culex tarsalis*, that carries rural St. Louis encephalitis and has similar epidemiology. Birds serve as maintenance and amplifying hosts, but its wintering-over mechanism is not understood.

The disease occurs in all age groups but preponderantly in children. Prodromal headache, fever, and malaise are followed in 2 or 3 days by focal and general neurologic signs, including convulsions in a high percentage of younger children. The disease is most severe in infants; death rates may be as high as 10% in this age group. That persistent learning and behavioral deficits may follow has been well documented.

The diagnosis should be considered for children exposed to mosquitoes during July, August, and September in the western United States and Canada. Increments in antibody titer between acute and convalescent sera confirm the diagnosis. No specific treatment is known, but an experimental vaccine is available for use among laboratory personnel.

EASTERN ENCEPHALITIS VIRUS

Eastern encephalitis, the most malignant of the arboviral encephalitides, most often affects and is most severe in children. The case-fatality rate is about 50%. The disease has a sudden onset and progresses rapidly, with symptoms including fever, headache, and convulsions (75% of patients). Unconsciousness may appear within 1 or 2 days, accompanied by cerebral edema and respiratory depression. Edema, especially of the face, may be prominent. The CSF often had elevated protein levels and a high mononuclear cell count, sometimes in the thousands. The convulsions, at times with focal abnormalities by electroencephalography (EEG) and tomography, may lead to an erroneous diagnosis of herpes simplex encephalitis. During the summer and early fall in the endemic areas of Massachusetts, New Jersey, Florida, other east coastal areas, upper New York state, and Michigan, serologic tests for eastern encephalitis virus should be done and found negative before patients are made to undergo brain biopsies. Diagnosis of eastern encephalitis is made on the basis of seroconversion demonstrated by neutralization tests with acute and convalescent sera. An experimental vaccine is now used in laboratory workers.

VENEZUELAN ENCEPHALITIS VIRUS

Venezuelan encephalitis virus infection usually results in a febrile illness and exceptionally in encephalitis. In Latin America, where large epidemics occur in horses and humans, about 4% of cases in patients less than 15 years of age result in encephalitis, with case fatality of about 20%. This disease is severe, characterized by convulsions, spastic paralyses, and cranial nerve disorders; fortunately, it is rare in the United States. Three cases in elderly persons exposed to swamp mosquitoes near Miami between 1967 and 1971, and 19 cases in persons exposed to mosquitoes during the 1971 outbreak of equine disease in Texas, have been recorded. There were no deaths or described sequelae. The 1971 outbreak was an extension of a large equine epidemic that originated in Central America and moved through Mexico to the United States. It was controlled by aerial spraying to kill mosquitoes and by vaccination of horses. The epidemic form of the disease has not been seen in the United States since. An antigenic variant is enzootic in swamp rodents and *Culex* (*Melanoconion*) spp. mosquitoes in Florida. Only persons who live in or enter the swamp areas for recreation are at risk.

POWASSAN ENCEPHALITIS VIRUS

Powassan encephalitis is a disease of all age groups characterized by headache and fever followed by pyramidal tract signs, tremors, progressive disorientation, coma, and sometimes death. Some patients give a history of exposure to tick bites; the disease occurs in the spring, summer, and early fall in rural areas of the northern United States and in Canada. About 25 cases have been recorded since 1958, when the virus was originally isolated from the brain of a child from Powassan, Ontario. No specific treatment or vaccine is yet available. This virus is serologically related to the tick-borne encephalitis virus of Europe and Asia.

ARBOVIRUS ENCEPHALITIS IMPORTED TO NORTH AMERICA

At least three major causes of mosquito-borne encephalitis and one of tick-borne encephalitis could potentially be found in travelers returning to the United States. Japanese encephalitis, closely related clinically and serologically to St. Louis encephalitis, should be suspected in those persons exposed in India, China, and Southeast Asia. The case-fatality rate may be as high as 40%. An inactivated vaccine produced in Japan is available experimentally through the Center for Disease Control for U.S. citizens going to live in high-risk areas. Rocio encephalitis occurs in the Sao Paulo region of Brazil and is associated with an acute encephalitis having cerebellar signs. The case-fatality rate is about 5%. Rift Valley fever is a disease of Africa. Transmission from sheep and cattle directly or by mosquitoes to humans in sub-Saharan Africa leads to fever, hemorrhagic disease, and blindness. Between 1977 and 1980 several thousand human cases were recorded in Egypt and some (fewer than 1%) had encephalitis that was usually not fatal. Tick-borne encephalitis virus transmitted by *Ixodes* ticks in eastern Europe and Soviet Asia is a relatively common cause of encephalitis in the spring and summer. The cases are forest-associated and carry a case-fatality rate of 1 to 5%.

HERPES SIMPLEX VIRUS

Herpes simplex virus is ubiquitous. Primary infection occurs early in life and is only rarely associated with primary encephalitis. More commonly, encephalitis is a secondary phenomenon when the virus is reactivated, presumably from the trigeminal ganglion, to produce severe, often focal necrotizing lesions in the brain. This condition may occur at any time of the year. The age distribution of encephalitis is bimodal: about one third of the patients are less than 20 years of age and two thirds are more than 40. This disease is treatable with adenine arabinoside or Acyclovir and must be diagnosed early for treatment to be most effective.

Patients present with fever, headache, lethargy, and malaise. The majority of patients show evidence of personality change, such as irritability or hallucinations. Localization to the temporal lobe may be indicated by temporal-lobe seizures, although seizures may also be generalized. CSF pleocytosis with slightly elevated protein levels may be found but are not diagnostic. Untreated, the disease course progresses rapidly from lethargy to coma, with a mortality of 60% or greater. A high percentage of survivors have permanent sequelae.

Electroencephalography may help localize a focus, especially if the procedure is done early in the disease. Computerized axial tomography may also be of some assistance in locating a lesion, although tomography is often negative in the first 5 days of illness.

Herpes simplex encephalitis is diagnosed by brain biopsy. The biopsy is done through a craniotomy, and, if diseased tissue is found (usually in the temporal lobe), biopsy is positive, which occurs in herpes simplex infections about 95% of the time. The brain tissue is examined histopathologically for Cowdry's type A inclusion bodies, by electronmicroscopy for virus particles, and by immunofluorescence for herpes simplex virus antigen. The brain tissue is inoculated into cell culture and may be inoculated intracerebrally in suckling mice to isolate and identify the virus.

If biopsy is contemplated, it should be done early, but not before available rapid diagnostic procedures for other causes of encephalitis, such as arboviruses and enteroviruses, have been done on serum and spinal fluid.

Treatment with adenine arabinoside should be started while one awaits the results of the biopsy. If the biopsy is negative, the treatment should be discontinued after 5 days—unless the clinical diagnosis indicates a typical herpes simplex virus infection and no other cause has been found. Adenine arabinoside is usually administered for a full course of 10 days if the virus is confirmed as the cause. A newer drug, Acyclovir, is also effective in herpes simplex encephalitis and is not associated to the same degree with complications such as fluid overload. In a combined multicenter trial of therapy, results with adenine arabinoside were excellent in patients less than 30 years of age who were treated before the onset of coma. In 3 of 182 patients, complications were attributed to the biopsy procedure, one with hernia and two with hemorrhage. All 3 patients survived.

RABIES VIRUS

Human rabies in the United States and Canada is rare, in most years between one and four cases being reported, usually from exposure to bites of bats, foxes, raccoons, or skunks. The disease is common, however, in many parts of the world, especially in South America, Asia, and Africa, where dogs and cats are the principal reservoir. The virus belongs to the family Rhabdoviridae and is a bullet-shaped particle containing single-stranded, negative-sense RNA. It buds from intracytoplasmic membranes and sometimes from the plasma membrane.

Prophylaxis is recommended for persons exposed to or bitten by rabid or suspected-rabid animals. For bites, the wound should be washed with soap and water or detergent, and the patient should receive 20 U/kg of human rabies immune globulin, half infiltrated around the wound and half intramuscularly. If human globulin is not available, then 40 U/kg of horse or other animal rabies immune antibody may be substituted, with usual precautions against hypersensitivity reactions. At the same time, immunization should be started with rabies human-diploid-cell vaccine given in a different site from that of the immune globulin. The vaccine is administered subcutaneously on days 0, 3, 7, 14, 30, and 90. Preexposure immunization on days 0, 7, and 28 is given to persons in high-risk professions or who travel to work or live in rabies-endemic areas.

Exposure to the bite of a rabid animal does not always mean that the patient will develop rabies, even if no prophylaxis is given. The attack rate depends on the age of the person, the species of biting animal, whether the bite went through the clothing, the severity of the wound, and the part of the body bitten. Probably far fewer than 50% of those bitten and not treated develop rabies.

The incubation period usually varies between 2 weeks and 3 months. Rarely does it extend to years. Rabies presents as an acute encephalitis with prodromal symptoms of fever, headache, anorexia, and malaise. The patient may demonstrate initial fasciculation, neuritic pain, or paresthesia at the original wound site. As the clinical course progresses, one or more of the following may be evident: excitability, anxiety, excess lacrimation and salivation, insomnia, convulsions, difficulty in swallowing, chest pain and tightness, and pharyngeal and laryngeal spasms precipitated by attempts to swallow water (hydrophobia) or by fanning the skin with air. Focal or ascending paralysis may be prominent. The patient is usually alert and rational, but coma may precede death, which usually is a consequence of cardiac or respiratory failure. The illness usually lasts about a week, but persons may survive much longer with vigorous medical care.

During life, rabies may be diagnosed from a history of exposure to a dog, cat, or rabid wild animal plus immunofluorescence-positive corneal smear or skin biopsy, or rising antibody titer with antibody in the CSF. The fluorescent focus inhibition test in cell culture is used for antibody determination. The patient should be held with full isolation precautions, and attending personnel should be vaccinated. There is no specific treatment after onset of illness; attempts at immunotherapy and use of interferon have not been successful.

At least three persons have survived rabies encephalitis, and other survivals can be anticipated. It is therefore important to maintain vigorous supportive treatment where indications exist, including maintenance of fluid and electrolyte balance, use of antibiotics, control of secretions, tracheostomy, and administration of oxygen, anticonvulsants, assisted respiration, adequate nutrition, and skin care.

LYMPHOCYTIC CHORIOMENINGITIS VIRUS

Lymphocytic choriomeningitis virus infects wild mice (*Mus musculus*) and laboratory rodents, which serve as the reservoirs. Virus is transmitted to humans through aerosol or fomites contaminated with rodent urine or other secretions. The virus contains two-segmented, single-stranded, negative-sense RNA. It replicates in the cytoplasm, where pleomorphic spherical or ovoid particles bud through the plasma membrane. The virions often contain cell ribosomes that appear like grains of sand, giving the family name Arenaviridae.

The disease in humans is usually inapparent, febrile, or systemic, but the febrile phase may be followed by choriomenigitis or encephalitis with fever, headache, stiff neck, and nausea. Localizing neurologic signs are unusual, and patients generally recover completely. The diagnosis is made serologically by complement-fixation or immunofluorescence test. The rare fatal cases have shown lymphocytic infiltration of the meninges, ependyma, and choroid plexus as well as perivascular cuffing in the brain. There is no specific treatment or vaccine for lymphocytic choriomeningitis. Prevention includes rodent control and monitoring of laboratory rodent colonies to eliminate infected animals.

DIAGNOSIS

Diagnosis may lead to curative treatment in herpes simplex encephalitis. More important, diagnosis may lead to life-saving public health preventive measures for poliomyelitis; the meningoencephalitides of measles, rubella, and mumps; rabies; and the arboviral encephalitides. For these diseases, preventive vaccines or vector control measures are available. The physician should report cases of encephalitis promptly, even if a diagnosis is not evident, because clustering of cases may signal an outbreak that can be controlled.

The differential diagnosis of viral CNS disease includes distinguishing from among postinfectious and postvaccinal encephalomyelitis, bacterial meningitides, such as those caused by Meningococcus, Staphylococcus, and Hemophilus; abscesses, parasitic diseases such as malaria, toxoplasmosis, trichinosis, amebiasis; Rickettsia, Chlamydia, Mycoplasma, Lyme disease, cerebrovascular accidents, neoplasms, serum sickness, toxic and metabolic encephalopathies, Reye's syndrome, multiple sclerosis, and Guillain-Barré syndrome.

A careful history is important. The diagnosis may become obvious if a sibling has mumps or if the patient was just vaccinated for one of the childhood diseases. Travel history may indicate exposure to an exotic virus, or a bite by a wild animal may mean exposure to rabies. Many other examples are possible. The enteroviruses and arboviruses have summer/fall seasonal activity and cause disease in specific age groups, factors that may help in the diagnosis. The physical examination may also be revealing. Flaccid paralysis leads to suspicion of poliomyelitis; temporal-lobe seizures with personality changes may signal the diagnosis of herpes simplex encephalitis. Generally, however, the signs and symptoms will be nonspecific.

The clinical laboratory is most useful to rule out nonviral causes of illness. The lumbar puncture should be done with a small-bore needle. If the pressure is greatly increased, only a small amount of fluid should be removed to prevent formation of a pressure cone. The presence of papilledema or evidence of a space-occupying lesion found on skull x-ray or computerized axial tomography should also dictate caution in performing the lumbar puncture. CSF should be examined for glucose, which may be reduced in tuberculous meningitis and rarely in lymphocytic choriomeningitis and mumps encephalitis. The CSF should be examined microscopically and by culture for bacteria. Most viral CNS infections will have a minimal to moderate increase in the number of CSF monocytes and the protein level, although in lymphocytic choriomeningitis and eastern encephalitis there may be greatly increased numbers of CSF monocytes.

To arrive at a specific diagnosis of viral CNS illness, one must either isolate the virus, demonstrate viral antigen or nucleic acid in tissues, or show evidence of recent antibody formation, such as IgM or a fourfold or greater rise or fall in titer. The virus laboratory is essential. Virus isolation should be attempted in cell culture from CSF, stool, and throat swab. If rabies is suspected, saliva should also be sampled. Isolating enteroviruses from stool may be possible for as long as 6 weeks after onset of illness, and enteroviruses and mumps virus are frequently isolated from the CSF.

Serologic tests are available for enteroviruses, arboviruses, childhood diseases, and rabies. The serologic diagnosis of herpes simplex virus is not helpful, because most of the population already has antibody and the levels may change in response to nonspecific stimuli. Herpes simplex encephalitis is diagnosed by biopsy of the temporal (sometimes parietal) lobe and examination of the brain for Cowdry type A intranuclear inclusions, usually found in oligodendroglia; for virus particles by electromicroscopy; and by demonstration of specific antigen by immunofluorescence. The virus is isolated and maintained in cell cultures and in suckling mice, but these procedures take more time. Biopsy of the skin and examination of corneal smears for rabies antigen may yield a diagnosis, but a negative test does not rule out rabies.

IgM is detected in serum and CSF for arboviruses and rubella virus using an IgM-capture, enzyme-linked immunosorbent assay (ELISA). This test is not in general use, but where available it often is diagnostic at the time of admission. Patients usually are not hospitalized until after the third day of CNS disease, which, in turn, is several days after the prodromal systemic infection. IgM antibodies in most cases have already been formed by the time the patient is first seen in the hospital. A specific presumptive diagnosis by serology otherwise is available only by testing acute and convalescent sera, and the result will not be of use in the acute-phase management of the patient. The IgM-capture ELISA is especially valuable for early diagnosis of arboviral encephalitis, and use of the assay may avert diagnostic brain biopsy when herpes simplex virus is erroneously suspected in these cases.

Inability to determine the cause of the infection is still one of the major factors inhibiting progress in treatment and control of acute viral CNS disease. More than half of the cases clinically resembling viral disease are not diagnosed (Table 6.1) because either the cause is not vigorously pursued or the causative agents have not yet been isolated. This need for cause determination is a challenge for the future.

Recommended Readings

Brinker KR, Monath TP: The acute disease. In Monath TP (ed): *St. Louis Encephalitis*, Washington DC, pp. 503–534. American Public Health Association, 1980.
Calisher CH, Thompson WH (eds): *California Serogroup Viruses*. New York, Alan R Liss, 1983.

Centers for Disease Control: *Annual Encephalitis Summary 1981.* Washington, DC, US Government Printing Office, 1982.

Evans AE (ed): *Viral Infections of Humans,* 2nd ed. New York, Plenum Medical, 1982.

Hattwick MA, Gregg MB: The disease in man. In Baer GM (ed): *The Natural History of Rabies,* vol 2, pp. 281–304. New York, Academic Press, 1975.

Johnson RT: *Viral Infections of the Nervous System.* New York, Plenum Medical, 1982.

Kono R. Miyamura K, Tajiri E, et al: Virological and serological studies of neurological complications of acute hemorrhagic conjunctivitis in Thailand, *J Infect Dis* 135: 706–713, 1977.

Monath TP: Central nervous system infections (acute). In Hsiung GD, Greene R (section eds): Section H: *Virology and Rickettsiology,* vol 1, part 1. In Seligson D (ed-in-chief): *Handbook Series in Clinical Laboratory Science.* West Palm Beach, FL, CRC Press, 1978.

Sabin AB: Poliomyelitis. In Braude AI (ed): *Medical Microbiology and Infectious Disease,* pp 1348–1365. Philadelphia, Saunders, 1981.

Whitley RJ, Soong S-J, Hirsch MS et al: Herpes simplex encephalitis. Vidarabine therapy and diagnostic problems. *N Engl J Med* 304:313–318, 1981.

7

Hepatitis and Diarrhea Viruses

SARAH H. CHEESEMAN and
NEIL R. BLACKLOW

This chapter combines discussions of the viruses of hepatitis and diarrheal disease not merely because their target organs are in the gastrointestinal system but also because their scientific histories share a unique feature. The discovery of these agents and elucidation of the pathogenesis, epidemiology, and immunology of the infections they cause occurred without in vitro cultivation of the human pathogens. Immunologic methods defined the viruses, and the necessary reagents were originally obtained from infected animals, patients with naturally occurring disease, and studies on human volunteers. Propagation in cultured cells has now been accomplished for hepatitis A virus and human rotavirus, but hepatitis B virus and Norwalklike viruses continue to defy efforts to cultivate them in the laboratory.

The agents that have been studied in detail and will be discussed here are hepatitis A and B viruses, rotavirus, and the Norwalklike viruses (e.g., Norwalk, Hawaii, Ditchling). The existence of one or more other causes of viral hepatitis, known as "non-A, non-B hepatitis," is certain. Other types of viruses have been seen in the stools of patients with diarrhea, but their medical importance in the clinical syndrome has not been determined. In addition, no satisfactory microbial cause has ever been established for many cases of apparently infectious gastroenteritis, which suggests the existence of still unknown agents or mechanisms.

VIRAL HEPATITIS

The Disease

CLINICAL SYNDROME

The clinical syndrome of hepatitis is characterized by fatigue, malaise, and anorexia, which may be so severe that the mere suggestion of food leads to revulsion and nausea. Smokers may note distaste for cigarettes. Disturbance of bilirubin excretion by hepatocytes into the biliary ducts leads to decreased bile pigments in the stool, which becomes clay-colored or chalky, and an increase in conjugated bilirubin in the circulation and the skin, with consequent jaundice. Pruritus may also result from bile-salt deposition in the skin. The excess bilirubin, filtered by the kidney, darkens the urine to the color of tea or cola. The severity of these signs and symptoms varies considerably, as does the presence and degree of fever and right-upper-quadrant abdominal fullness or pain. Derangement of liver function test results follows a hepatocellular pattern, with greatest elevations in levels of transaminases (serum glutamic oxaloacetic transaminase and glutamic pyruvate transaminase, also known as alanine aminotransferase). Conjugation of bilirubin is less affected than is its excretion; therefore, bilirubin, predominantly in the direct (conjugated) form, accumulates in the serum by what is believed to be a backwash mechanism. The alkaline phosphatase level may also be elevated, but if it is proportionately much more elevated than the transaminases, or if abdominal pain is severe, anatomic biliary obstruction or cholestatic jaundice produced by drugs or toxins must be suspected.

Hepatocyte dysfunction may occur, leading to failure of clotting factor and albumin synthesis and impairment of gluconeogenesis, which in turn results in fasting hypoglycemia. Prolonged prothrombin time and depressed serum albumin level are indicators of the gravity of the illness. Severe deficiency of clotting factors poses a risk of bleeding and presages liver failure. Accumulation of blood ammonia (a consequence of the diseased liver's inability to detoxify nitrogenous wastes through the urea cycle) and hepatic encephalopathy may ensue. This complication of viral hepatitis is rare but, when it occurs, often fatal. The syndrome is termed fulminant hepatitis and the pathologic correlate is *acute yellow atrophy*.

Patients with mild cases of viral hepatitis, in which the serum bilirubin level does not exceed 3, usually recover within 3 weeks, but those with more severe cases have prolonged recoveries often marked by clinical relapses.

DISCOVERY OF DIFFERENT FORMS

Variability in the clinical course, outcome, and source of hepatitis was first studied epidemiologically and led to the characterization of two forms, *infectious hepatitis* (hepatitis A) and *serum hepatitis* (hepatitis B or non-A, non-B). These were distinguished by exposure histories. Infectious hepatitis patients had histories of contact with jaundiced persons or of eating raw shellfish 15 to 60 days preceding onset. Serum hepatitis patients had a history of receiving transfusions or blood fractions, generally 45 to 180 days before the onset of illness, although one group reported intervals as short as 15 days. This disease was first recognized during World War II in military personnel who had been immunized against yellow fever. The vaccine contained live attenuated yellow fever virus (17D strain) and pooled human serum, which had been added as a stabilizing agent and in retrospect turned out to be the adventitious source of hepatitis infection.

The case fatality rate for serum hepatitis greatly exceeds that for infectious hepatitis, and progression to chronic liver disease seems limited to patients with serum hepatitis. Infectious hepatitis also differs clinically from serum hepatitis in having a greater frequency of fever, a more abrupt onset, and a predilection for young persons, with relative sparing of the elderly. During the period of the epidemiologic studies mentioned above (the 1960s), cases of hepatitis that could not be classified by exposure history resembled infectious hepatitis in clinical features and prognosis. Infectious hepatitis is now referred to as hepatitis A.

The difference between these two clinical patterns was confirmed by challenge experiments at Willowbrook State School, an institution for the mentally retarded in New York. Oral challenge with a serum pool derived from that institution, where hepatitis was endemic, produced the disease after a short incubation period (30 to 60 days); repeat challenge with the same inoculum intramuscularly produced a second bout of hepatitis after a longer period (38 to 146 days). Serum drawn from a subject a few days before onset of the first illness (MS-1) transmitted hepatitis with a short incubation period, whereas that obtained before the second illness (MS-2) transmitted hepatitis with a long incubation period. Hepatitis of the long-incubation type generally had a slower rise in the SGOT level and lasted longer than short-incubation disease. These studies also showed the transmissibility of both types of hepatitis to nonimmune persons in close contact with hepatitis patients in the setting of an institution for the severely retarded.

The conclusions of these studies were illuminated by a discovery in Australia. In 1965, Blumberg observed the formation of a precipitin line in agar gel between serum samples from Australian aborigines and samples from a patient with hemophilia who had received multiple transfusions. After much further work, the antigen in the aborigine serum was recognized as being related to viral hepatitis, particularly to that occurring after transfusions of whole blood. This new antigen was originally called Australian antigen and subsequently identified as the hepatitis B surface antigen.

Soon thereafter, it became possible to (1) verify that in most cases of serum hepatitis, the antigen discovered by Blumberg is present in the serum, (2) make a presumptive diagnosis of infectious hepatitis, (3) screen blood donors for hepatitis antigen, and thus reduce the rate of post-transfusion (serum) hepatitis, and (4) demonstrate that some cases of posttransfusion hepatitis were *not* related either to this antigen or to infectious hepatitis; these were therefore designated non-A, non-B hepatitis, a term that has persisted to the present.

Hepatitis B

VIRUS MARKERS

Hepatitis B, the current designation for the syndrome associated with Blumberg's Australia antigen, comprises most illness formerly categorized as serum hepatitis or homologous serum jaundice. The virus-host interactions in this infection are described predominantly in terms of antigens and their corresponding antibodies, because the virus has still not been shown to replicate in cell culture. The virus can be transmitted to chimpanzees, and much of our knowledge derives from studies in these animals. The hepatitis B virus (HBV) markers so far identified are hepatitis B surface antigen (HB_sAg), core antigen (HB_cAg), the enzyme DNA polymerase, the viral nucleic acid, and HB_e antigen (HB_eAg). Figure 7.1 depicts the sequential development and clearance of HBV markers in a typical acute infection, and Fig. 7.2 shows the pattern in a case that progressed to the carrier state and chronic liver disease. Each of the markers is discussed below.

Hepatitis B Surface Antigen and Antibody. Blumberg's discovery was aided by the immense quantity of antigen present in the serum of a carrier, 100 billion to 1 trillion particles per milliliter of blood. These particles can be seen by direct electron microscopy of serum. They occur in 22-nm spherical forms and as tubules 22 nm in diameter but varying in length. Dane described a 42-nm double-shelled particle that is now thought to be the virion itself. The outer coat of this particle is identical to the 22-nm-diameter particles, and sometimes tadpolelike forms are seen with the Dane particle as the head and a typical 22-nm tubular form as the tail. Thus, the 22-nm forms have been identified as excess viral surface coat material, or hepatitis B surface antigen (HB_sAg). This name has replaced the earlier terms, Australia antigen and hepatitis-associated antigen (HAA).

In acute hepatitis B infections, HB_sAg can be detected in the serum as early as 6 days after exposure and usually for about a month before liver function test results become abnormal (Fig. 7.1). In the days preceding overt hepatitis, a syndrome of fever, urticaria, arthralgias, and even frank arthritis (resembling serum sickness) may occur. This syndrome has been shown to coincide with the circulation of antigen-

Exposure

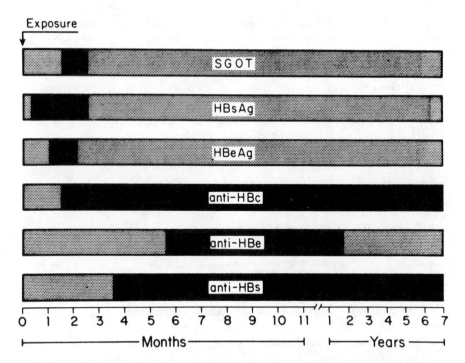

FIG. 7.1. Appearance and clearance of HBV markers and liver function abnormalities in a single patient with acute hepatitis B followed by recovery. The solid black portion of each bar represents the time during which that antigen or antibody was detectable in serum or, in the case of SGOT, elevated above normal. This case is somewhat unusual in that anti-HB$_e$ generally appears shortly after disappearance of HB$_e$Ag. (Reproduced by permission from Krugman S, Overly LR, Mushahwar IK, Ling C-M, Frosner GG, Deinhardt F: Viral hepatitis, type B: Studies on natural history and prevention reexamined. *N Engl J Med* 300:102, 1979.)

antibody complexes containing HB$_s$Ag and with consumption of complement.

The replacement of HB$_s$Ag by antibody to HB$_s$ in the circulation generally signals the change from infectious to immune status. The presence of one marker usually excludes the other, but a gap of months to more than a year may occur between the loss of HB$_s$Ag and the appearance of anti-HB$_s$. Anti-HB$_s$ correlates with firm immunity to rechallenge by HBV. However, not all persons progress from HB$_s$Ag-positive status to immune status. Those who do not become chronic carriers and may either be asymptomatic or have chronic liver disease of varying severity (chronic persistent or chronic active hepatitis). HB$_s$Ag is detectable in serum and in the cytoplasm of the hepatocytes in these cases. Three subtypes, or strains, of HB$_s$Ag have been identified, *adw*, *adr*, and *ayw*. These designations are primarily of use in epidemiologic studies.

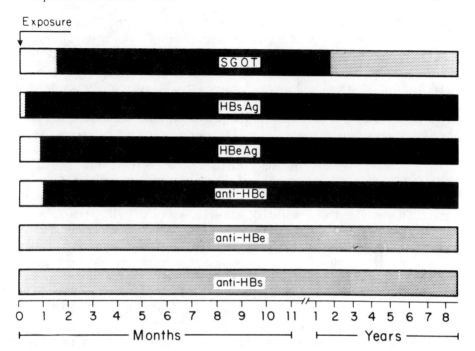

FIG. 7.2. Pattern of HBV markers and liver function abnormalities in a patient in whom hepatitis B progressed to chronic liver disease and the carrier state. The bars are explained in Fig. 7.1. Note that HB$_s$Ag and HB$_e$Ag persisted throughout 8 years of observation and that anti-HB$_s$ and anti-HB$_e$ never developed. (Reproduced by permission from Krugman et al, as in Fig. 7.1.)

Hepatitis B Core Antigen and Antibody. Hepatitis B core antigen activity corresponds to the inner nucleocapsid core of the Dane particle and is usually detected only in the nuclei of liver cells, although detergent treatment can unmask it in serum. Antibody to core antigen, anti-HB$_c$, is regularly found in the serum during acute hepatitis B infection, appearing 1 to 2 weeks before the rise in liver enzyme (transaminase) levels (Fig. 7.1). Persisting for the duration of HB$_s$ antigenemia and perhaps for life, anti-HB$_c$ may be the only serum marker detectable in the gap between the disappearance of HB$_s$Ag and the appearance of anti-HB$_s$. Thus, for purposes of determining the infection rate in a given population, any person whose serum contains any one of the three markers (HB$_s$Ag, anti-HB$_s$, or anti-HB$_c$) may be regarded as having had HBV infection. As with anti-HB$_s$, but not with the other markers associated with active infection, the prevalence of anti-HB$_c$ increases with age.

The presence of anti-HB$_c$ alone is a conundrum. On the one hand, it seems to be a sensitive epidemiologic tool, identifying certain (particularly elderly) persons who have no other markers of infection and, presumably, remote disease, now quiescent. On the other hand, anti-HB$_c$ as the sole marker of infection characterizes patients during the gap between circulating HB$_s$Ag and anti-HB$_s$, has been shown to denote infectivity of blood by transfusion, and signifies the presence of HB$_c$Ag in hepatocytes—all states of active infection. Recent studies have implicated "high titers" and IgM anti-HB$_c$ as evidence for continued viral replication, and a difference in quantity or type of antibody may explain the apparent paradox. Anti-HB$_c$ and anti-HB$_s$ often coexist. In one study, patients in whom the anti-HB$_s$ titer exceeded that of anti-HB$_c$ appeared to be asymptomatic carriers, whereas those in whom the anti-HB$_c$ level was relatively the greater had a 50% incidence of abnormal liver function tests. One viewpoint is that, in resolution of HBV infection associated with anti-HB$_s$, anti-HB$_c$ eventually disappears, but in the carrier state HB$_s$Ag may become undetectable while anti-HB$_c$ persists. The dynamics of declining anti-HB$_c$ and rising anti-HB$_s$ would then explain the value of the ratio between the respective titers in characterizing patient groups, as well as the observed increase in the prevalences of anti-HB$_s$ and anti-HB$_c$ with age. Anti-HB$_s$ is seen as the longest-lasting evidence of resolved infection and isolated anti-HB$_c$ as the last vestige of persistent infection.

DNA Polymerase.   During early stages of infection with hepatitis B virus, this viral enzyme is briefly detectable in the serum at about the time of appearance of anti-HB$_c$ (Fig. 7.1). It vanishes before liver function test values rise. In patients who progress to chronic liver disease, the enzyme reappears and may persist for a long time. DNA polymerase correlates strongly with circulating intact Dane particles and infectivity and has proved to be a sensitive laboratory indicator to follow during therapeutic trials directed against chronic HBV infection. Not present in asymptomatic carriers, the enzyme is not routinely assayed in most laboratories.

Hepatitis B Virus DNA.   The viral nucleic acid has recently been cloned by recombinant DNA technology to produce a probe for molecular hybridization studies.

Hepatitis B$_e$ Antigen and Antibody.   Magnius and Espmark first described the e antigen system in 1972; initially it was thought to be a host protein rather than a viral protein. It is now known to be a viral component located within the core of the Dane particle and expressed both in serum and infected hepatocytes. Virtually all patients develop e antigen at least transiently during acute hepatitis B, just shortly before the appearance of anti-HB$_c$. HB$_s$Ag carriers who retain HB$_e$Ag are more infectious than those who do not. This greater infectivity was first recognized in vertical transmission from mother to child and has since been confirmed for needlestick exposures, in which the risk of infection is roughly tenfold greater if the "donor" is HB$_e$Ag positive than if anti-HB$_e$ is present. Persistence of HB$_e$Ag for longer than 10 weeks during the acute illness forecasts a greater likelihood of chronic liver disease than its rapid clearance. HB$_e$Ag clearance and anti-HB$_e$ appearance correlate closely, with only about 4% of carriers bearing both markers simultaneously. Current detection methods leave fewer than 10% of cases unclassified with regard to the e antigen system, that is, lacking both the antigen and the antibody.

Delta (δ) Antigen and Antibody. The recently described δ antigen system presents an intriguing problem in interpretation. This antigen is found in hepatocytes of a large proportion of HB$_s$Ag carriers from southern Italy. Elsewhere, it occurs in carriers who are of Italian descent or are presumed to have had multiple HBV exposures, such as those with hematologic problems requiring frequent transfusions or clotting-factor replacement, homosexual men, and drug addicts. Like HB$_c$Ag, anti-δ appears in the serum when δ antigen is expressed in the liver, and δ antigen can be detected in detergent-treated serum. Delta appears simultaneously with HBV infection when serum containing HB$_s$Ag and anti-δ are given to previously uninfected chimpanzees. When the same mixed inoculum is given to chronic HB$_s$Ag carrier chimpanzees, δ appears in the liver and anti-δ in the serum after an abbreviated incubation period. Chimpanzees bearing anti-HB$_s$ and no δ markers seem to be immune to challenge with this inoculum. Delta antigen activity in serum seems to be associated with 35-nm spherical particles, although these particles have also been seen in preparations from subjects lacking δ markers. A ribonucleic acid smaller than the genome of any known animal virus has been associated with these particles. So far, δ markers have never been found in the absence of HB$_s$Ag. Delta is now known to be a defective virus that requires hepatitis B helper functions and circulates in an HA$_s$Ag coat. Delta infection is associated with increased risks of fulminant hepatitis and chronic liver disease, although its presence seems to dampen the expression of other HB markers. Anti-δ may coexist with anti-HB$_e$ in serum, and HBV DNA has been identified in such serums.

EPIDEMIOLOGY AND TRANSMISSION

A disease transmitted only by transfusion should be rare, but the quantity of HB$_s$Ag present in the blood is so great that even minute amounts such as those transferred by cutaneous needle puncture can result in disease in as many as 11% of susceptible persons. Thus, apparently nonparenteral transmission, such as was seen at Willowbrook State School (see above), could represent either minimal blood contact, as from a cut or scrape, or the existence of another mode of spread. Sexual partners frequently transmit the disease, and HB$_s$Ag can be found in semen, in which the presence of microscopic traces of blood may be the source of infection. Nonetheless, many groups in which overt parenteral transmission is unlikely have exceedingly high seroprevalence of HB$_s$Ag, such as the populations of Africa and the Far East. In those

TABLE 7.1   Prevalence of Markers of Hepatitis B Infection in the United States

Population	HB$_s$Ag (%)	HB$_e$Ag (% of those with HB$_s$Ag)	Any Marker of HBV Infection (%)
Volunteer blood donors, NY	0.3–0.4	4–9	3–7
Institutionalized mentally retarded	10–20	4–27	35–80
Male homosexuals	5–6	56–61	35–80
Drug addicts	3–7	35	60–80
Chinese Americans	9	4–24	81
Southeast Asian refugees	13	NA[a]	70–85
Dialysis patients	3–5	71	20–80
Medical personnel No or infrequent blood contacts	0.3	NA	3–10
Frequent blood contacts (includes dentists and dialysis staff)	1–2	(dialysis staff: 30)	15–30
Prisoners (males)	1–8	NA	10–80
Staff of institutions for mentally retarded	1	NA	10–25

[a] NA = not available.
Source: Data from Szmuness W, Harley EJ, Ikram H, Stevens CE: Sociodemographic aspects of the epidemiology of hepatitis B. In Vyas GN, Cohen SN, Schmid R (eds): *Viral Hepatitis*, Philadelphia: Franklin Institute Press, 1981; Szmuness W, Neurath AR, Stevens CE, Strick N, Harley EJ: Prevalence of hepatitis B "e" antigen and its antibody in various HB$_s$Ag carrier populations. *Am J Epidemiol.* 113: 113–121, 1981, and *Morbidity and Mortality Weekly Report* 31: 318, 1982.

settings both horizontal and vertical transmission occur, with the mother transmitting the virus to her child at the time of delivery, presumably by surface contamination with maternal blood. The presence of HB$_s$Ag in cord blood is not predictive of transmission, implying that the route is not breakdown of the maternal-fetal placental barrier.

Table 7.1, compiled from several sources, shows the rates of seropositivity for the U.S. population groups having a higher than usual incidence of HBV infection.

RELATION TO PRIMARY HEPATOCELLULAR CARCINOMA

In the parts of the world where HB$_s$ antigenemia is common, primary hepatocellular carcinoma is the most frequent malignancy. The relative risk for this tumor among HB$_s$Ag-positive Chinese government employees is about 200 times that for HB$_s$Ag-negative employees. The HBV genome has been identified in some of these tumors, and HB$_s$Ag titers seem to be higher among patients with this malignancy than among asymptomatic carriers. The presence of cirrhosis appears to be an important precondition to the development of hepatocellular carcinoma, and alcohol may be an important contributing factor along with HBV. A recent report presented evidence that HBV has some characteristics of the retrovirus group.

PROPHYLAXIS

The morbidity and mortality of the acute illness, as well as the risk of chronic liver disease, have spurred efforts toward developing prophylaxis against HBV infection. By far the most widespread and effective measure has been avoidance of exposure by rejecting blood for transfusion from HB$_s$Ag-positive donors. Since this policy was instituted, the rate of posttransfusion hepatitis has fallen by 80 to 85%, and the proportion attributable to hepatitis B virus has dropped to 10%. Also basic to hepatitis B prevention is proper handling of instruments and devices contaminated by human blood or tissues. Where feasible, as with needles and syringes, disposable materials should be used. Simple disinfection, as with alcohol or quaternary ammonium compounds (e.g., benzalkonium chloride), does not reliably inactivate HBV. When materials must be reused, therefore, vigorous mechanical washing followed by autoclaving, or the use of 0.5% hypochlorite, glutaraldehyde solutions, or gas sterilization with ethylene oxide is recommended.

Selection of blood donors having high titers of anti-HB$_s$ and preparation of immunoglobulin fractions from their plasma yields human hyperimmune globulin against HBV, known as HBIG. In several trials, HBIG was shown to decrease the risk to the spouse of a hepatitis B patient and to the infant born to an HB$_s$Ag-carrying mother. In studies comparing HBIG with standard immune serum globulin as protection against percutaneous exposures from HB$_s$Ag-positive patients, the concomitant presence of some anti-HB$_s$ and low levels of HB$_s$Ag in the lots of immune serum globulin used as controls confounded the evaluation of efficacy. HBIG generally prevented clinical hepatitis, although not completely so.

Following the early clues from Willowbrook State School, where boiled HB$_s$Ag-positive serum was found to protect against later hepatitis B infection, a vaccine was prepared from extensively purified formalin-inactivated HB$_s$Ag-positive serum. This subunit vaccine, tested in homosexual men with high risk of contracting HBV infection, has been shown to induce high titers of anti-HB$_s$ in 96%, with a protective efficacy of 92% in the population as a whole and virtually 100% among those with brisk antibody responses. The incidence of hepatitis B began to decline in the vaccinated group after the second month, that is, still within the incubation period for exposures that took place before vaccine administration. This vaccine may, therefore, be useful prophylactically even after exposure. The vaccine is immunogenic in infants, in whom it has been used as prophylaxis against vertically transmitted HBV infection. A trial in medical personnel demonstrated safety and immunogenicity, but the number of HBV infections in the control group was too low for evaluation of efficacy. Populations being considered for vaccine use include members of the high-risk groups listed in Table 7.1 who lack evidence of HBV infection, as well as persons with an identifiable recent exposure, in whom perhaps both HBIG and vaccine may be used. Patients likely to need multiple transfusions by virtue of embarking on cytotoxic chemotherapy or a diagnosis of a clotting disorder or red-blood-cell dyscrasia might benefit greatly from this vaccine, if it proves to be immunogenic in them. Three intramuscular injections are required for maximal efficacy, and side effects in the short term have differed little from those reported with placebo. Long-term effects are still unknown, as is the ultimate place of this vaccine in the medical armamentarium; an alternative preparation consisting of HB_sAg produced by recombinant DNA is also under trial.

Hepatitis A

Specific recognition of hepatitis A became possible in 1973, when Feinstone, Kapikian, and Purcell demonstrated the presence of 27-nm particles in stool samples from patients with infectious hepatitis. When mixed with serum from convalescent patients, the particles were seen by electron microscopy to be specifically aggregated (immune electron microscopy). This method is too cumbersome to be widely applicable, and only recently have diagnostic tools become available to the practicing physician. Commercial radioimmunoassay tests for IgM antibody specific for hepatitis A virus permit the diagnosis of acute infections, and tests for IgG antibody can be used to define immunity. Recently, the hepatitis A virus (HAV) has been grown in cell culture.

THE VIRION
  Electron microscopy reveals HAV to be a small structure lacking an envelope, morphologically similar to the parvoviruses. Unlike the parvoviruses, however, the nucleic acid of HAV is RNA, and the virus is now thought to be more closely related to the picornaviruses. Virus growth in cultures of fetal monkey cell lines, hepatoma cells, or human fibroblasts produces no cytopathic effect, but intracellular HAV can be detected by radioimmunoassay. Eventually, after adaptation to cultured cells, some virus is released into the medium. Most investigators have found a very long eclipse time before new virus is produced. Cell-culture–grown HAV, used to study viral structure, reveals a genome of molecular weight 2.8×10^6 daltons, consisting of linear single-stranded RNA and four major virion polypeptides.

VIRUS MARKERS IN CLINICAL ILLNESS
  HAV is excreted in the stool for about 2 weeks before the onset of symptoms or elevation in liver function test findings. Once symptoms appear, HAV can be detected in the feces of only 50% of patients dur-

ing the first week of illness and of 25% during the second. Viremia occurs briefly 3 to 7 days before the onset of jaundice, but transmission through serum, which can be produced experimentally, does not occur naturally and only with great rarity through blood transfusions. HAV has neither a gastrointestinal nor a serum carrier state. Antibody to HAV is produced early and is detectable by the time of clinical illness; IgG anti-HAV apparently persists for life.

Serologic studies in the United States show rising prevalence of IgG anti-HAV with age, with the rate of rise faster in groups of lower socioeconomic status. About 70% have antibody by age 50, and are thus immune to hepatitis A, which explains the previously noted rarity of this disease among the elderly.

EPIDEMIOLOGY AND TRANSMISSION

HAV is transmitted by the fecal-oral route, and contagion is greatest in the preicteric phase, correlating with the presence of virus in the stool. Many asymptomatic, or "anicteric," infections occur; adults are more likely than children to develop icterus. In crowded conditions with poor sanitation, nearly universal infection occurs in early life but with little clinical illness. This situation holds in the tropics, and travelers from more-developed countries who stray off the beaten path are at risk of acquiring hepatitis A. This experience is duplicated in the United States in day-care centers, where fecal-oral exchange among young children is facilitated, especially among those under age 2 who are still in diapers. The children are usually found to have asymptomatic infection, but their teachers and parents often become ill.

Food-borne and waterborne outbreaks of hepatitis A do occur. A particularly famous waterborne outbreak affecting a college football team was traced to backflow of groundwater to the drinking fountain on a practice field after a transient drop in water pressure. Food-borne outbreaks are a particular problem, as the infected foodhandler will shed virus in feces for several weeks before the onset of symptoms and may contaminate many food products during that time. Preventing these outbreaks requires good sanitary facilities and personal hygiene practices—most important, scrupulous handwashing at all times.

In the home, as in restaurants, contacts will have already been exposed before the patient is known to be sick. Patients requiring hospitalization for hepatitis A are usually well into the illness and unlikely to be excreting much virus, so no particular precautions are needed to protect personnel and other patients. Nonetheless, stool from any patient should be considered potentially infectious for hepatitis A, as well as many other pathogens, and handled with appropriate concern. In addition, patients admitted with hepatitis should be considered as potentially having hepatitis B or hepatitis non-A, non-B until proved otherwise; blood and needle precautions should be taken until the diagnosis of hepatitis A is confirmed by the laboratory.

PROPHYLAXIS

Once exposure has occurred, or in anticipation of it (as in travel to endemic regions), normal serum immunoglobulin (IG) will prevent clinical illness in a sizable proportion of those at risk. Household contacts of hepatitis A cases should receive immune globulin. It may not be cost-effective to screen for susceptibility in this setting. Travelers and others considering IG in advance of exposure might well be tested for anti-HAV before administration. If anti-HAV is present, IG is not needed. Prophylaxis is warranted because of the morbidity of the illness. Among employees at Willowbrook State School, the average loss of time from work due to hepatitis was 60 days and there is the possibility, albeit rare, of fulminant disease (as described earlier). An attenuated live virus vaccine has been developed that has successfully protected marmosets when challenged with wild-type HAV.

Non-A, Non-B Hepatitis

Currently about 90% of posttransfusion hepatitis must be classified as non-A, non-B. Before a case is ascribed to the putative non-A, non-B agent(s), cytomegalovirus and Epstein-Barr virus infection should be excluded. The term *non-A, non-B* hepatitis acknowledges our ignorance. Nonetheless, some information regarding possible causes has accumulated. Clusters of non-A, non-B disease have been traced to particular lots of clotting-factor concentrates, with production of disease and 27-nm particles in the liver when the implicated lots were injected into chimpanzees. Short-incubation-period non-A, non-B hepatitis has been documented, verifying the observation in 1965 that serum hepatitis appeared as early as 15 days after transfusion. Several new antigen-antibody systems have been identified by immunodiffusion methods similar to Blumberg's in the serum of patients with non-A, non-B hepatitis and by immunofluorescence in hepatocytes of those developing chronic liver disease. So far, however, uniform recognition and classification of these systems have not been achieved.

Standard IG given before large volumes of transfused blood (average 12.3 units per patient) greatly reduced the frequency of icteric hepatitis and its progression to chronic liver disease in cardiac surgery patients. This effect suggests that the normal blood donors from whose serum IG is prepared must have antibody to non-A, non-B agent(s), and thus that nonparenteral spread of the agent or agents may occur. In fact, reports of outbreaks of infectious-hepatitislike illness not related to known hepatitis viruses and presumed to be non-A, non-B in origin are appearing with increasing frequency.

Another approach to preventing non-A, non-B hepatitis after transfusions is not to use units of blood taken from donors having elevated levels of serum alanine aminotransferase. Difficulties with the standardization of alanine aminotransferase assays, the problem of rejection of donors who may not be infectious, and the generation of undue concern about disease among such donors have hindered the widespread adoption of this suggestion.

Other Viruses Causing Hepatitis

Cytomegalovirus and *Epstein-Barr virus* are discussed more fully in Chapter 10. Here it must suffice to say that the spectrum of acute illness caused by primary infections ranges from classic infectious mononucleosis, with elevation of transaminase levels as an incidental finding, to a syndrome in which the signs and symptoms of hepatitis predominate. Liver-function abnormalities, and perhaps even chronic hepatitis, are also features of the cytomegalovirus syndrome after organ transplantation.

Outside the United States and in caring for travelers, one should recall the importance of viruses other than the traditional hepatitis agents discussed above. The name *yellow fever* derives from the classic syndrome of fulminant hepatitis produced by a mosquito-borne togavirus. After a 3-day period of fever, chills, headache, backache, and vomiting, the clinical symptoms abate briefly before the onset of the fever, jaundice, and gastrointestinal bleeding that characterize the illness. Milder cases may occur in which an influenzalike syndrome is the only manifestation. This disease was eradicated from the Panama Canal Zone by mosquito control as a result of pioneering research by Walter Reed, but the virus persists in primate reservoirs in jungle areas (sylvatic yellow fever) in South America and sub-Saharan Africa. From these areas, yellow fever is periodically reintroduced into populated areas. Attenuated-live-virus vaccine is used for travelers to jungle or epidemic areas, and mosquito control is used to halt the progress of urban outbreaks.

GASTROENTERITIS VIRUSES

Clinical Syndromes

Only a small proportion of acute gastroenteritis or diarrheal illness can be attributed to known pathogenic bacteria and protozoa; therefore, it has long been assumed that viruses cause much of this disease. Nonetheless, it was not until the early 1970s that specific viral agents were identified as the cause of the two major clinical forms of acute infectious nonbacterial gastroenteritis. These syndromes arise from alterations in the small-bowel mucosa, with malabsorption of fat and xylose and deficiencies in brush-border enzymes such as disaccharidases. The syndromes are both characterized by watery diarrhea, without either blood or leukocytes in the stool, and by fluid and electrolyte imbalances. The two syndromes are distinguished by their clinical patterns and age predilection.

The first is represented in outbreaks of illness with abrupt onset, rapid spread, and spontaneous resolution in 24 to 48 hours. Either vomiting or diarrhea, or both, may occur, along with abdominal cramps, headache, low-grade fever, malaise, and myalgias, but the patients' distress comes mainly from the severity of gastrointestinal symptoms. This disease affects adults and school-age children, and has well earned such names as *winter-vomiting disease* and *epidemic collapse*. It is unpleasant but self-limiting. Several agents have been identified in the stools of patients from such outbreaks by use of convalescent serum and the technique of immune electron microscopy to aggregate the sparse 27-nm virus particles. To capture the virus for laboratory investigation, it has been necessary to challenge human volunteers with bacteria-free filtrates of these stools. Because Norwalk virus was the first to be identified, and studies of it have progressed the farthest, it will be taken as the paradigm for discussion.

The second type of clinical syndrome consists of 5 to 8 days of fever, diarrhea, and vomiting, preponderantly in infants 6 months to 2 years of age. The consequences of this prolonged gastrointestinal disturbance are dehydration and electrolyte imbalance, which frequently occasion hospitalization for persons in developed countries and threaten death where appropriate fluid therapy is not available. The 70-nm rotavirus particle was first recognized by electron microscopy of duodenal biopsy specimens from children with this syndrome in Australia, and this finding has been widely confirmed in diarrheal stool samples examined throughout the world.

Norwalk and Norwalklike Viruses

These agents bear the names of the locations of the outbreaks from which they derive; they are listed in Table 7.2 along with indications as to their antigenic relatedness. An epidemic in Norwalk, Ohio, affected half of the students at an elementary school. A filtrate of stool from a secondarily infected teacher in that school failed to produce disease in animals or detectable virus in cell culture. However, the same filtrate transmitted the syndrome to about 50% of the adult volunteers to whom it was fed. Stools and serums from these volunteers have become the reagents for solid-phase radioimmunoassays that detect the virus in stool and a corresponding antibody response in serum. Clinical observations and small-intestine biopsies of the same volunteers have taught us what we know about the pathophysiology of the infection. To date, Norwalk virus has not been propagated in the laboratory, nor has it produced disease in experimental animals.

TABLE 7.2 Norwalklike Viruses

Designation/Location	Antigenic Relatedness[a]	Disease Induction in Volunteers
Norwalk/United States	Distinct[b,c]	Yes
Hawaii/United States	Distinct[b,c]	Yes
Ditchling/England	Distinct[c]	Not done
Cockle/England	? Distinct[c]	Not done
Montgomery County/ United States	Related to Norwalk[b,c]	Yes
W/England	Related to Ditchling[c]	Yes
Parramatta/Australia	? Distinct[c]	Not done
Colorado/United States	Not known	Yes
Small round virus/Japan	? Distinct[c]	Not done
Marin County/United States	Not known	Not done

[a] The Norwalk, Hawaii, and Ditchling strains are antigenically distinct from one another and can be regarded as reference serotypes. Viruses listed as "? Distinct" are unrelated to one or two of the reference serotypes but have not been compared with all three.

[b] Antigenic relations were studied with cross-challenge experiments in volunteers.

[c] Antigenic relations were studied by immune electron microscopy.

Source: Reprinted from Blacklow NR, Cukor G: Viral gastroenteritis. *N Engl J Med* 304:397–406, 1981.

THE VIRION

Norwalk virus is a round, 27-nm particle that lacks an envelope. It resembles the DNA-containing parvoviruses but also the RNA-containing caliciviruses and cannot be classified definitively at this time. Its infectivity resists ether, acid, and heat. Despite the apparently small number of virions shed in stool, it gives rise to a high secondary attack rate, perhaps because the virus also occurs in vomitus and may be aerosolized from that source.

SEROLOGIC RESPONSE AND IMMUNITY

Antibody titers to Norwalk virus rise after recovery from symptomatic illness. An IgM response occurs, and detectable IgM antibody to Norwalk agent may persist as long as 20 weeks after the episode. Beyond these events, the immunologic response to this infection is most unusual. Some volunteers lack serum antibody and repeatedly resist challenge with infectious inocula without developing either illness or antibody. Others either have or develop antibody and respond to repeated Norwalk challenge with characteristic symptoms. With each episode, they develop Norwalk-specific IgM and show boosts in their IgG titers. Even local gut antibody fails to give protection and, in fact, correlates with susceptibility. Some short-term resistance to clinical illness occurs after infection, however. Thus, rechallenge after 4 to 8 weeks did not produce illness in three of four volunteers who had had repeat symptoms when the interval between challenges was 27 to 42 months.

The basis for the apparent absolute resistance to infection in some persons is still unknown.

EPIDEMIOLOGY

Norwalk virus has been implicated in 34% of epidemics of viral gastroenteritis, occurring in a large number of settings, including food-borne and waterborne outbreaks. Some food-borne outbreaks involved inadequately cooked shellfish and thus have incriminated contaminated water. Nursery outbreaks have not been recognized.

The seroprevalence of Norwalk antibody resembles somewhat that of hepatitis A. After the disappearance of maternal antibody, most persons remain seronegative until early adolescence. Thereafter, antibody prevalence rises steadily with age, until about two thirds of adults have the antibody. In underdeveloped countries, antibody develops during childhood, although still not in infancy.

PROPHYLAXIS

Beyond enteric precautions, little can be offered in the way of prophylactic measures for Norwalklike virus infections. The peculiar immune response leaves little expectation for the success of a vaccine, except perhaps in the very short term. Research into the biology of intrinsic resistance would seem to be the most promising and intriguing approach.

Rotavirus

Rotavirus is a major cause of diarrhea in infants and young children. Studies of this virus have been greatly aided by the existence of animal counterparts antigenically related to human rotavirus and useful as models for the human disease. In fact, most animal species have a rotavirus that is endemic and causes diarrhea among their young. These animal viruses, some of which grow in cell culture, have been used to prepare reagents for solid-phase immunoassays. In fact, enzyme-linked immunosorbent assay (ELISA) using antibody raised against a simian rotavirus is now available commercially for detecting human rotavirus in diarrheal stools.

In 1981, human rotavirus from stool was finally propagated in cell culture. The key to this achievement apparently was imitation of the intestinal milieu by including trypsin in the medium and the use of roller culture tubes. The trypsin has been shown to "uncoat" the virus, releasing its nucleic acid for penetration into the cells.

THE VIRION

The structure of rotavirus is better understood than that of any of the other viruses discussed in this chapter. It gains its name from its cross-sectional resemblance to a wheel (*rota,* in Latin), with the rim being the double-shelled outer capsid and the hub (a wide one), the inner core. These are connected by 32 short spokes, or capsomers. Unlike the spokes of a wheel, the capsomers actually form the capsid, so they should, perhaps, be described as more like thumbtacks in shape. The rotavirus genome consists of 11 segments of double-stranded RNA.

Virus is shed in the feces during acute illness and may persist for up to 8 days and sometimes longer. It withstands heat and ether treatment and exposure to weak acids but loses infectivity at a pH below 3, unlike the enteroviruses and Norwalk virus. At least four serotypes of rotavirus are known.

SEROLOGIC RESPONSE AND IMMUNITY

Serum antibody levels rise after rotavirus infection, but the presence of antibody does not necessarily imply resistance to infection. Infection does confer at least short-term, strain-specific immunity, however, and reinfection in childhood probably involves different serotypes of rotavirus. Both strain-specific serum antibody and secretory IgA in jejunal fluid contribute to clinical immunity in adults experimentally exposed to the virus. Secretory IgA to rotavirus is also found in human milk, and the titer rises during maternal infection.

EPIDEMIOLOGY

Half of all cases of infantile diarrhea requiring hospital admission are due to rotavirus. In temperate climates, most rotavirus illness occurs during the winter months. Spread from infected children to other patients and personnel in the hospital occurs with considerable frequency. Although adults are less likely to have symptomatic rotavirus infection than young children are, they certainly can become ill. Further, adult contacts may have asymptomatic infection, with virus shedding and the attendant risk of transmission to a susceptible child. The problem of nosocomial infection requires enteric precautions for all children with diarrhea and particular care (preferably the use of gloves) in handling diapers. The occurrence of viral transmission may necessitate cohorting children by diagnostic categories: those having gastroenteritis, those exposed to gastroenteritis at home or in hospital, and those newly admitted for unrelated disease.

PROPHYLAXIS

Children less than 4 months old have a low incidence of rotavirus gastroenteritis, suggesting a protective effect of maternal antibody and the potential efficacy of a vaccine. If a preparation could be developed that was immunogenic and protective when given within the first few months of life, it would reduce morbidity in the United States and mortality in the Third World. Natural infection occurring in the first two weeks of life with a rotavirus strain prevalent in a hospital nursery in Melbourne, Australia, seems to be asymptomatic and provided some protection against severe clinical illness in the ensuing three years. At present, breast-feeding is a practical and effective means of reducing (but not eliminating) the incidence of rotavirus disease. A bovine rotavirus is currently being tested as a live vaccine in young children, and in initial studies has stimulated immunity to naturally occurring illness.

Viral Gastroenteritis Agents of Uncertain Medical Importance

Enteric adenoviruses, recognized by electron microscopy in diarrheal stool specimens and not cultivatable in most cell culture systems, are associated with gastroenteritis in young children. These agents are serologically distinct from previously described strains of adenovirus that are readily grown in cell culture.

Coronaviruses cause diarrheal illness in newborn animals whose mothers lack breast-milk antibody to the coronavirus of their species. Particles similar in appearance have been reported in stools from babies with neonatal necrotizing enterocolitis and acute infantile gastroenteritis, but these associations remain to be confirmed. Coronaviruses have been found as frequently in the feces of asymptomatic adults as in stools from affected persons during outbreaks of diarrhea.

Caliciviruses have been convincingly linked to several outbreaks of gastroenteritis in children. Perhaps they are related to the Norwalk virus.

Astroviruses have been found in the stools of children with diarrhea, but efforts to transmit the syndrome in serial fashion by administering filtrates of these stools to adult volunteers have failed.

Clarifying the importance of these potential agents of gastroenteritis awaits the proper combination of epidemiologic, immunologic, and virologic techniques. New approaches to cultivating unusual viruses in cell or organ culture and to recognizing the replication of noncytopathic agents may be necessary. In the absence of in vitro propagation of the viruses, reagents for immunoassays must be produced in susceptible animal models or human volunteers. Whatever the source of materials, such assays require stringent controls. Nonetheless, the progress made so far in elucidating the diseases described in this chapter demonstrates the power of immunologic tools and the challenges that still face the medical scientist.

Recommended Readings

Alter HJ: Hepatitis B: A tribute to nondirected medical research. *Seminars in Liver Disease* 1:2–6, 1981.
Beasley RP, Lin C-C, Hwang L-Y, Chien C-S: Hepatocellular carcinoma and hepatitis B virus: A prospective study of 22,707 men in Taiwan. *Lancet* 2:1129–1133, 1981.

Bishop RF, Davidson GP, Holmes IH, Ruck BJ: Virus particles in epithelial cells of duodenal mucosa from children with acute nonbacterial gastroenteritis. *Lancet* 2:149–151, 1974.

Blumberg BS, Alter HJ, Visnich S: A "new" antigen in leukemia sera. *JAMA* 191:541–546, 1965.

Brandt CD, Kim HW, Yolken RH et al: Comparative epidemiology of two rotavirus serotypes and other viral agents associated with pediatric gastroenteritis. *Am J Epidemiol* 110:243–254, 1979.

Cukor G, Blacklow NR: Human viral gastroenteritis. *Microbiol Rev* 48:157–179, 1984.

Dolin R, Blacklow NR, DuPont H et al: Transmission of acute infectious nonbacterial gastroenteritis to volunteers by oral administration of stool filtrates. *J Infect Dis* 123:307–312, 1971.

Feinstone SM, Kapikian AZ, Purcell RH: Hepatitis A: Detection by immune electron microscopy of a virus-like antigen associated with acute illness. *Science* 182:1026–1028, 1973.

Grady GF, Chalmers TC, Boston Inter-Hospital Liver Group: Viral hepatitis in a group of Boston hospitals: I. A retrospective study of 1675 patients, II. A prospective controlled epidemiologic study. *N Engl J Med* 272:657–666, 1965.

Greenberg HB, Valdesuso J, Yolken RH et al: Role of Norwalk virus in outbreaks of nonbacterial gastroenteritis. *J Infect Dis* 139:564–568, 1979.

Hadler SC, DeMonzon M, Ponzetto A, et al: Delta virus infection and severe hepatitis: An epidemic in the Yucpa Indians of Venezuela. *Ann Intern Med 100:339–344, 1984.*

Kapikian AZ, Kim HW, Wyatt RG et al: Human reovirus-like agent as the major pathogen associated with "winter" gastroenteritis in hospitalized infants and young children. *N Engl J Med* 294:965–972, 1976.

Kapikian AZ, Wyatt RG, Dolin R, Thornhill TS, Kalica AR, Chanock RM: Visualization by immune electron microscopy of a 27nm particle associated with acute infectious nonbacterial gastroenteritis. *J Virol* 10:1075–1081, 1972.

Krugman S, Giles JP, Hammond J: Infectious hepatitis: Evidence for two distinctive clinical, epidemiological, and immunological types of infection. *JAMA* 200:365–373, 1967.

Krugman S, Ward R, Giles JP, Bodansky O, Jacobs AM: Infectious hepatitis: Detection of virus during incubation period and in clinically inapparent infection. *N Engl J Med* 261:729–734, 1959.

Morse LJ, Bryan JA, Hurley JP, Murphy JF, O'Brien TF, Wacker WEC: An epidemiologic report of the Holy Cross College football team hepatitis outbreak. *J Environ Health* 34:470–476, 1972.

Rizzetto M, Gerin JL, Purcell RH: Delta antigen: Evidence for a variant of hepatitis B virus or a non-A, non-B hepatitis agent? In Pollard M (ed): *Perspectives in Virology XI.* New York, Alan R Liss, 1981.

Robinson WS: The enigma of non-A, non-B hepatitis. *J Infect Dis* 145:387–395, 1982.

Stevens CE, Neurath RA, Beasley RP, Szmuness W: HB$_e$Ag and Anti-HB$_e$ detection by radioimmunoassay: Correlation with vertical transmission of hepatitis B virus in Taiwan. *J Med Virol* 3:237–241, 1979.

Szmuness W, Harley EJ, Ikram H, Stevens CE: Sociodemographic aspects of the epidemiology of hepatitis B. In Vyas GN, Cohen SN, Schmid R (eds): *Viral Hepatitis.* Philadelphia, Franklin Institute Press, 1981.

Szmuness W, Stevens CE, Zang EA, Harley EJ, Kellner A: A controlled clinical trial of the efficacy of the hepatitis B vaccine (Heptavax B): A final report. *Hepatology* 1:377–385, 1981.

8

Viral Infection of the Fetus and Newborn Infant

JOHN F. MODLIN

The fetus and newborn infant are uniquely susceptible to several viruses that cause only mild or asymptomatic disease in the mature human host. Susceptibility to infection relates both to the mechanism of transmission and to the immaturity of the fetal or neonatal immune system. The manifestations of congenital infection are frequently the consequence of viral infection during critical stages of organogenesis and organ maturation in the developing host. In addition, the persistent infections established by some viruses may have serious pathologic effects months to years after birth.

With rare exception, the mother is the source of infection for the fetus or neonate; thus, viral infection is said to be *vertically* transmitted. Viral infections so transmitted can be classified according to the timing of the maternal infection and the mode of transmission to the fetus or neonate. *Intrauterine,* or congenital, infections are those acquired by the developing fetus in utero. *Perinatal* viral infections are transmitted from the mother to the infant immediately before, during, or shortly after delivery. Because the causative agents, the pathophysiology, and the clinical course of intrauterine and perinatal viral infections differ considerably, they are considered separately in this chapter.

The TORCH acronym (T = toxoplasmosis, O = other, R = rubella, C = cytomegalovirus, H = herpes simplex virus) was conceived in the early 1970s as a guide to the diagnostic evaluation of infants with suspected congenital infection. In concept, the acronym is useful because of overlapping characteristics of congenital toxoplasmosis, rubella, and cytomegalovirus (CMV) infections. However, laboratory services for TORCH serologies have been greatly misused and are often not essential for diagnosis. The inclusion of herpes simplex virus (HSV) is ill-conceived, because HSV rarely causes intrauterine infection. Furthermore, several agents not included in the TORCH acronym are also important causes of neonatal viral infection. For these reasons, classifying neonatal infections as either intrauterine or perinatal is preferred. A diagnostic approach to each type of neonatal infection is suggested here.

INTRAUTERINE VIRAL INFECTIONS

Although many viral agents have been implicated as causes of intrauterine infection and fetal malformation, only CMV and rubella virus are proven causes of intrauterine infection. Substantial clinical evidence supports a role for varicella-zoster virus (VZV) as being a rare cause of fetal malformation. *Toxoplasma gondii,* a protozoan parasite, is an important nonviral cause of intrauterine infection that often must be considered in the evaluation of the infant with suspected intrauterine infection. Although maternal viral illness is a common event during pregnancy, many viral infections are limited to a localized site, such as the respiratory epithelium, gastrointestinal epithelium, or skin. Only viruses that produce a maternal viremia are capable of infecting the placentofetal unit.

The outcome of placental or fetal infection after maternal viremia depends on many factors, including the gestational age at the time of maternal infection. Early in gestation, transplacental viral infection may cause the fetus to die and later be reabsorbed or aborted, or the infection may produce defects during critical stages of organogenesis. Later in pregnancy, intrauterine infection may result in retardation of intrauterine growth, stillbirth, or premature labor and delivery. Fetal infection at any stage of gestation characteristically persists well beyond delivery. Infants with severe congenital viral infections will have generalized manifestions of "viral sepsis" at birth and high postnatal mortality. Many infants, including most of those with congenital CMV infections, will be asymptomatic at birth. Asymptomatic infants have an increased risk of central nervous system (CNS) developmental defects later in childhood.

Evaluation of Suspected Congenital Infection

Because most infants with congenital infection are asymptomatic at birth, the presence of infection is generally unrecognized. Occasionally maternal illness during gestation will lead to work-up of congenital infection. However, the overwhelming majority of maternal infections with agents that cause intrauterine diseases are asymptomatic. The minority of congenitally infected infants who have apparent abnormalities during the neonatal period often have nonspecific symptoms. This is especially true for infants with congenital CMV infection or congenital toxoplasmosis, in which the presence of petechiae, hepatosplenomegaly, and thrombocytopenia is characteristic of infection with either agent. However, a careful examination and consideration of the clinical, lab-

TABLE 8.1   Distinctive Clinical Features of Congenital Infections Caused by Cytomegalovirus, Rubella Virus, and *Toxoplasma gondii*

Rubella	Cataracts, congenital heart disease (especially patent ductus arteriosus and peripheral pulmonic stenosis), microphthalmia, long bone lesions
Cytomegalovirus	Microcephaly, pneumonitis, periventricular cerebral calcification
T. gondii	Hydrocephalus, chorioretinitis, scattered cerebral calcifications

oratory, and roentgenographic findings will often suggest a specific cause. Those features that help differentiate congenital infections with CMV, rubella, and toxoplasma are listed in Table 8.1.

Nonspecific laboratory tests are rarely of value. A total IgM antibody level of >20 mg/dl is more often found in babies with congenital infections than in uninfected babies. However, the test is both insensitive (50% false-negative rate) and nonspecific (60 to 80% false-positive rate). The presence of atypical lymphocytes on the peripheral blood film suggests CMV infection. Calcifications shown on the skull roentgenograms strongly suggest CMV or toxoplasma infection. Histopathologic methods are occasionally helpful. A supravital stain (Giemsa, Wright's) of urine sediment will demonstrate renal epithelial cells containing nuclear inclusions in about 50% of infants having CMV in the urine.

At present, the *optimal method* for identifying the specific cause of intrauterine infection depends on the agent being considered. These methods are shown in Table 8.2 and discussed below in more detail. Submission of sera for testing against the battery of TORCH agents is

TABLE 8.2   Methods of Choice for the Laboratory Diagnosis of Specific Intrauterine Infections

	1st Choice	2nd Choice
Cytomegalovirus	Isolation of CMV from urine throat swab or blood in cell culture. FAG stains of cells in urine	Demonstration of stable or rising CMV antibody titer, *or* demonstration of CMV-specific IgM antibody
Rubella virus	Isolation of rubella virus from urine, throat swab, or blood, *or* demonstration of rubella virus IgM antibody	Demonstration of stable or rising titer of antibody to rubella virus
T. gondii	Demonstration of IgM antibody to *T. gondii*	Demonstration of a stable or rising titer of antibody to *T. gondii*

of secondary importance. A follow-up serum specimen when the infant is 2 to 4 months of age must be tested before the routine TORCH serologies can be interpreted; in the interim, the diagnosis is usually made by other methods. Nonetheless, TORCH may be of value in situations in which other tests are not available.

Cytomegalovirus

Human CMV is a member of the herpesvirus family. The universal nature of the CMV-human host relationship is shown by the fact that 50 to 80% of adults in the U.S. have antibody to CMV. Infants, young children, and sexually active young adults have the highest rate of primary infection. Although the exact mode of person-to-person spread is unknown, intimate contact with CMV-containing oropharyngeal or genital secretions apparently is necessary for transmission. CMV is also transmitted by blood transfusion. Primary infection in the normal host is usually asymptomatic. Young adults will sometimes have a mononucleosis syndrome with a negative heterophile test during primary CMV infection. After primary infection, the virus persists in a latent state in the host, although the site of persistence and the mechanism of latency are unknown. Reactivation of CMV may occur later in life during periods of immunosuppression, such as in pregnancy and during cancer chemotherapy and organ transplantation.

CMV INFECTIONS IN PREGNANCY AND RISK OF INTRAUTERINE TRANSMISSION

Asymptomatic excretion of CMV can be found in the urine of 2 to 7% and cervical secretions of 3 to 15% of pregnant women. In general, youth, low parity, and lower socioeconomic status are associated with maternal CMV infection. Cervical CMV excretion rates also increase from 0 to 3% during the first trimester to 8 to 15% at term. Most CMV infections during pregnancy represent reactivation rather than primary infection. No maternal symptoms have been associated with gestational CMV infection.

That intrauterine transmission occurs with both primary and reactivation CMV infections during pregnancy is now firmly established. The overall rate of intrauterine CMV infection is 0.5 to 2.5% of all live births, as defined by the presence of CMV viruria at birth. Recent studies in Birmingham, Alabama, have confirmed that a mother with primary CMV infection is more likely to give birth to an infected infant than a mother with reactivation of CMV. These studies place the risk of intrauterine transmission at approximately 50% after primary infection, whereas the estimated risk of intrauterine infection after maternal reactivation of CMV infection is less than 10%. However, because reactivation of CMV infection during pregnancy is far more common than primary infection, most congenitally infected infants are born to women having recurrent CMV infections.

CLINICAL MANIFESTATIONS

Only 5 to 10% of congenitally infected infants appear abnormal at birth. The most common clinical manifestations are low birth weight, hepatomegaly, splenomegaly, jaundice, and petechiae, each occuring in most symptomatic infants. Pneumonitis is present in 30 to 40% of cases. Central nervous system and ocular defects are slightly less common in infants with overt disease. These defects consist primarily of microcephaly and chorioretinitis, although microophthalmia and cataracts are sometimes observed. Laboratory studies often demonstrate atypical lymphocytosis, thrombocytopenia, and mildly elevated transaminase levels. Characteristic periventricular intracerebral calcifications are noted on skull x-rays in about one third of the cases. With the possible exception of the irreversible effects of CNS damage, frank teratogenic effects do not result from infection with CMV.

Few infants with cytomegalic inclusion disease die in the postnatal period. Pneumonitis may produce a life-threatening illness, but this and other non-CNS manifestations gradually clear within the first few weeks of life. Unfortunately, residual CNS defects remain in more than 80% of infants with overt congenital CMV disease, regardless of whether or not CNS involvement is apparent in the neonatal period. These defects include microcephaly, mental retardation, seizures, hearing loss, and visual deficits. The prognosis for most infants who have no neonatal manifestations is far better. Nonetheless, diminished intelligence and hearing loss occur considerably more often in these infants than in uninfected siblings.

The treatment of congenitally infected infants consists of only supportive care. Many antiviral agents, including adenine arabinoside, acyclovir, and interferon, have been administered to infected infants without altering the course of disease. Some agents transiently reduce CMV excretion of CMV in the urine of infected neonates.

PATHOPHYSIOLOGY

Fetal infection presumably occurs as a result of maternal CMV viremia. The placenta may provide a relative barrier to fetal infection, because only a proportion of infants born to mothers with primary CMV infections become infected and placental CMV infection in the absence of fetal infection is known to occur. Once introduced, CMV is disseminated hematogenously through fetal tissues. Histologic and virologic evidence of CMV replication has been demonstrated in various tissues, including those of the lung, liver, reticuloendothelial system, ventricular ependyma, and the organ of Corti. Ultimately, healing is accompanied by complete resolution of the inflammatory changes in non-neural tissues. However, residual fibrosis and calcification may persist within the brain and sensory organs.

At birth, infectious CMV can be recovered from many sites and sources, including oropharyngeal secretions, cerebrospinal fluid (CSF), buffy coat cells, and urine. Longitudinal studies of CMV excretion in urine have shown high titers of infectious virus ($\geq 10^4$ TCID/ml) at birth and persistent excretion over several years. The reasons for prolonged viral replication are not fully understood. Human CMV infection is known to be immunosuppressive a priori. Congenitally infected infants express a variety of deficient T-lymphocyte-dependent responses, and the degree of immunosuppression correlates with the presence of clinical disease and the quality of virus excreted by the infant. B-lymphocyte responses are less affected. Not only are high levels of CMV-specific IgG, IgM, and IgA present at birth, but ongoing CMV replication in the face of humoral antibody leads to the formation of circulating immune complexes. Rheumatoid factor is also frequently present in the circulation. Whether the potentially immunoreactive immune complexes and rheumatoid factor play a role in the pathogenesis of congenital CMV disease is unknown.

DIAGNOSIS

The standard diagnostic procedure is isolation of CMV from the newborn infant. Urine is the specimen most frequently submitted to the virology laboratory, but CMV is often isolated from oropharyngeal secretions and the buffy coat of preservative-free, heparinized blood. Isolation of CMV in the first 3 weeks of life is diagnostic of intrauterine CMV infection; beyond that age, intrauterine infection cannot be distinguished by virus isolation alone from infection contracted at birth. A simple and rapid technique for detecting CMV virions in urine by electron microscopy has been described; this method identifies about 80% of urine specimens that contain infectious CMV. Simple staining of a centrifuged urine sediment with Wright's or Giemsa stain for inclusion-bearing renal epithelial cells is a less sensitive, but an extremely specific, rapid, and simple means of detecting the presence of CMV viruria.

Serologic methods are more complicated and less useful than CMV isolation. The diagnosis can be made by demonstrating a stable or rising titer of CMV antibody during the first 3 or 4 months of life. This method necessitates that sera taken in the first few days of life and also at 2 to 4 months of age be tested. Conventional serologic methods (CF, IHA, IFA) measure both IgM and IgG antibody. Although not widely available, the CMV-specific radioimmunoassay for IgM antibody has proved to be a sensitive and specific means of diagnosing congenital CMV infection.

PREVENTION

Preventing primary maternal CMV infections is virtually impossible because of the ubiquitous nature of the virus and virtual absence of specific symptoms. It is prudent for pregnant women to avoid direct contact with congenitally infected infants and other patients known to be excreting CMV, although the value of isolating infected patients is unknown.

Rubella Virus

Rubella virus is an RNA togavirus that causes a mild exanthemous illness known as rubella, or German measles. The primary medical and public health importance of rubella infection concerns the ability of the virus to infect the fetus in utero when rubella occurs during pregnancy. Otherwise, the effects of postnatal rubella infection are largely inconsequential.

The introduction in 1969 of live, attenuated rubella virus vaccine has greatly altered the epidemiology of rubella. Before this from 75 to 85% of persons (except in geographically isolated populations) had rubella infection by late adulthood. Infection frequently happened during extensive epidemics, which occurred every 6 to 9 years. Although rubella virus infects susceptible persons of all ages, the highest attack rates occurred in school-aged children. Widespread use of rubella vaccine has reduced rubella attack rates in all age groups, but most strikingly in young children. Consequently, most cases now occur among adolescents and young adults, many of whom are of childbearing age. Institutional outbreaks among high school students, college students, and hospital employees have been a prominent form of transmission of rubella in recent years.

RUBELLA INFECTIONS IN PREGNANCY AND RISK OF INTRAUTERINE TRANSMISSION

Although subclinical reinfection occurs in persons with both natural and vaccine-induced immunity, such reinfections do not affect the fetus. From 15 to 25% of women of childbearing age are seronegative, and thus capable of transmitting rubella virus to the developing fetus when primary rubella infection occurs during pregnancy. The risk of maternal rubella relates directly to the number of rubella infections occurring in the community.

When maternal infection does occur, the risk of transmission to the fetus depends largely on the gestational stage at the time of infection. During the first 8 weeks of gestation, the risk of fetal infection is 50 to 80%. Thereafter, the risk falls rapidly, so that by 16 weeks of gestation, intrauterine transmission occurs in fewer than 10% of cases.

CLINICAL FEATURES OF CONGENITAL RUBELLA SYNDROME

Most infected infants will have manifestations detectable at birth, although 20% or more may at first appear normal. The classic triad of defects attributable to congenital rubella are sensorineural hearing loss, cataracts, and congenital heart disease. The hearing loss, which usually results from damage to the organ of Corti, may be responsible for the delay in language development in the early years of life. The cataracts are typically found in the central, posterior portion of the lens; they may occur unilaterally or bilaterally. Patent ductus arteriosus is the most common type of congenital heart disease, but peripheral pulmonary artery stenosis, atrial septal defects and ventricular septal defects also occur. In addition, low birth weight due to intrauterine growth retardation, chorioretinitis, hepatosplenomegaly, radiographic lesions in long bones, and thrombocytopenia are common features of the congenital rubella syndrome. Maternal rubella during the first 12 weeks incurs the risk of congenital heart disease, while hearing loss and chorioretinitis may be caused by maternal rubella contracted at any time during the first 16 weeks of gestation.

Additional abnormalities may become apparent beyond the neonatal period, regardless of whether symptoms were present initally. A "late-onset" rubella syndrome has been described as beginning at 3 to 6 months of age and consisting of failure to thrive, interstitial pneumonia, skin rash, lymphocytic meningitis, thrombocytopenia, and neurologic deterioration. Hypogammaglobulinemia and circulating immune complexes have also been described in infants with congenital rubella. Later in life, endocrine abnormalities such as diabetes mellitus and lymphocytic thyroiditis are well known, and a progressive panencephalitis similar to subacute sclerosing panencephalitis has been described.

Only about 15% of congenital rubella patients die in the first decade of life. The prognosis for survivors largely depends on the degree of CNS involvement and the developmental effects of the visual and auditory deficits. The long-term follow-up studies of Cooper and others have shown that two thirds of congenital rubella patients have varying degrees of hearing loss, and half are mentally retarded.

PATHOPHYSIOLOGY

Pathologic evidence of placental involvement is often found following maternal rubella, mostly as small, necrotic foci within the endothelium of small placental blood vessels. The placenta may act as a relatively effective protective barrier because placental lesions may be found in the absence of fetal infection, particularly when maternal rubella occurs after 20 weeks' gestation. Infected fetuses examined at autopsy have multiple foci of vascular endothelial necrosis throughout many organs. In vitro studies suggest that rubella virus may inhibit cell division and growth and cause chromosomal breaks in cultured leukocytes. These mechanisms, and probably others, are responsible for the embryopathic effects of persistent rubella virus infection of the fetus. Rubella virus infection of the congenitally infected newborn infant is generalized; virus has been recovered from urine, oropharyngeal swabs, feces, CSF, blood, and bone marrow. Active viral replication persists for weeks to months; virus has been recovered from the urine of infants with congenital rubella up to 19 months of age.

IgG, IgM, and occasionally IgA antibodies to rubella virus are present in the serum of infected infants at birth, following which antibody levels decline more rapidly than in individuals who contract rubella later in life. Hemagglutination-inhibition (HI) antibody to rubella virus is no longer detectable in 15 to 20% of congenitally infected children beyond the age of 4 years. Both hypogammaglobulinemia and hypergammaglobulinemia have been described in the congenital rubella syndrome. The humoral immune response to antigens such as tetanus and diphtheria toxoid is sluggish. Deficient T-cell response to nonspecific mitogens and to purified rubella antigen occur in the more severely affected congenital rubella patients. Children with congenital rubella who have lost detectable rubella HI antibodies respond poorly, or not at all, to attenuated rubella vaccine.

DIAGNOSIS AND MANAGEMENT OF MATERNAL RUBELLA

It has become standard obstetrical practice to determine the immune status of pregnant women serologically, often at the first prenatal visit. A history of rubella, or German measles, is unreliable unless diagnosed by laboratory methods. Fetuses of seropositive women are not at risk.

The following steps are recommended for pregnant women who are seronegative, or of unknown immune status, when they are exposed to a known or suspected case of rubella: (1) A serum specimen should be obtained and split, one half being sent to the laboratory for immediate determination of rubella antibody. (2) If this specimen is positive for antibody, the patient can be reassured; if she is seronegative, a second specimen should be obtained in about 3 weeks, regardless of whether the woman develops symptoms of rubella. (3) The second serum specimen should be tested for rubella antibody *simultaneously* with the remainder of the first specimen. Seroconversion from negative to positive antibody status is diagnostic of primary rubella infection.

The rubella HI test remains the standard for diagnosis of postnatal rubella infection. For routine testing, both acute and convalescent sera are required. Alternatively, demonstration of the presence of IgM antibodies to rubella virus is diagnostic of acute infection occurring within the preceding 4 to 6 months; however, this test is of limited availability. In recent years, a number of rapid, inexpensive serologic tests for determining a person's rubella-immune status have been introduced, none of which, however, allows a distinction to be made between IgM and IgG antibody. These commercially available tests are therefore not appropriate for diagnosis of acute or recent infection.

DIAGNOSIS AND MANAGEMENT OF CONGENITAL RUBELLA SYNDROME

Diagnosis of maternal rubella during pregnancy demands a thorough clinical and laboratory evaluation of the infant at birth. The diagnosis of congenital rubella rests on either recovery of rubella virus from the newborn infant or serologic evidence of intrauterine rubella virus infection. The latter is best accomplished by demonstration of the presence of IgM antibodies to rubella virus in cord serum or serum taken from the infant during the first weeks of life. A stable or rising level of rubella HI antibody during the first 2 to 4 months of life is also diagnostic of congenital rubella syndrome.

There is no specific therapy for infants congenitally infected with rubella virus. Most children with rubella syndrome have neurodevelopmental or perceptual deficits, or both, that require early intervention and attention throughout childhood.

PERINATAL VIRAL INFECTIONS

Several viruses are known to be transmissible from mother to infant during the perinatal period, namely, CMV, HSV, VZV, enteroviruses, and hepatitis B virus. In addition, there is clinical evidence that influenza, mumps, measles, and adenoviral infection may occasionally be contracted by the neonate. The mechanism of transmission and outcome of neonatal infection vary widely among these agents. Transmission may occur at any time from a few days before to a few days after parturition, although some viruses, particularly HSV, are transmitted principally during labor and vaginal delivery. Characteristically, infected infants appear well at birth, manifesting clinical evidence of infection only after a variable incubation period. HSV, VZV and some of the enteroviruses can cause severe, frequently fatal, disease in the neonate. On the other hand, perinatal CMV infection is rarely, if ever, responsible for overt disease. Similarly, hepatitis B virus rarely causes clinical hepatitis in newborns, who may, however, become chronic carriers of HBV.

Evaluation of Suspected Perinatal Viral Infections

A carefully obtained history of all past maternal infections is important. This record should include information not only on symptoms of recent illness, but also on any history of or exposure to genital herpes or varicella-zoster. Women who have acute hepatitis B infection in middle to late pregnancy are at high risk for transmitting HBV to their newborns during labor and delivery. Geographical, social, and occupational characteristics help to define the risk that a pregnant woman may be an asymptomatic carrier of HBV. These factors are discussed in greater detail later.

In general, the diagnosis of perinatal viral infection depends on isolation and identification of the suspected agent from the infected neonate. With the exception of HBV infections, serologic tests are only minimally useful because they require that both acute and convalescent specimens be examined.

Herpes Simplex Virus

Two related, but biologically separate, types of herpes simplex virus (HSV) cause human infection. HSV type 1 (HSV-1) is transmitted predominantly by orolabial contact, although various types of sexual exposure and other means of direct contact are responsible for spread of some HSV-1 infections (e.g., whitlow). Infection with HSV-1 is so common that 50 to 80% of the general population have the corresponding antibody, although only a few show clinical symptoms. Nonvertical transmission of HSV-2 occurs exclusively through sexual contact, the risk of genital infection depending largely on social factors, including the degree of sexual promiscuity. In general, 10 to 30% of women of childbearing age have specific antibody to HSV-2 antibody. Both types of HSV show a propensity to remain latent in regional sensory nerve ganglia after primary infection and to cause recurrent mucocutaneous lesions throughout life. Recurrent genital HSV infections are predominantly due to HSV-2, whereas recurrent orolabial lesions (cold sores) are almost exclusively caused by HSV-1. Both types are rare causes of more serious disease in normal persons (e.g., encephalitis) or in immunocompromised patients.

MATERNAL HSV INFECTIONS IN PREGNANCY AND VERTICAL TRANSMISSION

Although intrauterine transmission of HSV has been documented, the major risk to the newborn comes from infection acquired during the perinatal period. The maternal genital tract is the source of most perinatal HSV infections for infants who become infected during vaginal delivery. Neonatal infection may occur after either primary, recurrent, or asymptomatic genital herpes infection in the mother. Recent data have shown that 80% of women with a history of genital herpes will have symptomatic recurrences sometime during pregnancy, most having more than one recurrence during a single gestation. There are no definitive data regarding the rate of asymptomatic shedding of HSV from the maternal genital tract at term, although combined information from several recent studies suggests that a rate of 1% might be expected among women with a history of genital herpes.

Primary maternal genital HSV infection is suspected to represent a greater risk to the infant at term than recurrent or asymptomatic infection. During a primary infection virus is shed in higher titer for longer periods than in recurrent disease. If symptoms of primary herpes (i.e., a first attack) develop late in pregnancy, there may not be enough time for IgG antibody to develop and cross the placenta to provide protection for the infant at term. However, there are no data to substantiate the possibility that primary infection during pregnancy poses any greater inherent risk to the infant than recurrent maternal disease, which accounts for the majority of cases. The actual risk of disease occurring in a newborn infant born vaginally to a woman with active recurrent lesions at term is unknown, but considered to be significant. Therefore, cesarean delivery is usually performed to prevent exposure of the parturent infant to maternal lesions. This leaves those who are asymptomatic but still shedding HSV as an important source of neonatal infection. In a recently published study of the natural history of neonatal herpes simplex, it was reported that 70% of infants found to be infected had come from mothers who were asymptomatic at the time of delivery.

CLINICAL MANIFESTATIONS OF NEONATAL INFECTIONS WITH HSV

Whether maternal genital herpes predisposes the mother to premature labor and delivery or whether premature or otherwise compromised infants are inherently more susceptible to infection than normal term infants is an unanswered question. In any event, neonates found to be infected with HSV are more likely to have been premature or to have suffered other perinatal complications. Although there have been a few reports of clinically evident herpes during the first day or two of life, the majority of infants infected at birth do not develop overt disease until they are 1 to 3 weeks of age. In 80% of cases, neonatal herpes presents first with skin lesions; and 70% of those initially showing cutaneous herpes will develop progressive, systemic disease as a result of hematogenous dissemination of virus to liver, spleen, brain, lungs, and kidneys. In others, the infection remains localized to skin and mucous membranes or to the eye. About 20% of infected infants never develop skin lesions, but are found to have incurred disseminated disease or infection localized to the CNS.

The skin lesions of neonatal herpes occur as isolated or clustered vesicles surrounded by erythema, which turn rapidly into shallow encrusted ulcerations. The lesions appear most often in anatomic sites usually exposed during parturition (face, scalp, perineum) or at sites of trauma caused by application of scalp electrodes or forceps. In most instances, the cutaneous eruption tends to remain localized. In a few cases, however, extensive skin involvement resembling epidermolysis bullosa has been reported. The first manifestations of disseminated herpes simplex often suggest bacterial sepsis. The infant is lethargic, feeds poorly, and later develops apnea and progressive jaundice. Within 48 hours of onset, frank hepatitis, pneumonia, encephalitis, thrombocytopenia, or diffuse intravascular coagulation may supervene. Seizures, hypotension, and intractable bleeding presage a fatal outcome.

Some infants slightly older (i.e., 2 to 4 weeks) than those with disseminated disease (2 to 14 days) will present with signs and symptoms that are limited to the CNS. After a brief period of general lethargy and irritability, CNS involvement is heralded by the onset of focal motor seizures that may become repetitive and respond poorly to anticonvulsant therapy. Opisthotonos, high-pitched cry, diminished reflexes, hy-

pothermia, and apnea all signify severe CNS disease. Low-grade pleocytosis with a predominance of lymphocytes is usually found in the CSF.
The electroencephalogram generally shows abnormal periodic slow wave
complexes that tend to recur throughout the record and indicate diffuse involvement of the brain.

PATHOPHYSIOLOGY

Although transplacental infection with HSV is well documented, the
dominant source of perinatal infection is the maternal genital tract during the perinatal period, most neonatal infections being acquired during the infant's passage through the birth canal. The major evidence
supporting this mechanism of transmission is the protection apparently
provided through delivery by cesarean section. Premature rupture of
the membranes, however, exposes the fetus to virus that may ascend
from the lower genital tract.

In the immunocompetent human host, cell-mediated immune mechanisms are of central importance in recovery from infection with HSV.
In the newborn infant, immunologic immaturity is the most critical factor in determining the extent of disease following perinatal infection
with HSV. In this connection, in vitro studies have shown neonatal lymphocytes to have relatively deficient immune responses. The role of humoral antibody in modulating neonatal infection is uncertain. Women
with recurrent genital herpes usually have high titers of IgG antibody,
which can be measured in cord blood. There is some clinical evidence
to suggest that the presence of this transplacentally acquired antibody,
though not fully protective against infection, may be associated with
disease of reduced severity in the newborn infant.

DIAGNOSIS

Cells scraped from the base of a mucocutaneous lesion can be mounted
and stained rapidly with Giemsa or Wright to reveal the multinucleated
cells and intranuclear inclusions characteristic of herpes virus infection
(Tzanck prep). Specific HSV antigen can be identified in frozen or appropriately fixed tissues by immunofluorescence or immunoperoxidase
methods. Definitive diagnosis usually rests on the recovery of HSV in
cell cultures inoculated with vesicle fluid or scrapings from mucocutaneous lesions. In infants with disseminated infection, the virus can be
recovered from throat swabs or from samples of urine, feces, blood, or
CSF. Diagnosis may be difficult in infants in whom infection is limited
to the CNS. Recent clinical experience suggests that in these cases specimens from peripheral sites, blood, and CSF are unlikely to contain infectious HSV and that a definitive diagnosis can be made only through
brain biopsy.

Infants who survive neonatal herpes can be expected to produce detectable antibody, but only after an interval of 10 to 14 days following
onset. Serologic studies, therefore, rarely help in the initial diagnosis or

managment of the disease. Acute and convalescent serum samples, however, can be used toward serologic confirmation of the diagnosis, particularly when attempts at virus isolation have failed or appropriate cell culture facilities are not readily available.

PREVENTION AND TREATMENT

Retrospective clinical data indicate that the risk of infection is about 50% for infants delivered vaginally to women with active genital herpes, and that this risk is reduced to less than 10% with delivery by cesarean section. Although these data have not been confirmed in controlled studies, it has become standard practice to recommend section at term for women with clinically evident genital herpes, although the benefit may be lessened if the membranes have been ruptured more than 4 hours before delivery. In pregnant women with a history of genital herpes, and in the absence of clinically apparent diseases, it is generally recommended that, beginning at the 34th to 36th week of gestation, weekly cultures for HSV be taken of the labia and cervix in order to detect asymptomatic shedding of virus. In the absence of symptoms, the route of delivery is determined at the onset of labor by results of the most recent viral culture. Although the risk seems to be very low, postnatal transmission from maternal extragenital herpes (i.e., cold sores) as well as from nonmaternal sources is well documented. Nevertheless, a mother with either genital or oral herpetic lesions need not be restricted in her contact with her infant, provided precautions are taken to prevent its direct exposure to sites of active disease. Breast-feeding is permissible, except in the presence of herpetic lesions on the breast.

The recent development of antiviral agents that are active against HSV has somewhat improved the prognosis of perinatally infected infants. A national collaborative study has established that adenine arabinoside (vidarabine), in daily intravenous doses of 15 mg/kg body weight, reduces mortality and improves the outcome of disease in those infants who survive primary infection. Acyclovir [9 (2-hydroxyethoxymethyl)-guanine] is another promising agent, which shows enhanced in vitro activity against HSV. Although it has been proved to be clinically effective in other types of severe HSV infections in adults, controlled clinical studies in neonates are incomplete. The intravenous daily dose is 750 mg/m^2/day. With either vidarabrine or acyclovir, therapy should be instituted as early in the course of the disease as possible.

PROGNOSIS

The outcome in untreated cases of neonatal herpes is grim, with severe neurologic impairment being the rule among survivors. The mortality rates for infants with disseminated disease or with encephalitis alone are 75 and 50%, respectively. Even when disease appears to remain localized to the skin, the mucous membranes, or the eyes, about a third of the infants will develop serious sensory or other neurologic deficits.

Early therapy with vidarabine reduces mortality of disseminated disease to 40% and that of encephalitis to 10%. However, many survivors are left with serious developmental handicaps.

Enteroviruses

The enteroviruses are small, nonenveloped, RNA viruses that belong to the family of picornaviruses. Sixty-seven distinct serotypes are known to cause human disease. Traditionally the enteroviruses have been divided into four classes, according to certain biologic characteristics: polioviruses, Coxsackie A viruses, Coxsackie B viruses, and echo viruses. Recently, it has become conventional to designate newly discovered human serotypes simply as enterovirus, for example, enterovirus 71, without regard to the aforementioned classification.

The enteroviruses are common human pathogens. In temperate climates, enteroviral diseases occur most frequently in summer and fall; but they may be encountered at any time of year. Usually, a few serotypes will circulate seasonally in a community, causing sporadic human illness. At intervals of about 5 to 8 years, a single serotype may predominate and infect a large proportion of the population to cause localized outbreaks of the disease. Clinical attack rates are inversely related to age.

Only a small fraction of human enteroviral infections are symptomatic, although in young children clinical illness is likely to be more evident than in adults. The most common clinical manifestations include undifferentiated febrile illnesses, often with pharyngitis, exanthem, and myalgia. Many enteroviral serotypes cause aseptic meningitis. A number of less common clinical manifestations have generally been associated with specific enteroviruses, such as poliomyelitis with poliovirus, herpangina with Coxsackie A virus, pleurodynia and myocarditis with Coxsackie B virus, and hemorrhagic conjunctivitis with enterovirus type 70. When maternal enterovirus infection occurs in the perinatal period, the virus may be transmitted to the fetus or neonate, sometimes with devastating results. In the United States, before poliomyelitis was brought under control, maternal paralytic poliomyelitis caused a 50% mortality among infants infected perinatally with poliovirus. Coxsackie B virus (serotypes 1 through 5) and echo viruses, especially type 11, continue to cause severe, and frequently fatal neonatal infections.

MATERNAL ENTEROVIRUS INFECTION AND PERINATAL TRANSMISSION

The clinical manifestations of maternal enteroviral infection in pregnancy depend on both host factors and the specific type of enterovirus involved. Clinical studies of pregnant women with poliovirus infections and more recent studies in animals suggest that pregnancy itself may augment the severity of enteroviral infection. In pregnant mice, the most severe infections occur during the late stages of gestation.

In pregnant women, Coxsackie B virus infections are associated with several clinical manifestations including fever, pharyngitis, and myalgia. Specific symptoms have not been associated with echo viral infection contracted during pregnancy. However, during a recent outbreak of illness caused by echo virus type 11, about half the women who gave birth to infected infants had had a syndrome of fever and lower abdominal pain that, in some instances, was misdiagnosed as abruptio placentae.

If teratogenic effects result from enteroviral infection in early pregnancy, they are infrequent. Retrospective serologic studies have suggested that the risk of congenital heart disease increases slightly in infants of mothers who contract infections with either Coxsackie B3 or B4 viruses during pregnancy. Undoubtedly the neonate is most threatened when maternal infection occurs during the peripartum period. In reported cases of perinatal enterovirus transmission, the onset of symptoms in the mother occurred in some cases up to 5 days before delivery as well as in the intrapartum period; other cases were asymptomatic. An unexpectedly high number of infants delivered by cesarean section, undertaken in some cases because of maternal illness, have been found to be infected with an enterovirus. This finding is strong evidence that transmission of virus to the fetus or infant does not depend on passage through the birth canal or contact with the maternal perineum.

CLINICAL MANIFESTATIONS

In most perinatally infected infants, onset of symptoms occurs at 2 to 6 days of age, and rarely within the first 24 hours after birth. The infected infant presents with fever, irritability, lethargy, poor feeding, or jaundice, symptoms that may initially suggest bacterial sepsis. Within 24 to 48 hours, however, the infant's condition generally deteriorates, with evidence of diffuse visceral involvement.

The dominant features of Coxsackie B virus infection in the neonate are myocarditis and meningoencephalitis. Myocarditis is heralded by the sudden onset of congestive heart failure, with arrhythmias, tachypnea, tachycardia, hepatomegaly, cardiomegaly, and elevated enzymes indicative of cardiomyopathy. Signs of meningoencephalitis may range from irritability and poor feeding to seizures and coma. The CSF usually shows a pleocytosis with a predominance of lymphocytes. In addition, some infants with Coxsackie B virus infection have been reported to have hepatitis, pancreatitis, and adrenal necrosis. The course of neonatal Coxsackie B virus infection is highly variable. During nursery outbreaks, death rates of 30 to 60% have been reported. In survivors, the noncardiac manifestations often improve within days, but residual myocardial damage in some infants may lead to persistent cardiac failure.

Neonatal echo viral infection has been less frequently reported than Coxsackie viral infection. Although the features of echo viral neonatal infections may overlap somewhat with those of Coxsackie B viral infection, the cardinal finding has been overwhelming hepatic necrosis. Diffuse, uncontrollable bleeding secondary to hepatic failure and consumption coagulopathy is often a dramatic feature. Intraventricular hemorrhage leads to seizures and coma, and renal medullary hemorrhages are found in infants with renal failure. Severe hypotension and uncontrollable bleeding are terminal events. At postmortem examination, there is hepatic and often adrenal necrosis, but little evidence of inflammatory changes in other organs. The mortality from perinatally acquired echo viral hepatitis is 80% or greater.

In addition to these syndromes, many milder cases of disease in young infants have been reported to be caused by a variety of enteroviruses, including some serotypes of Coxsackie A virus. These self-limited infections have been responsible for exanthems, aseptic meningitis, lethargy, poor feeding, and nonspecific febrile illness.

PATHOPHYSIOLOGY

Neonatal enteroviral infections are acquired from both maternal and nonmaternal sources. The mechanism of vertical transmission is unknown. However, because some infants have virologically proved infections at birth, infection clearly can take place in utero before delivery. Studies in experimental animals suggest that transplacental infection occurs during maternal viremia but that the newborn may not manifest infection until several days after birth. Ascending infection from the maternal genital tract and intrapartum infection are two other possible routes of vertical transmission. Transplacental antibody from the mother does not seem to prevent perinatal infection, but it is probably important in modifying or preventing serious disease. Reports of outbreaks in newborn nurseries indicate that occasionally enteroviral disease is transmitted postnatally, and in some cases, nosocomial transmission from infected staff members has been implicated.

Severe neonatal infection is characterized by rapid, systemic spread of virus, by means of an overwhelming viremia. Virus can usually be recovered from multiple clinical specimens, including oropharyngeal secretions, stool, urine, blood, and CSF. At postmortem examination, only a few target organs (e.g., heart, liver) may show evidence of active inflammation, but virus can be recovered from many organs.

The reasons for the innately high susceptibility of the neonate to enteroviral infection and for its failure to control such infection are under active laboratory investigation. Studies in mice and in cultured primate cells show that neonatal tissues have a greater cell-binding affinity for some enterovirus serotypes. Furthermore, the immaturity of the macrophage function in the newborn may also be important, because of the important role that the macrophages have been known to play in the immune response.

DIAGNOSIS

The laboratory diagnosis of neonatal enteroviral infection depends on isolation and identification of virus in cell culture or in newborn mice. Because of the potential for nosocomial spread of enteroviral disease, identifying the source of infection may be important. Oropharyngeal and fecal specimens for virus isolation should be obtained from the mother, whether or not she has had symptoms of viral infection.

MANAGEMENT

There is no established means for preventing neonatal enteroviral infections, nor is there a specific antiviral therapy. Management consists of supportive care. Infected infants shed virus from multiple sites and must be isolated from other newborn infants, with full infection control precautions.

Varicella-Zoster Virus

Varicella-zoster virus (VZV) is a common cause of human infection. Primary infection occurs predominantly in childhood in the form of varicella (chickenpox). Reactivation of latent VZV later in life is the cause of herpes zoster (shingles), a syndrome of cutaneous vesicular eruption and neurogenic pain, generally within a single dermatome. Except in the immunocompromised host, VZV infections are self-limited and of little long-term consequence, although severe postherpetic neuralgia may be both incapacitating and intractable. Maternal VZV infection that occurs before the third trimester carries a low degree of risk for causing congenital defects in utero. The major risk to the neonate occurs when maternal varicella presents during the perinatal period.

MATERNAL VARICELLA AND RISK OF PERINATAL TRANSMISSION

Because 95% of persons living in the United States have antibody to VZV by 15 years of age, varicella in pregnancy is rare. Women coming from tropical countries, particularly tropical islands, are less likely to have had childhood varicella and thus are at greater risk of becoming infected during pregnancy. About 25% of women who develop varicella within the last 3 weeks of pregnancy will transmit infection to their offspring. Although the risk of neonatal infection seems to vary little with the timing of maternal illness, the risk of severe illness and mortality varies considerably. Data collected by the Centers for Disease Control show that the case-fatality rate for newborns of mothers who develop the characteristic rash within 4 days before delivery is about 30%, compared with no mortality for infants of mothers with onset of rash more than 4 days before delivery.

Evidently, the longer the period between onset of rash and delivery the greater the opportunity for development of maternal IgG antibody to VZV and hence passive protection of the fetus. Development of chickenpox in the mother up to 2 days *after* delivery must also be considered a high risk to the newborn unprotected infant, although there are no published data to substantiate this supposition. Herpes zoster in late gestational women is not considered to be a threat to the fetus or newborn, both because communicability of VZV is low and antibody is always present to provide passive transplacental protection.

CLINICAL MANIFESTATIONS OF NEONATAL VARICELLA

Infants with perinatal VZV infection may already have the characteristic rash at birth, or may develop it at any time up to 10 days of age, depending on the chronology of the maternal infection and hence the degree of passive protection of the fetus effected by transplacental maternal antibody. In general, infants who develop the rash within the first 5 days of life suffer milder disease than those with later onset of the rash. The clinical spectrum of neonatal varicella varies from mild illness to overwhelming systemic disease, characterized by extensive skin lesions and pneumonia, the usual cause of death. Postmortem examination, reported in only a few fatal cases, has shown scattered foci of necrosis in many organs, including skin, lungs, liver, spleen, adrenal glands, kidneys, pancreas, and gastrointestinal tract.

PATHOPHYSIOLOGY

The frequency with which varicella skin lesions have been reported to be present at birth confirms the impression that VZV may be transmitted across the placenta, probably in the course of maternal viremia. Infants who develop varicella during the first 10 days postpartum are also likely to have acquired their infection in utero. Relatively little is known, however, about the pathophysiology of perinatal varicella, because the number of cases reported (50) is limited and there is as yet no animal model in which to analyze the infection. As already noted, clinical observations suggest strongly that the extent and severity of neonatal disease depend on many factors, chief among which are the timing of maternal varicella and the concomitant immune response in relation to delivery and appearance of disease in the infant, as well as the limitations of the neonate's own immune defenses.

MANAGEMENT AND PREVENTION OF NEONATAL VARICELLA

Varicella-zoster immune globulin (VZIG) has been shown to prevent or modify varicella infections in immunocompromised children when given within 3 days of exposure. Although no similar data exist regarding protection of perinatally exposed infants, it is now recommended that these infants be given VZIG at birth if their mothers develop the rash of varicella between 4 days before and 2 days after birth. In addition, there are two licensed antiviral drugs that are active against VZV in vitro (acyclovir, vidarabine). Although no data exist for their use in neonatal varicella, one of these drugs should be given to infants with visceral involvement.

Hepatitis B Virus

About 20 years have elapsed since the first descriptions of "Australia" antigen in the serum of Australian aborigines and its association with long-incubation (serum) hepatitis. In the ensuing explosion of knowledge about the virology, epidemiology, and pathophysiology of hepatitis B, the virus has been identified as a 42-nm particle containing dsDNA, reverse transcriptase, and capsid proteins that include the surface antigen (HB_sAg), the original Australia antigen. The nature of the public health problem of hepatitis B varies with geopolitical and socioeconomic factors. In southeast Asia and sub-Saharan Africa, HBV infects virtually the entire population, and 15 to 20% of persons become chronic carriers of HBV. In these individuals, HB_sAg persists in the serum as a result of continued replication of HBV in the liver. Chronic carriers of HBV are at increased risk of developing chronic hepatitis, cirrhosis, and malignant hepatomas. The high carrier rate in these population groups is thought to result from perinatal vertical transmission of the virus.

In the United States and other developed countries, the problem of infection with HBV differs considerably. The overall prevalence of antibody to HBV is much lower, that is, 5 to 10%, and hepatitis is largely confined to persons who are at risk of acquiring virus percutaneously from blood or blood products or through sexual contact with infected individuals. The high-risk categories include hemophiliacs and others who require transfusion with large quantities of blood (e.g., cardiac surgical patients), intravenous drug abusers, medical personnel sustaining accidental trauma from needles or other instruments contaminated with blood or blood products (especially surgeons, dentists, surgical nurses, hemodialysis and laboratory personnel), homosexual males, and heterosexual contacts of chronic HBV carriers. Risk of primary infection also carries the secondary risk of the chronic carrier state.

The overall risk of developing the chronic carrier state after acute HBV infection is about 6%, but this figure is higher among persons infected as young children or during periods of immunosuppression. The overall chronic carrier rate in the United States is therefore extremely low, that is, less that 1%. The carrier rate is much higher among the aforementioned increased-risk groups, as well as among immigrants from parts of the world where HBV infection is endemic.

MATERNAL HBV INFECTION AND RISK OF PERINATAL TRANSMISSION

The potential for vertical transmission occurs either when the mother has acute HBV infection during pregnancy or when the mother is a chronic carrier of HBV. In the former case, the primary risk to the newborn exists when onset of maternal hepatitis occurs during the last trimester, or shortly after delivery. These infants have more than a 75% chance of becoming infected. In contrast, the risk of perinatal transmission is close to zero when maternal hepatitis occurs early enough in pregnancy and antibody to HB_sAg develops before parturition.

Numerically, more infants are at risk from mothers who are chronic carriers than from mothers who have acute infection in late pregnancy. Among chronic carrier mothers, the incidence of vertical transmission depends greatly on whether or not HB_eAg is present in the mother at delivery. Women who are HB_eAg-positive have more than a 90% chance of transmitting HSV to their newborn infants, whereas the probability of infection among infants of HB_eAg-negative mothers is less than 5%. About 20% of women of childbearing age who are chronic carriers are HB_eAg-positive.

CLINICAL MANIFESTATIONS OF PERINATAL HBV INFECTION

Because the newborn infant is not usually exposed to maternal HBV until parturition, markers for HBV infection do not appear in the infant's blood until after an incubation period of 2 to 3 months. Although acute, icteric hepatitis is known to occur in some perinatally infected infants, 80 to 90% remain asymptomatic. Some clinically well infants who have been carefully followed have developed only mild hepatic enlargement and modest elevation of serum transaminases.

Although some infected infants resolve their infections with expression of serum antibody to HB_sAg (HB_sAb), most remain HB_sAg-positive and when retested at least 6 months later are found still to lack antibody. These infants thus meet the definition of chronic carrier status, which persists for many years, perhaps for life. As yet, there are no firm data on which to predict the consequences of the chronic carrier state that follows infection acquired during infancy. However, most workers in the field strongly suspect these individuals have high risk for developing chronic active hepatitis, cirrhosis, and primary hepatocellular carcinoma (hepatoma).

PATHOPHYSIOLOGY

The presence of HB_eAg indicates the presence of the infectious HBV virion (Dane particle) in maternal blood and a high probability that HBV will be transmitted to the newborn at birth. The exact mechanism of transmission is unknown, but it is likely to involve exposure of the infant to maternal blood during labor or delivery. HB_sAg has been found in cord blood, amniotic fluid, and newborn gastric aspirates, but these findings seem to correlate poorly with neonatal infection. HB_sAg is also found in the breast milk of mothers who are chronic carriers, but breast-feeding does not increase the risk of perinatal infection.

The serologic events that accompany neonatal infection have not been well characterized. Longitudinal studies of adult human volunteers show that HB_sAg appears in serum 8 to 15 weeks after exposure, followed shortly by evidence of the circulating Dane particle. The elevation of serum transaminase levels that follows is accompanied by the appearance of serum antibody to the HBV core antigen (HB_cAb). Persons with self-limited infection tend to develop clinical illness, but they rapidly clear HB_sAg from the serum and later develop antibodies to HB_sAg. Subclinical hepatitis is the rule among those who are destined to have persistent infection and to become chronic carriers of HB_sAg. Chronic carriers do not make antibody to HB_sAg, but they tend to have persistently high levels of HB_cAg.

DIAGNOSIS

The diagnosis of hepatitis B depends on demonstration of one or more serologic markers for HBV infection. The presence of HB_sAg in serum indicates current infection. Persistent infection has been defined as the presence of HB_sAg in serum specimens taken at least 6 months apart. HB_sAb is indicative of past infection and permanent immunity. Patients with permanent natural immunity also have low levels of HB_cAg.

MANAGEMENT AND PREVENTION OF PERINATAL HBV INFECTION

Recent studies in populations in which HBV infection is endemic have shown that perinatal transmission can be substantially reduced by administration of hepatitis B immune globulin (HBIG) to the newborn infant. Regimens that include adminstration of HBIG immediately after birth and one or two subsequent doses are 75% effective in preventing the chronic carrier state. Although the optimum schedule for HBIG administration remains to be defined, the U.S. Public Health Service Advisory Committee on Immunization Practices (ACIP) recommends a dose within 30 minutes of birth, and subsequently at 3 and 6 months of age. Close to two thirds of passively immunized infants develop antibody to both HB_sAg and HB_cAg, indicating that subclinical HBV infection has occurred, followed by development of permanent immunity to HBV (so-called passive-active immunization). The remaining one third of passively immunized infants are protected from infection with HBV, but become susceptible within 3 to 4 months, as HBIG is catabolized. Later, these infants are found to have no HBV serologic markers, and therefore are not immune and are at risk of contracting primary HBV infection.

Studies in Taiwan have suggested that a significant number of passively protected infants become infected within the first 3 years of life. For this reason, it has been recommended that perinatal HBIG administration be complemented with a course of HBV vaccine (HB_sAg purified from pooled plasma of chronic carriers) beginning at 3 months of age, in order to ensure active immunity for all infants at risk. In addition to these steps, infants of HB_sAg-positive mothers must be followed up in order to assess the effect of these preventive measures and to determine the status of the infants with respect to HBV. This information can be derived by testing sera for HB_sAg and antibody to HB_sAg and HB_cAg at 12 months of age.

Perinatally infected infants are rarely infectious at birth and do not need to be isolated. Routine blood and needle precautions, however, are necessary for the mother, who is infectious. There is no need to prevent full mother-infant contact, including breast-feeding. Temporary proscription of breast-feeding may be indicated, however, if the mother's nipples are cracked or bleeding.

Recommended Readings

Alford CA: Chronic intrauterine infection. In Galasso GJ, Merigan TC, Buchanan RA (eds): *Antiviral Agents and Viral Diseases of Man.* New York, Raven Press, 1984.

Alford CA, Pass RF: Epidemiology of chronic congenital and perinatal infections of man. *Clin Perinatal* 8:397, 1981.

American Academy of Pediatrics Committee on Infectious Diseases: Expanded guidelines for use of varicella-zoster immune globulin. *Pediatrics* 72:886, 1983.

Arvin AM, Yeager AS, Bruhn FW, et al: Neonatal herpes simplex infection in the absence of mucocutaneous lesions. *J Pediatrics* 100:215, 1982.

Beasley RP, Hwang L-Y, Lin CC, et al: Hepatitis B immune globulin (HBIG) efficacy in the interpretation of perinatal transmission of hepatitis B virus carrier state. *Lancet* 2:388, 1981.

Beasley RP, Hwang L-Y: Postnatal infectivity of hepatitis B surface antigen-carrier mothers. *J Infect Dis* 147:185, 1983.

Brunell PA, Kotchmar GS: Zoster in infancy: Failure to maintain virus latency following intrauterine infection. *J Pediatrics* 98:71, 1981.

Chin J: Prevention of chronic hepatitis B virus infection from mothers to infants in the United States. *Pediatrics* 71:289, 1983.

Cooper LZ: Congenital rubella in the United States. in: Krugman S, Gershon A (eds): Proceedings of the International Conference on Rubella Immunization. *Am J Dis Child* 118:3, 1969.

Griffiths PD, Stagno S, Pass RF, et al: Infection with cytomegalovirus during pregnancy: Specific IgM antibodies as a marker of recent primary infection. *J Infect Dis* 145:647, 1982.

Hanshaw JB, Scheiner AP, Moxley AW, et al: School failure and deafness after "silent" congenital cytomegalovirus infection. *N Engl J Med* 295:468, 1976.

Kaplan MH, Klein SW, McGhee J, et al: Coxsackie B infections in infants less than three months of age: A serious childhood illness. *Rev Infect Dis* 5:1019, 1983.

Kibricks S: Herpes simplex infection at term: What to do with mothers, newborns and nursery personnel. *JAMA* 243:157, 1980.

Krajden S, Middleton PJ: Enterovirus infections in the neonate. *Clin Pediatrics* 22:87, 1983.

Kumar ML, Nankervis GA, Jacobs IB, et al: Congenital and postnatally acquired cytomegalovirus infections: Long-term follow-up. *J Pediatrics* 104:675, 1984.

Lee FK, Nahmias AJ, Stagno S: Rapid diagnosis of cytomegalovirus infection in infants by electron microscopy. *N Engl J Med* 299:1266, 1978.

Leland D, French MLV, Kleiman MB, et al: The use of TORCH titers. *Pediatrics* 72:41, 1983.

Light IJ: Post-natal acquisition of herpes simplex virus by the newborn infant: A review of the literature. *Pediatrics* 63:480, 1979.

Meyers JD: Congenital varicella in term infants: Risk reconsidered. *J Infect Dis* 129:215, 1974.

Modlin JF: Fatal echovirus 11 disease in premature neonates. *Pediatrics* 66:775, 1980.

Nahmias AJ: The TORCH complex. *Hosp Practice,* May 1974, p 65.

Osborn JE: Cytomegalovirus: Pathogenicity, immunology, and vaccine initiatives. *J Infect Dis* 143:618, 1981.

Overall JC, Glasgow LA: Virus infections of the fetus and newborn infant. *J Pediatrics* 77:315, 1970.

Reesnik HW, Reerink-Brongers EE, Lafeber-Schut BJT, et al: Prevention of chronic Hb$_s$Ag carrier state in infants of HB$_s$Ag-positive mothers by hepatitis B immunoglobulin. *Lancet* 2:436, 1979.

Remington JS, Klein JO (eds): *Infectious Diseases of the Fetus and Newborn Infant,* 2nd ed. W.B. Saunders, Co., Philadelphia, 1983.

Schweitzer IL: Vertical transmission of the hepatitis B surface antigen. *Am J Med Sci* 270:287, 1973.

Stagno S, Reynolds DW, Pass RF, et al: Breast milk and the risk of cytomegalovirus infection. *N Engl J Med* 302:1073, 1980.

Stagno S, Pass RF, Dworsky ME, et al: Congenital cytomegalovirus infection. *N Engl J Med* 306:945, 1982.

Whitley RJ, Nahmias AJ, Visintine AM, et al: The natural history of herpes simplex virus infection of mothers and newborns. *Pediatrics* 66:489, 1980.

Whitley RJ, Yeager A, Kortus P, et al: Neonatal herpes simplex infection: Follow-up evaluation of vidarabine therapy. *Pediatrics* 72:778, 1983.

Williamson AP: The varicella-zoster virus in the etiology of severe congenital defects. *Clin Pediatrics* 14:553, 1975.

Yeager AS, Grumet FC, Hafleigh HB, et al: Prevention of transfusion-acquired cytomegalovirus infections in newborn infants. *J Pediatrics* 98:281, 1981.

Yeager AS, Palumbo PE, Malachowski N, et al: Sequelae of maternally derived cytomegalovirus infections in premature infants. *J Pediatrics* 102:918, 1983.

9

Childhood Exanthemas

JOHN F. MODLIN

The epidemiology of measles and mumps, along with that of rubella, has undergone dramatic changes during the past 2 decades because of the development of safe and effective live, attenuated viral vaccines (see Chapter 14). Before the early 1960s, infection with the viruses of measles and mumps was a virtually universal experience; and only one in five children reached adulthood without also having had rubella. Widespread immunization with measles-mumps-rubella (MMR) vaccine, and lasting adult immunity after naturally acquired infection, have ensured a commensurately high incidence of passive transplacental protection for neonates. Despite the dramatic decline in the incidence of measles (from 200 to 500 cases in 1963 to 1.3 cases in 1981 per 100,000 population) sporadic cases still occur, generally in older persons who have escaped exposure or immunization in childhood. In underdeveloped countries, moreover, all three diseases remain largely uncontrolled because the vaccines are not generally available. Measles in particular continues to be a major cause of significant morbidity and mortality among young children in parts of Asia and Africa. Even though neither measles nor mumps has been clearly implicated in fetal or neonatal disease, a brief discussion of both will be added here to supplement the earlier discussion of rubella (see Chapter 8).

MEASLES (RUBEOLA)

Agent

The agent of measles is a pleomorphic enveloped virus, the genome of which is contained in a single, unsegmented linear strand of RNA in helical configuration with protein. Because its genome is unsegmented, and hence incapable of recombination, measles virus has remained genetically stable over the years, and there is therefore only one serotype. The envelope contains lipid and at least two glycoproteins, the hemagglutinin (H) and the fusion (F) factor; there is no neuraminidase. The H protein is responsible for attachment of the virion to the host-cell membrane; the F protein mediates fusion of viral envelope with cell membrane to allow penetration of the viral genome into the susceptible cell. The F protein also mediates cell-to-cell fusion with formation of syncytia characteristic of measles virus cytopathology (see Chapters 1 and 3).

Humans are the only natural host for measles virus. The agents of canine distemper and rinderpest are antigenically related to measles virus; these three agents comprise the genus Morbillivirus in the family Paramyxoviridae.

Transmission, Pathophysiology, and Clinical Manifestations

A person who has contracted measles is highly infectious from the prodromal stage, that is, the 3 to 4 days preceding the onset of symptoms, until 2 to 3 days after the appearance of the typical exanthem. Infection is spread to nonimmune persons by contact with virus-laden respiratory secretions and aerosols created by coughing and sneezing. Humans are the only natural reservoir.

During the incubation period, that is, the 9 to 11 days between exposure and onset of first symptoms, viruses replicate in the mucosa of the upper respiratory tract and adjacent lymphoid tissue, and spread hematogenously to reticuloendothelial tissue throughout the body. During a second period of viremia, the virus spreads to the lower respiratory tract, skin, and other target organs, coinciding with the onset of illness.

The first clinical manifestations are fever, cough, coryza, and conjunctivitis, increasing in intensity during the ensuing 3 to 5 days. During this prodromal stage, small, punctate bluish white lesions that are diagnostic of measles appear on the buccal mucosa.

The typical rash of rubeola consists of discrete maculopapular lesions that appear first on the neck and face 3 to 4 days after the onset of the prodromata, and then spread caudally over trunk and extremities, progressing to confluence on the face, neck, and upper trunk. Symptoms and signs resolve during the next 5 to 6 days, and, in the absence of complications, subside completely within 10 days. Dry cough may persist for several weeks.

Diagnosis

The experienced physician can usually make the correct diagnosis on clinical grounds alone. However, it is advisable to confirm the diagnosis retrospectively by isolation of virus from the respiratory tract (throat washing or nasal swab) or serologically by examining paired serum samples for complement fixing or for the presence of HI antibody (see Chapter 13).

Complications and Mortality

Otitis media due to common bacterial pathogens (e.g., *Streptococcus pneumoniae, Haemophilus influenzae, Staphylococcus aureus, Streptococcus pyogenes*) may occur in up to 15% of all cases of measles in previously normal children. Less frequently, croup or bronchiolitis and giant-cell pneumonia may be part of the measles syndrome, particulary in young children and those with chronic cardiopulmonary disorders. Pneumonia may also be caused by any of the secondary bacterial pathogens just mentioned. Respiratory complications account for most measles-associated deaths, particularly in children under 5 years of age.

The incidence of encephalitis in association with acute measles is about 1 per 1000 cases, being somewhat higher in children of school age, but otherwise independent of the severity of underlying disease. CNS symptoms generally occur during early convalescence from typical measles, 4 to 7 days after the appearance of the rash. In these cases, fever recurs, accompanied by headache, lethargy, irritability, and clouding of the sensorium. In up to half the cases of encephalitis, generalized seizures or coma may supervene and persist for days to weeks. In 90% of cases, there is CSF pleocytosis, with 10 to 500 leukocytes per cubic millimeter, lymphocytes predominating, and an abnormal encephalogram. The mortality attending measles encephalitis is 10 to 15%; 25 to 30% of those who survive have significant and permanent neurologic sequelae.

Subacute sclerosing panencephalitis (SSPE) is a rare late complication, which may not appear until years after an unremarkable episode of measles, and which for unknown reasons affects predominantly white, male children from rural areas. Early symptoms of SSPE include personality changes, decline in intellectual powers, impaired physical coordination, and disturbance of sleep and vision, followed by the appearance of myoclonic reflexes in one or more extremities. Although remission may occur in some patients, the course of SSPE progresses inexorably through major loss of mental function, extrapyramidal and cranial nerve dysfunction, seizures, coma, and inanition, to death generally 6 months to 3 years after onset. During life, the diagnosis is substantiated by the finding of characteristic paroxysmal bursts of activity in the EEG and greatly elevated levels of CSF immunoglobulin, most of which is antibody to measles virus. Brain tissue taken by biopsy or at autopsy shows characteristic histopathology. (For more detailed discussion of the pathogenesis of SSPE, see Chapter 11.)

Measles in the Altered Host

Some degree of immunosuppression occurs transiently during otherwise uncomplicated measles, as evidenced by anergy to tuberculin or purified protein derivative (PPD), a phenomenon recognized many years ago. However, in a host already immunodeficient because of malignancy, immunosuppressive chemotherapy, or any of the congenital immunodeficiency syndromes, measles may be particularly severe and virus may be shed from the respiratory tract for unusually prolonged periods. Giant-cell pneumonia, which may occur even in the absence of characteristic exanthem, is the principal cause of the high mortality of measles in the immunocompromised patient. Malnutrition, particularly in underdeveloped countries, predisposes to severe measles, especially in young children. In central Africa, the reported mortality among children hospitalized because of measles is 5 to 25%. In most of these children, who may also suffer from complications such as croup and pneumonia, gastroenteritis, ulcerative stomatitis, purulent keratoconjunctivitis, and extensive epidermal desquamation, prolonged measles may in turn aggravate preexisting protein-calorie malnutrition. In parts of Asia and Africa, where health standards are relatively low and vaccines have not been available, measles continues to be a major cause of morbidity and mortality among young children.

Although immunization during pregnancy is not recommended, inadvertent administration of live, attenuated virus vaccine in early pregnancy carries no known risk to the fetus.

Epidemiology

This information is covered in Chapter 15.

MUMPS

Before the introduction of live, attenuated mumps virus vaccine in 1968, mumps was endemic in most populations. Clinical disease occurred usually during the winter and spring months, and predominantly among children of school age, who then were the chief source of infection for susceptible younger children and adults. Virtually universal immunization of infants with MMR vaccine has dramatically reduced the incidence of mumps, which is today a relatively rare disease.

Agent

The virus of mumps is in the genus Paramyxovirus of the family Paramyxoviridae. The helical nucleocapsid comprises the RNA genome in single-stranded, unsegmented linear form, in association with core (N) protein (sometimes referred to as "soluble" or S antigen), antibodies to which are not protective. As with the other paramyxoviruses, hemagglutinin (H) and neuraminidase (N) activities are associated with a single-envelope glycoprotein (HN protein, sometimes referred to as "virion" or V antigen). Being on the outer surface of the virion, HN antigen is the major protective antigen of the virus (i.e., elicits protective antibody). Because its genome is unsegmented, mumps virus is genetically stable. Hence there is only one serotype, which shows some slight antigenic cross-reactivity with type 1 parainfluenza (Sendai) virus.

Pathophysiology

Mumps is highly contagious. Persons with subclinical infection, that is, lacking notable clinical signs, may nonetheless be a source of infection for susceptibles. Transmission occurs through droplets from respiratory secretions containing the virus. The period of infectivity, during which mumps virus can be recovered from respiratory secretions (e.g., throat washings), saliva, or urine, extends from 2 to 3 days before onset through the first 3 to 5 days of clinical illness. The incubation period, from exposure and infection to the development of signs and symptoms, is usually 16 to 18 days, during which the virus replicates in the upper respiratory tract and regional lymph nodes and, by direct extension, may invade the salivary glands (parotid, submaxillary, and sublingual). Parotitis, the most common clinical manifestation, may be unilateral or sequentially bilateral, and usually persists for 7 to 10 days. During the incubation period, virus will have been disseminated hematogenously to infect glandular epithelium of various organs, notably gonads, pancreas, thyroid, and mammary glands and, in some cases, the central nervous system, even in the absence of recognized parotitis. Signs and symptoms frequently reflect these sites of secondary viral infection. Fever, headache, anorexia, and abdominal pain are general accompaniments.

Antibodies to mumps virus can be detected within 3 days of the onset of symptoms, by either HI or CF. Antibody measured in the CF test using "soluble" (i.e., nucleoprotein) antigen, generally appears before that detected in CF tests with the "viral" [i.e., envelope (HN)] antigen, and is usually the first to decline during convalescence. Antibody to viral antigen persists indefinitely and limits the hematogenous spread of virus both during the primary attack and after subsequent reinfections, which can be assumed to recur on reexposure. Reinfections remain asymptomatic, but doubtless contribute to the maintenance of immunity. Cellular immune responses, including the early effects of interferon, have important roles in resolving the primary systemic infection.

Complications

Before the introduction of mumps vaccine, mumps virus was the most common identifiable cause of aseptic meningitis, particularly in the absence of recognized parotitis. In these cases, mumps can be diagnosed relatively early in the disease by demonstration of a fourfold or greater rise in antibody to the soluble (nucleoprotein) antigen of mumps virus. In community outbreaks of mumps parotitis, 5 to 10% of all cases are accompanied by symptoms or signs of CNS involvement, most commonly meningitis. A small proportion of these patients will develop signs of encephalitis, such as delirium, seizures, and coma. In the usual case of meningoencephalitis due to mumps virus, the CSF is characterized by moderate and predominantly lymphocytic pleocytosis, normal or slightly elevated protein, and normal or slightly low glucose, the latter in comparison with simultaneously determined blood glucose level. The prognosis is uniformly good; neurologic sequelae are rare, with one exception. Permanent eighth-nerve deafness is associated with mumps in up to 4% of cases; and in most cases it is fortunately unilateral. The pathogenesis of this complication is unknown.

Epididymo-orchitis occurs in about 25% of postpubertal males who contract mumps; and in 40% of these it is bilateral. Acute edema increases the pressure within the indistensible tunica vaginalis, leading to some degree of atrophy, but rarely if ever to complete sterility even when both testes are affected. Up to 30% of adult women who contract mumps may develop mastitis, and a smaller number suffer abdominal signs and symptoms that suggest acute oophritis. In most cases of mumps, some degree of transient renal involvement is evidenced by microscopic hematuria.

Interstitial nephritis with reduced glomerular filtration rate has been reported in a few cases. Other rare complications in adult cases of mumps include arthritis, thryoiditis, and myocarditis. Because of its predilection for glandular tissue, mumps has been suspected of being the cause of juvenile diabetes mellitus, but this association has never been clearly substantiated. Acute pancreatitis, however, does occasionally occur, substantiated by transiently elevated levels of serum and urinary amylase. Mumps must be distinguished from other disorders accompanied by swelling of the face of neck, such as cervical lymphadenitis, and from parotitis caused by other viruses (e.g., enteroviruses, influenza and parainfluenza viruses, cytomegalovirus) and pyogenic bacteria. In the latter cases, there is usually purulent drainage from Stenson's ducts, from which bacterial pathogens can frequently be cultured. Noninfectious causes of parotid enlargement include Sjögren's syndrome and leukemia.

RUBELLA VIRUS

The close association of rubella with birth defects was first noted in 1941 by Gregg in Australia, who described a syndrome of congenital cataracts and heart lesions in infants born after an epidemic of rubella. The primary medical and public health importance of rubella relates to the ability of the virus to cross the placenta, and damage the fetus. "German measles," or rubella, is caused by a positive-stranded enveloped icosahedral virus, which was first isolated in primary human amnion cells in 1962 by Weller and Neva and later adapted to growth in other cells of human origin. In cell culture, rubella virus causes relatively little cytopathology, and infected cells do not hemadsorb. The presence of virus can be demonstrated by interference as well as by immune fluorescence microscopy. (See also Chapter 15 for further discussion of rubella cytopathology and interference.)

Because of its molecular biology, the virus is now classified with the togaviruses, although it is atypical, since humans are its only natural host. There is only one serotype. Complement fixation or hemagglutination (with chick hatching erythrocytes) inhibition (HI) are used in serodiagnosis. Except for fetal rubella, the disease is of little consequence when acquired postnatally. The usual route of transmission is the respiratory route. The incubation period ranges from 14 to 21 days and in older patients is accompanied by coryza, sore throat, headache, minimal fever, and swelling of posterior cervical, occipital, and postauricular lymph glands. The maculopapular exanthem, which is sometimes pruritic, appears 1 to 4 days later, first on the face and spreading distally, and fading within 3 to 4 days. About 40% of patients do not develop a rash after infection, which may therefore go unrecognized unless diagnosed serologically in retrospect. Antibody appears in the serum on the second or third day of the rash, peaking 4 to 6 weeks later. Immunity to reinfection is lifelong, and is to be equated with the presence of serum IgG HI antibody at any level. In 20 to 40% of patients, rubella is accompanied by arthralgia of varying severity, beginning as the rash fades, and persisting for 1 to 2 weeks. Arthralgia may be particularly severe in older adults, particularly women, who contract rubella for the first time.

Until recently, rubella was seen regularly, with peak incidence during the spring. Before the introduction of vaccine in 1969, large epidemics occurred in the United States every 6 to 9 years with lower incidence of disease during the intervening years. The last nationwide epidemic occurred in 1963–1964, in which 12 million cases of rubella and 11,000 cases of congenital rubella syndrome were reported. With widespread use of the vaccine, the overall incidence of rubella cases has diminished by more than 90% and the epidemic disease is no longer observed, although localized outbreaks still occur occasionally in institutional settings.

The age-specific incidence of reported rubella cases has also changed with the wide application of immunization. Accordingly, the proportion of cases occurring in teenagers and young adults is considerably greater than before the vaccine became available. This change is reflected in the fact that 15 to 20% of women of childbearing age have no detectable antibody to the virus.

Recommended Readings

Axnick NW, Shavell SM, Witte JJ: Benefits due to immunization against measles. *Pub Health Rep* 84:673–680, 1969.

Barkin RM: Measles mortality. *Am J Dis Child* 129:307–309, 1975.

Cooper L: Rubella: A preventable cause of birth defects, in: *Birth Defects: Original Article Series.* vol IV (7), National Foundation-March of Dimes, New York, 1968, pp 23–25.

Centers for Disease Control: Current status of rubella in the U.S. *J Infect Dis* 142:776, 1980.

Dudgeon JA: Congenital rubella. *J Pediatrics* 87:1078, 1975.

Enders JF, Katz SL, Milovanovic MJ, et al: Studies on an attenuated measles virus vaccine. *N Engl J Med* 263:153, 1960.

Gordon JE, Ingalls TH: Modern measles. *Am J Med Sci* 228:334, 1954.

Grand MG, Wyll SA, Gehlbach SH, et al: Clincal reactions following rubella vaccinations. *JAMA* 220:1569, 1972.

Herrmann KL, Halsted SB, Brandling-Bennett AD, et al: Rubella immunization: persistence of antibody 4 years after a large-scale field trial. *JAMA* 235:2201, 1976.

Hillemann MR, Buynak EB, Weibel RE, et al: Development and evaluation of the Moraten measles virus vaccine. *JAMA* 206:587–590, 1968.

Hinman HR, Brandling-Bennett AD, Bernier RH, et al: Current features of measles in the United States: Feasibility of measles elimination. *Epidemiol Rev* 2:153, 1980.

Hinman HR: Measles and rubella in adolescents and young adults. *Hosp Prac* October 1982, p 137.

Horstmann D: Controlling rubella: Problems and perspectives. *Ann Intern Med* 83:412, 1975.

Horstmann DMS, Liebhaber H, LeBouvier GL, et al: Rubella reinfection of vaccinated and naturally immune persons exposed in an epidemic. *N Engl J Med* 283:771–778, 1970.

Krugman S: Present status of measles and rubella immunization in the U.S. *J Pediat* 90:1–12, 1977.

Landrigan PJ, Witte JJ: Neurological disorders following live measles virus vaccination. *JAMA* 233:1459–1462, 1973.

Mann JW, Preblud SR, Hoffman RE, et al: Assessing risks of rubella infection during pregnancy. *JAMA* 245:1647–1652, 1981.

Centers for Disease Control: *Measles Surveillance,* Report No. 11, Issue 1, September, 1982.

Measles Prevention (ACIP Statement): *MMWR* 31:217, 1982.

Modlin JF, Jabbour JT, Witte JJ, Halsey MA: Measles, measles vaccine and SSPE. *Pediatrics* 59:20–29, 1977.

Modlin JK, Herrmann KL, Brandling-Bennett AD, et al: Risk of congenital abnormality after inadvertent rubella vaccination of pregnant women. *N Engl J Med* 294:972, 1976.

Murphy MD, Brunell PA, Lievens AW, Shehab ZM: Effect of early immunization on antibody response to reimmunization with measles vaccine as demonstrated by enzyme-linked immunosorbent assay (ELISA). *Pediatrics* 74:90, 1984.

Panum PL: Measles in the Faroe Islands. Bibliothek for Laeger 1:270–344, 1847 (translated in *Medical Classics* 3:829–886).

Robinson RG, Dudenhoeffer FE, Holroyd HJ, et al: Rubella immunity in older children, teenagers and young adults: A comparison of immunity in those previously immunized with those unimmunized. *J Pediatrics* 101:188, 1982.

Rubella Prevention (ACIP statement): *MMWR* 30:37–42.

Schwartz AJF: Preliminary tests of a highly attenuated measles vaccine. *Am J Dis Child* 103:216–219, 1962.

Tardieu M, Grospierre B, Durandy A, et al: Circulating immune complexes containing antigens in late-onset rubella syndrome. *J Pediatrics* 97:370, 1980.

10

Herpesvirus Infections

JOSEPH S. PAGANO

Herpesviruses are among the most important viruses medically and the most interesting virologically. New drugs to treat herpetic infections and improved methods of diagnosis have focused attention on these viruses. The rapidly unfolding molecular biology of these complex viruses, coupled with beginning insights into how they cause infection and pathologic changes, add to the interest in them. The herpesviruses are not only medically consequential but inescapable; almost everyone eventually becomes infected by the several members of this group of viruses.

The herpesviruses cause a wide spectrum of pathogenic effects that may occur at any time of life, following either primary infection or reactivated, latent infection. The frequency and importance of both of these types of acute infections with the herpesviruses have increased in recent years as a result of modern immunosuppressive therapies, which provide the setting for reactivated infection, and from changes in sexual mores that have contributed to increased numbers of primary infections with the sexually transmitted herpesviruses.

Although the herpesvirus group includes more than 70 agents that infect many species of animals, only 5 commonly infect human beings: herpes simplex virus (HSV) types 1 and 2, varicella-zoster virus (VZV), cytomegalovirus (CMV), and Epstein-Barr virus (EBV). All are similar in morphology. They also have common biologic properties such as a propensity toward production of latent infection. At the cellular level they are all capable of producing multinucleated giant cells, recognized in cell monolayers as syncytia. However, with the exception of HSV 1 and 2, which are genetically similar but not identical, the herpesviruses are distinct at the level of the genetic information encoded in their genomes. The diseases produced by the herpesvirus group cover a wide spectrum, which may share some pathologic and clinical features.

Infection with herpesviruses usually occurs during infancy, after which, in most cases, the virus becomes latent but can be reactivated at any time by natural or iatrogenic stimuli. The clinical manifestations of the reactivated infection may differ strikingly from those accompanying primary infection. Several herpesviruses, such as HSV, CMV, and EBV, have an additional importance based on their ability to transform infected cells in vitro, and EBV is associated with at least two human malignancies, Burkitt's lymphoma and nasopharyngeal carcinoma.

BIOLOGIC FEATURES

Most herpesvirus infections are probably silent, although the ratio of inapparent to symptomatic primary infections differs greatly with each virus. For example, most CMV infections are asymptomatic, whereas virtually all primary VZV infections (viz., chickenpox), whether in the susceptible child or adult, produce at least a few scattered skin lesions (pocks). Many EBV infections are silent, whereas probably most HSV (types 1 and 2) infections do produce symptoms initially. HSV, however, can cause asymptomatic primary infection. Herpesvirus infections tend to be persistent, which means that, in contrast to latent infection, active viral replication continues with production and shedding of infectious virus, usually without accompanying symptoms. CMV and EBV are especially disposed to produce persistent infection. CMV may be shed in the urine for many months or for years, and EBV from the oropharynx for long periods.

In addition to episodic reactivation of latent virus causing renewed virus excretion with or without symptoms, repeated exogenous superinfections may also occur, particularly with HSV-2 and CMV in the genital areas and occasionally with HSV-1 infections in and around the oral cavity. New exogenously acquired superinfections due to EBV or VZV, however, while they may occur, are believed to be of little significance. This striking biologic difference among the members of the herpesvirus group may be explained on the basis of immunologic factors and pathogenetic features underlying infection with each type of virus. However, following either symptomatic or asymptomatic infection, each of the herpesviruses enters a latent phase that persists for life, interrupted only by unpredictable episodic virus reactivation and disease, sometimes of major proportions. The latency state represents a remarkable balance between host and virus that is of great interest from a virologic point of view, but the periods of imbalance, when active virus replication is taking place, are specifically medically consequential. These striking biologic features of viral behavior constitute one of the hallmarks of herpesviruses, which are unique in this respect. From the standpoint of immunization and treatment, all these phases in the life history of herpesviruses in human beings call for different methods of approach.

VIRUS STRUCTURE AND ANTIGENICITY

Herpesviruses, particularly those infecting human beings, are virtually identical morphologically. HSV, VZV, CMV, and EBV virions are indistinguishable by electron microscopy. These are large viruses, the intact viral particle having a diameter of 150 to 200 nm and comprising a core of double-stranded DNA and protein surrounded by a symmetrical protein capsid composed of a specific number of capsomers that display icosahedral symmetry. The nucleocapsid, which develops in the cell nucleus, in turn is covered by an envelope derived from the nuclear membrane of the host cell. The envelope contains not only lipids and lipoproteins but also virus-coded glycoproteins and is essential for virus infectivity.

Herpesvirions contain many polypeptides, most of which are defined at present only in terms of molecular weight or antigenic identity. Some of the polypeptides are phosphoproteins. Several of these viruses, notably HSV, CMV, and EBV, have a protein kinase activity that is intimately associated with the virion and perhaps contributes to the structure. These viruses also code for virus-specific enzymes, particularly DNA polymerases, thymidine kinases, and nucleases, which differ from cellular enzymes. The viral enzymes are produced in the infected cell, but for the most part are not included in the virions.

The genomes of herpesviruses are linear and double-stranded and have a molecular weight anywhere from 100×10^6 daltons (EBV) to 150×10^6 daltons (CMV); indeed the human CMV genome is one of the largest viral genomes known. At their termini and situated internally within the genome, stretches of repeated sequences are found in addition to the sequences unique to each strain of herpesvirus. The specific arrangements of these common, repeated sequences differ for each type of herpesvirus, but overall they follow a generally similar distribution pattern. The significance of these novel sequence arrangements is unknown, but they may be involved in the generation of defective genomes that, in turn, may facilitate latent infection. Defective herpesvirions sometimes contain fewer than the full complement of sequences that make up the complete viral genome. For example, CMV particles containing only 100×10^6 daltons are common. On the other hand, defective virions that contain unit-length genomes also exist; defectiveness comes about because reiterated common sequences are substituted for unique sequences.

The EBV genome exists intracellularly in a physical form called the episome or plasmid, which contains no virion structural protein. EBV plasmids are found in the nuclei of latently infected cells and assume a nucleosomal structure identical to that of cellular chromatin. The plasmids are replicated in cells as if they were cellular constituents, synthesis evidently being catalyzed by host-cell DNA polymerase rather than by virus-specific DNA polymerase. Closed, circular genomes of this sort, which theoretically might arise because of the terminal homology of most types of herpesviral genomes, have thus far been detected only in cells infected with either EBV or *Herpesvirus saimiri,* a virus of lower primates.

All of the herpesviruses except for HSV types 1 and 2 are genetically and antigenically distinct. HSV types 1 and 2 share about 50% of their genome content, with the homologous regions being discontinuous. No homology has been detected among the other members of the herpesvirus group, although several simian EBV-like agents have been found to bear varying degrees of homology to human EBV.

The homology of various viral genomes to one another can be assessed by digestion with restriction endonucleases. Such nucleases recognize specific sites on different genomes and cleave the genome into specific DNA fragments that have characteristic electrophoretic patterns. These patterns directly reflect and precisely identify discrete differences among virus strains otherwise difficult to detect. In general, members of the herpesvirus group are diverse, with no two isolates being exactly the same when examined by this sensitive technique. Restriction enzyme analysis is therefore an extremely useful epidemiologic and clinical tool for identifying specific strains of viruses and tracing their transmission, especially with HSV and CMV.

Antigenically most of the human herpesviruses differ from one another. Only HSV types 1 and 2 are antigenically cross-reactive, their relatedness being ascribed to possession of shared polypeptides. On the other hand, the two viruses are sufficiently distinct to be usually, but not invariably, recognized by reactivity with type-specific antisera. Considerable variation exists among strains of the same type, however, and individual HSV isolates may not fall clearly into either the type 1 or type 2 class.

VZV isolates appear to be antigenically similar or identical, in accord with the limited degree of variation among VZV isolates disclosed by restriction endonuclease analyses. VZV and HSV antigens do not cross-react in vitro, although HSV infection may evoke a transient heterologous antibody response to VZV.

Human strains of CMV are unrelated genetically to other herpesviruses but have some degree of DNA homology to simian CMV. Like HSV, CMV isolates show great heterogeneity when analyzed by restriction endonuclease digestion. Furthermore, a range of antigenic diversity among CMV isolates can be demonstrated by neutralization kinetics or cross-complement-fixation analyses. However, no clear-cut subgrouping of strains can be established by antigenic analysis.

A number of antigenic determinants have been recognized in EBV that are not cross-reactive with other human herpesviruses, again reflecting the lack of nucleic acid homology among members of the herpesvirus group. The viral capsid antigen (VCA) of EBV is a structural part of the virion. Any person who has been infected with EBV will develop antibodies to VCA that persist for life. Cells abortively infected with EBV do not necessarily produce mature virus, but are found to contain EBV early antigen (EA). In patients, the appearance of EA antibodies is usually indicative of acute infection, either primary or reactivated, and also accompanies EBV-associated malignancies. Cells latently infected with EBV contain the EBV nuclear antigen (EBNA), the function of which is unknown. Persons who have been infected with EBV eventually develop EBNA antibodies, the presence of which is taken as an indication of past, but currently inactive, infection with EBV. The presence of antibodies to EBNA does not indicate an EBV-associated malignancy inasmuch as this antigen is also produced during the normal course of benign EBV infection and during infectious mononucleosis. In short, the patterns of the several antibody responses to EBV antigens follow characteristic temporal sequences that are useful diagnostically.

PATHOGENESIS

Herpesviruses can infect many cell types, the range of susceptible cell types being greater in infants than in adults. Primary infection, the latent state, and reactivated infection with the same virus in the same individual each may involve different cell types, and each herpesvirus strain has its own spectrum of cellular interactions. Underlying these general features of cellular susceptibility to viral infection is the predilection of herpesviruses for epithelial cells. Signs and symptoms of disease may result directly from viral replication and cell death, but disease and tissue damage may also result from the presumably normal cellular immune responses to virus-infected cells.

Both HSV types 1 and 2 infect primarily epithelial cells, type 1 virus classically causing lesions in the oropharynx (herpes labialis) and type 2 causing lesions in the genital mucosa and adjacent skin (herpes genitalis). Site- or organ-specific localization of infection is typical of herpesviruses, but it is not absolute, and disseminated infections may occur in both infants and immunosuppressed adults. In such patients, the virus is found in the respiratory and gastrointestinal epithelium, widely dispersed in the skin, and in parenchymal organs such as the liver. Primary infection also occurs in the respiratory epithelium, but in persons with normal immune mechanisms such infections are inapparent. After primary infection in oral or genital sites, virus may spread to adjacent nervous tissue. HSV-1 establishes latent infection in the trigeminal ganglia and type 2 in sacral ganglia. Following various stimuli (e.g., ultraviolet light, fatigue, unrelated infection, pregnancy), virus latent in ganglionic sites causes lesions to appear at peripheral locations in the skin and mucosa. Such are the characteristic patterns of both labial and genital herpetic infections.

Although latent ganglionic infection is entirely silent, HSV infection elsewhere in the central nervous system (CNS) is symptomatic. In the United States herpes simplex virus type 1 is the leading cause of sporadic encephalitis. The pathogenesis of herpes simplex encephalitis is uncertain, but the CNS is believed to become involved secondarily, perhaps from primary infection of the respiratory tract after which a cell-associated viremia may carry the virus to the CNS. Another possibility is that the virus may reach the brain directly from the trigeminal ganglion. It is likely that encephalitis may result not only from primary infection at any age but also from reactivated infection, since latent HSV infection is almost universal in the older age group. HSV-2 rarely causes encephalitis, but it is not infrequently associated with benign, "aseptic" meningitis. In newborn infants, disseminated HSV-2 infection is usually acquired directly from genital lesions of the mother. Not only primary but also secondary infections with HSV can be, and probably often are, asymptomatic but nonetheless evoke an immune response.

With VZV, the agent causing both chickenpox and herpes zoster ("shingles"), the respiratory epithelium is the primary site of asymptomatic infection and viral replication. In nonimmune adults exposed to VZV, serious penumonitis may result. Cell-associated virus then spreads hematogenously to the skin; during viremic spread, other organs, such as the liver, may become infected but generally asymptomatically, although functioning of those organs may be transiently impaired. Spread to parenchymal organs is also characteristic of severe reactivated infection in immunosuppressed persons. In the skin lesions, large amounts of virus replicate freely during the vesicular stage and can be detected by direct electron microscopic examination or culture of vesicular fluid. However, the virus is transmitted from person to person by respiratory droplet aerosols and usually not from infected vesicles, so that contagion also occurs before infection becomes apparent. A noteworthy consequence of virus dissemination to the skin is the silent infection of peripheral sensory nerve ganglia that occurs in several parts of the body after primary exposure to VZV. The virus persists for life in these nerve sites and may be reactivated segmentally to cause shingles, which is followed during regression of the lesions by severe, sometimes intractable, postherpetic neuralgia.

CMV replicates in a number of epithelial tissues, including respiratory tract, salivary glands, and kidney tubules; from all of these sites virus may be shed for prolonged periods. CMV is also common in cervical secretions, especially late in pregnancy. During CMV mononucleosis, the virus can be found in the semen, where it may persist after it disappears from other sites. CMV is detectable in the semen of up to 25% of asymptomatic homosexual men. CMV-infected lymphocytes and polymorphonuclear leukocytes disseminate the virus through the bloodstream to distant sites. Age has pronounced effects on the cell types that are susceptible to infection. If infection occurs in utero, there is a major involvement and destruction of fetal CNS tissue, whereas post-natal CMV infection rarely causes encephalitis. CMV infections occurring after birth are usually asymptomatic, but may produce the "mononucleosis syndrome" with transient hepatitis. CMV also appears to be able to infect the gastrointestinal tract, but the pathologic significance of such infections is not established.

CMV infections provide striking examples of the fact that the symptoms and signs produced on virus reactivation differ from those produced at the time of primary infection. For example, CMV seems to cause a severe and often fatal interstitial pneumonitis in immunosuppressed recipients of bone-marrow transplants, whereas overt pneumonia during primary CMV infection is extremely rare. On the other hand, fever, leukopenia, and mild hepatitis and the much more rarely seen chorioretinitis occur not only in primary infection but also in reactivated infection in immunosuppressed renal allograft recipients. In such patients, primary infection is generally more severe than reactivated latent infection. Finally, CMV has recently been detected in association with endemic Kaposi's sarcoma, both in Africa and in male homosexuals in the United States.

Epstein-Barr virus probably replicates in epithelial cells of the oropharynx and parotid glands, B lymphocytes becoming secondarily infected. Although circulating B lymphocytes do not seem to support virus replication in vivo, they harbor the episomal or plasmid form of the EBV genome. In this latent state, infection is disseminated hematogenously to various organs and tissues. An apparent disparity exists between the widespread symptoms that may arise during EBV infection on the one hand and the lack of direct evidence of virus infection and replication at the sites of disease on the other. During the course of recognized infectious mononucleosis, for example, symptoms point to the involvement of many organs and sites, including the throat, lymphatic tissue, the liver, and both the peripheral and central nervous systems. However, the virus itself has not been demonstrated in any of these sites except for the parotids and the oropharynx and surrounding lymphatic tissue. The involvement of the other sites may be a consequence of immunologic responses to latently infected B lymphocytes.

African Burkitt's lymphoma is a malignancy of B lymphocytes with peculiar patterns of incidence and geographic distribution that suggested to Dr. Denis Burkitt that this cancer might be caused by a transmissible agent. Although the tumor characteristically localizes in the jaw, it also involves other organs, in particular the ovaries and testes. Burkitt's lymphoma is closely associated with EBV infection; the tumor cells contain latent Epstein-Barr virus DNA in episomal form. The American form of the disease is rare, it does not have the epidemiologic features of African Burkitt's lymphoma, and the tumor cells only occasionally contain the viral DNA. In Africa, EBV infection takes place early in life, with about 90% of children becoming infected during the first year, almost all asymptomatically. This early exposure to the virus may be a factor in the development of Burkitt's lymphoma, which is predictable on the basis of persistently high EBV antibody titers starting from infancy. Other factors, not as yet ascertained, are probably needed to explain the high incidence of this malignancy in Africa.

Another human malignancy, undifferentiated carcinoma of the posterior nasopharynx, is also closely associated with EBV infection. The association is pathogenetically interesting inasmuch as the cell type involved in nasopharyngeal carcinoma (NPC) is epithelial, which is consistent with the notion that the primary target cell for EBV replication is epithelial. Certain ethnic groups, especially Chinese living in Canton and selected populations living in northern Africa, have an extraordinarily high incidence of NPC. The disease is also common in east and west Africa and among Alaskan Eskimos; in all these groups NPC can be considered to be endemic. In addition, a sporadic form of the disease is found worldwide in various populations. In both the endemic and sporadic form, there is a strong association with EBV infection, which may well be its primary cause, but certain human genotypes are suspected to be predisposed to the development of NPC.

LABORATORY DIAGNOSIS

Herpesvirus infections cause cytologic changes that can be identified by light microscopy, including multinucleated giant cells and intranuclear inclusion bodies of the Cowdry A type. These changes are relatively nonspecific but have some utility when only fixed cytologic and pathologic specimens are available.

Virus isolation is the preferred method for diagnosing herpetic infections. HSV is easily recovered and propagated in many standard cell lines in which the cytopathic effects produced by the virus are obvious within 24 to 48 hours after inoculation. VZV is more difficult to recover from specimens, and the cytopathic effects develop more slowly. CMV can be cultured from the urine, cervix, semen, oropharynx, and, at times, from blood leukocytes. The cytopathic effects produced by each of these three viruses are not distinctive enough to permit positive identification, which is accomplished either by neutralization of cytopathic effect with specific antisera to the various viruses or demonstration by fluorescence microscopy of specific viral antigens in fresh tissue cells (e.g., cells recovered from urine or vesicular fluid), or by cultured cells inoculated with clinical specimens. The detection of EBV is more difficult because it cannot be cultured in standard cell lines. Its presence in oropharyngeal secretions, including those from the parotid gland, is demonstrable through transformation assays with human cord blood lymphocytes. The lymphocytes, after exposure to EBV-containing fluids, proliferate indefinitely in culture, and the cells are found by immunofluorescence to contain a new EBV-specific antigen, EBV nuclear antigen (EBNA), as well as EBV DNA by hybridization analysis.

Serologic methods for detection of antibodies to the herpesviruses are numerous. For all except EBV, the standard serologic test is complement fixation (CF) with antigens of known viral specificity. These tests have some utility as an indication of past infection, but they are relatively insensitive. Complement-fixing antibodies to VZV and sometimes to CMV are relatively short-lived, but they may reappear after reactivation or reinfection. Herpes simplex virus types 1 and 2 antibodies can be distinguished to some extent by neutralization tests. Appropriately prepared and absorbed type-specific HSV antisera are used to distinguish virus types. CMV antibodies can also be detected by indirect fluorescence microscopy, using the patient's serum as primary reagent. These tests are more reliable and correlate well with CMV neutralization, a test that is not routinely available in clinical diagnostic laboratories. The detection of IgM antibody to CMV as well as to EBV is indicative of recent acute infection.

Up to 80% of cases of infectious mononucleosis (IM) due to EBV are characterized by the appearance of heterophile antibodies, which are also referred to as sheep cell agglutinins, non-Forssmann, IM, or Paul-Bunnell antibodies. These IM heterophile antibodies are distinct from universally present Forrsmann antibodies, and do not cross-react with any known EBV antigens. Although the origin of heterophile antibodies is obscure, they are useful in providing the basis for a presumptive diagnosis of IM due to EBV. Specific evidence of EBV infection comes from detection of antibodies to viral capsid antigens (VCA), early antigens (EA), and EBV nuclear antigen (EBNA). The VCA antibodies persist for life and do not indicate when infection occurred, whereas antibodies to EA arise shortly after primary or reactivated infection and fall to low or undetectable levels within a few months. On the other hand, antibodies to EBNA do not appear in the circulation until 1 to 3 months after infection but thereafter persist indefinitely. It is therefore possible not only to diagnose EBV infection with great specificity and sensitivity, but also to establish generally the chronology of infection.

A number of improved and more sensitive methods for rapid diagnosis of herpesvirus infections are coming into use or are under development. The tests emphasize techniques that demonstrate viral antigens directly in fresh tissue specimens. In addition to direct and indirect immunofluorescence tests, radioimmunoassays and enzyme-linked antibody assays are now becoming available. Detection of new antigens that arise from virus-altered cell membranes may become a useful technique, for example, in the diagnosis of HSV encephalitis. Finally, nucleic acid hybridization techniques should become more generally available in diagnostic laboratories in the next few years.

One major problem in diagnosis of herpesvirus infections is the fact that any of these may reactivate, and replication and excretion may be renewed concurrently with some other disease process. Indeed, infection with one herpesvirus may lead to reactivation of another member of this group of viruses. In short, the isolation of one of these viruses from a patient does not necessarily prove that it caused the given condition.

IMMUNITY

Both humoral and cell-mediated mechanisms contribute to the immune responses characterizing infection with each of the herpesviruses. The relative importance of the respective immune systems differs from one virus to the other, and the impact of cell-mediated immunity on primary and reactivated infection clearly differs from that of humoral immunity. Exogenous reinfection with EBV or VZV is rare, if it occurs at all, whereas repeated reinfection may occur with CMV or either type of HSV. This dichotomous behavior within the herpesvirus group is due to the protective capacity of circulating antibodies to EBV and VZV, whereas antibodies arising from previous infection with CMV or HSV do not prevent reinfection. Reinfection is permitted because of the ability of CMV and HSV to establish infection through direct mucosal transfer of virus in local areas that are sheltered from immunologic response mechanisms. In such instances of direct local transmission, a viremic phase of the infection is circumvented, and the opportunity never arises for circulating antibodies to affect the course of infection.

Cell-mediated immune mechanisms have several roles in the pathogenesis of herpetic infections. They prevent or limit reactivated infection, and they probably are also important in terminating primary infection. On the other hand, depression or dysfunction of cellular immune responses permits or promotes the reactivation of herpesviruses. As already noted, cells of the immune system, including macrophages, also may have an important role in disseminating virus, which, being intracellular, remains unneutralized by antibody. With HSV, VZV, and CMV, the reactivated infection may produce serious disease; however, reactivated EBV infection is usually asymptomatic. Reactivation of these viruses occurs despite the presence of high titers of viral antibodies, which leads to the inference that changes in the cellular immune system play the predominant role in permitting reactivation as well as latent infection.

Other factors that are apparently not immunologic can also predispose to reactivation of herpesvirus infection. For example, recurrent herpes labialis seems to be triggered by a variety of stimuli, including environmental and endocrinologic factors. The factors that provoke reactivation seem to act as local triggers, but whether they do so in the trigeminal ganglion (where the virus is latent) or somehow by feedback from the superficial site (where the virus replication in the lesion occurs) is not known. Reactivating stimuli include ultraviolet light, stress, fever, and local mechanical irritation. Pregnancy also seems to increase susceptibility to reactivation of HSV and CMV; whether this effect is due to the endocrine changes or to immunologic changes that occur during pregnancy is unknown.

The limitation on primary infection imposed by cellular immune mechanisms is well illustrated in EBV infection, in which virus-specific cytotoxic T lymphocytosis (atypical lymphocyte response) arises as a response to the proliferation of EBV-infected B lymphocytes in the circulation. EBV infection during the course of infectious mononucleosis also evokes a nonspecific "killer" lymphocyte response. An important further step in this type of response is illustrated by infection with CMV that may lead to a disequilibrium of the normal ratio of suppressor-to-helper T-lymphocyte subsets. Thus, herpesvirus infections through combinations of such effects may cause both profound immuno-suppression and immunostimulation.

EPIDEMIOLOGY

Within the herpesvirus group, the epidemiology of infections differs from one virus to another, but there are also common epidemiologic features for different subsets of the viruses. For example, herpesvirus simplex type 2 and genital strains of cytomegaloviruses are transmitted sexually, whereas the other members of the herpesvirus group apparently are not. Common to all the herpesviruses is their universal presence in the general population, infection usually occurring before middle life. However, in underdeveloped countries and lower socioeconomic environments, infection occurs early in life, within the first year for some of the viruses. In Kenya and Uganda, for example, Epstein-Barr virus infects 90% of children by the end of the first year, whereas in the United States only 40% of students entering college, who are mostly from upper socioeconomic groups, have antibodies to EBV. Early infection with EBV is asymptomatic or produces nondescript symptoms, and therefore the syndrome of infectious mononucleosis is virtually unknown in underdeveloped countries and lower socioeconomic groups and is rare in infants and young children.

The epidemiology of Epstein-Barr virus infection in relation to Burkitt's lymphoma offers a striking example of the clues that epidemiologic patterns may provide as to the cause of a malignancy. This unusual lymphoma occurs in the so-called lymphoma belt of Africa in a pattern that is characterized by a striking age-related incidence, restricted geographic distribution at lower altitudes, and case clustering. These features suggest that a transmissible agent (EBV?) is involved in its causation. The endemic form of Burkitt's lymphoma occurs only in Africa; in the United States, where the tumor is much rarer, no such epidemiologic pattern has been revealed, nor is the disease invariably associated with demonstrable EBV infection.

Cytomegalovirus is the only herpesvirus whose capacity to infect in utero is well documented. It also differs from the other herpesviruses in causing perinatal infections with relatively high frequency. Later in life, the incidence of CMV infection follows a pattern similar to EBV infection, with a clear relation to age and socioeconomic status. Primary infections with HSV-1 and VZV are largely afflictions of childhood, whereas HSV-2 and genital CMV infections begin in adolescence and reach their peak in early adulthood through sexual transmission. The incidence of both of these infections relates directly to the frequency and diversity of sexual activity. For example, up to 50% of pregnant teenagers have active cervical CMV infection. Both CMV and HSV-2 infections have a very high frequency in male homosexuals.

HSV-1 is spread primarily by exchange of oral secretions and by direct contact with infected skin or mucous membranes. Spread by means other than person-to-person contact has not been proved. Type 1 virus tends to infect the oropharynx and type 2 virus the genitalia, but this distribution is unpredictable in the individual case. The existence of active herpetic lesions (HSV-1 or -2) in the vagina or cervix may lead to disseminated herpetic infections transmitted to the newborn during delivery. With the availability of effective antiherpetic therapy, cesarean section should soon no longer be necessary, provided that therapy is started early enough to suppress virus replication completely.

The most contagious of herpesvirus infections—and one of the most contagious of all human diseases—is varicella. Although vesicular fluid is infectious, transmission occurs primarily by droplet infection through the respiratory tract and does not require direct contact. VSV-infected aerosols have been documented to travel considerable distances in hospital settings. Infection is also transmitted during the 3-week incubation period, well before the rash appears. Active lesions of herpes zoster are sometimes a source of infection for susceptible children and adults. Few adults, however, are susceptible to exogenous VZV because of the solid immunity that follows childhood infection, which in most persons occurs within the first decade of life. Typically, nonimmune adults who contract VZV infections are those who are habitually exposed to children, such as schoolteachers.

Cytomegalovirus like HSV is spread directly from person to person through secretions such as saliva, vaginal secretions, and semen. As with HSV-2, cervical CMV infection is a source of fetal or neonatal infection. Unlike HSV-2, however, CMV can infect the fetus in utero early in pregnancy, apparently by transplacental transfer of virus, with resultant severe fetal damage. The exact mode of CMV transmission in utero—whether viremic or by direct spread from the infected cervix—remains unknown. Because the presence of antibodies to CMV prevents neither reactivation of infection nor reinfection with exogenous virus, a mother who gives birth to one infant with congenital CMV infection may later have another infant who is also infected in utero. Although the same strain of virus may be involved, infection in succeeding infants is less likely to be symptomatic. CMV can also be transmitted by breast-feeding, because the virus is found in milk. Finally, posttransfusion CMV mononucleosis, as distinct from EBV infectious mononucleosis, is an infrequent but well-described syndrome.

The transmission of EBV probably requires close contact and exchange of oral secretions. Saliva and parotid secretions contain infectious virus that may persist for many months or even years. The incubation period and duration of contagiousness is ill defined; estimates are weeks or more.

CONCLUSION: PREVENTION AND TREATMENT

The only established preventive measure for any of the herpesvirus infections is the use of varicella hyperimmune globulin to prevent or modify VZV infection, particularly in leukemic children and immunosuppressed adults threatened with infection. Experimental vaccines against CMV and VZV composed of live, presumably attenuated viruses are under study. Such vaccines present potential problems inasmuch as the living vaccine virus is expected to persist in a latent form in the same manner as wild-type virus acquired through natural infection. Consequently, even though the primary infection caused by the vaccine virus may be silent, later reactivation of the virus might cause overt disease. Inactivated vaccines, which consist of disrupted virus from which the DNA has been removed, are also being tested experimentally. Such viral subunit preparations have the advantage that they cannot produce latent infection, but their capacity to induce durable immunity is uncertain, particularly as they evoke primarily humoral and little cell-mediated immunity. The use of live EBV preparations as vaccines seems to be precluded, inasmuch as the virus is a suspected oncogenic agent. No experimental EBV vaccines are yet available.

Several licensed antiviral drugs and experimental agents are now available for treatment of HSV infections. Iododeoxyuridine (IUDR) and trifluorothymidine are useful for treatment of early ocular keratitis. Adenine arabinoside (ARA-A, vidarabine), when used for treatment of HSV encephalitis, reduces mortality, but the incidence of severe complications is still considerable, especially if coma has supervened before treatment can be initiated. Children with HSV encephalitis tend to have a more favorable response to treatment with ARA-A, even if they are comatose when first treated. Acyclovir [9-(2-hydroxyethoxymethyl)-guanine] is a new antiviral agent that is the most effective thus far approved for treatment of herpesviral infections. It is useful for treatment of HSV-1 or -2, whether in the form of local lesions or disseminated disease. Acyclovir has recently been shown to be more effective and safer than ARA-A for treatment of herpes simplex encephalitis. The drug is also effective in disseminated VZV infections in immunocompromised hosts. In vitro, acyclovir effectively halts replication of EBV, but it has little effect on CMV either in vitro or in vivo human infections. Recent studies with acyclovir give some indication that it may have an effect on EBV infection. In clinical trials, the main side effects have been transient, reversible rises in blood urea nitrogen and serum creatinine levels, which respond to hydration, and pain at the site of local injections of the drug.

The mechanism of this drug is of particular interest, inasmuch as in cultured HSV-infected cells, acyclovir is activated by specific phosphorylation by the HSV-coded thymidine kinase. The acyclovir monophosphate thus produced is converted by cellular enzymes to di- and triphosphate forms of the drug. The triphosphorylated acyclovir is active intracellularly at the site of infection and halts herpes simplex virus DNA replication by interfering with the action of the virus-specified DNA polymerase. The drug is highly specific and virtually nontoxic for normal, uninfected host cells. In in vitro systems, the drug is ineffective against the latent, episomal form of the Epstein-Barr virus, although it does check replication of the linear form of the EBV genomes that go to make up virus.

Other drugs under development and still experimental include phosphonoformic acid, which operates by interference with herpesvirus-specified DNA polymerases and in vitro is generally effective against herpesviruses, inasmuch as all herpesviruses appear to code for a novel DNA polymerase. E-5-(2-bromovinyl)deoxyuridine (BVDU) also seems to be an effective drug in vitro and is soon to undergo clinical trials. Other new agents include 1-(2-deoxy-2-fluoro-β-D-arabinofuranosyl)-5-methyluracil (FIAC), 1-(2-deoxy-2-fluoro-β-D-arabinofuranosyl)-5-iodocytosine (FMAU), and 9-(1, 3 dihydroxy-2-propoxymethyl) guanine (DHPG), which are promising antiherpetic agents in vitro. Some of these newer experimental drugs are effective against cytomegalovirus in vitro.

Recommended Readings

Guinan ME et al: The course of untreated recurrent genital herpes simplex infection in 27 women. *N Engl J Med* 297:69, 1977.

Huang E-S, Pagano JS: Comparative diagnosis of cytomegaloviruses: New approach. In Kurstak E, Kurstak C (eds): *Comparative Diagnosis of Viral Diseases*, Vol 1, Human and Related Viruses, Part A. New York, Academic Press, 1977.

Kaplan AS: *The Herpesviruses.* New York, Academic Press, 1973.

Pagano JS, Nedrud JG: Latency of the Epstein-Barr virus and cytomegalovirus. In Nahmias AJ, Dowdle WR, Schinazi RF (eds): *The Human Herpesviruses, An Interdisciplinary Perspective*. Part IV, pp 206–218. New York, Elsevier-North Holland, 1980.

Pagano JS, Datta AK: Perspectives on interactions of Acyclovir with Epstein-Barr and other herpesviruses. *Am J Med* 73:18–26, 1982.

Spruance SL, Overall JC, Kern ER, et al: Natural history of recurrent herpes simplex labialis: implications for antiviral therapy. *N Engl J Med* 297:69, 1977.

Sixbey J, Pagano JS: New perspectives on the Epstein-Barr virus in the pathogenesis of lymphoproliferative disorders. In Remington JS, Swartz MN (eds):*Current Clinical Topics in Infectious Diseases*, Vol 5. New York, McGraw-Hill, 1984.

11

New and Emerging Viral Diseases

NEAL NATHANSON and
FRANCISCO GONZALEZ-SCARANO

For many years, viruses have been recognized as the cause of a number of acute infectious diseases. During the last 10 years, however, certain chronic diseases have been linked to persistent infection with some of these same conventional and well-characterized infectious agents. Also described have been a number of "slow virus diseases" either caused by less conventional viruses or for which no viral etiology is known.

SPONGIFORM ENCEPHALOPATHIES

The spongiform encephalopathies (Table 11.1) are human and animal diseases that are classified together because they share similar neuropathologic features. All of these diseases have been transmitted experimentally by inoculation of affected brain tissue into susceptible hosts (primates, sheep, and rodents). The putative causal agents have not been purified, and their structure and composition are therefore unknown. Their established properties, however, suggest that they differ from any of the presently known classes of viroids, infectious single-stranded RNA, and viruses. Plant viroids (small single-stranded RNAs) have been proposed as models for the spongiform agents.

The pathologic features of this group of diseases consist of spongiform change and status spongiosus. The former, considered pathognomic, is characterized by vacuolization within the neuropil, with neuronal loss. Late in the course of the disease, vacuolization is accompanied by reactive astrocytosis, and it is then called status spongiosus (Fig. 11.1). No inflammatory infiltrate occurs in either parenchyma or membranes in any of the spongiform encephalopathies. In Creutzfeldt-Jacob dis-

TABLE 11.1   Spongiform Encephalopathies of Animals and Humans

Disease	Agent	Natural Host	Distribution
Kuru	Kuru	Humans	Fore peoples, New Guinea
Creutzfeldt-Jacob disease	C-J	Humans	Worldwide
Scrapie	Scrapie	Sheep	Worldwide
Transmissible mink encephalopathy	TME	?Sheep[a]	Mink ranches

[a] Probably represents scrapie transmitted by consumption of contaminated sheep offal.

ease, the cortex, basal ganglia, and cerebellum are affected severely, whereas in kuru the cerebellum is the site of the most severe lesions.

Kuru

Discovered after World War II in the peoples of the Fore linguistic group in the highlands of eastern New Guinea, kuru is apparently confined to that area of the world. It is a progressive neurologic disease that begins with ataxia and ends, usually within 1 year, with complete motor incapacity.

The suspicion that a transmissible agent might cause kuru derived from Hadlow's observation that the disease greatly resembled scrapie, a disease of sheep, whose symptoms include nervousness and excitability and, later, tremor, ataxia, and wasting. Kuru was initially transmitted to chimpanzees (and later to other subhuman primates) by the intracerebral inoculation of affected brain. Inoculated animals develop symptoms after intervals of 10 to 80 months; symptoms resemble those in humans, and the pathologic lesions (confined to the brain) are similar.

Transmission in the primitive Stone-Age-like culture of the Fore probably occurred during ritual endocannibalism, in which the relatives of a recently deceased person consumed partially cooked tissues of the body. Since this practice was usually limited to women and children, it probably accounted for the fact that relatively few cases of kuru occurred in men.

The practice of cannibalism was discouraged by the Australian trust officials and was terminated for the most part in the late 1950s. Since that time (Fig. 11.2), the incidence of kuru has gradually declined, and by 1980 the disease had virtually disappeared. Although kuru was considered an exotic disease limited to an isolated primitive people, the studies by Gajdusek and others have dramatized the potential importance of transmissible agents in certain chronic diseases. For instance, the declining incidence of kuru made it possible for researchers to de-

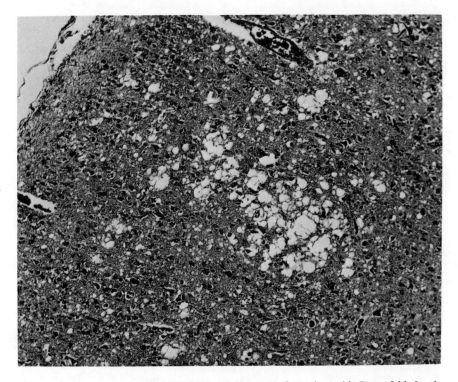

FIG. 11.1. Photomicrograph from the cerebral cortex of a patient with Creutzfeldt-Jacob disease to show vacuolization, which is the salient characteristic of the spongiform encephalopathies (×100). (courtesy of Dr. Donald Gilden).

termine its distribution and the variations in the incubation period, which probably ranges from 2 to 20 years (Fig. 11.3).

Creutzfeldt-Jacob (C-J) Disease

A subacute, progressive, and invariably fatal disease, with symptoms and lesions confined to the central nervous system, Creutzfeldt-Jacob disease is found worldwide, although it is rare, with an annual incidence of about 0.5 to 1 per million.

Occurring in middle age, Creutzfeldt-Jacob disease affects both sexes with about the same frequency. Its overriding feature is a global dementia, which progresses much more rapidly than the more common senile or presenile dementias. Gait abnormalities and visual disturbances are also frequent, as are disturbances of consciousness and seizures. Death usually occurs within 1 year of onset.

Most cases of Creutzfeldt-Jacob disease are isolated, and a major enigma is the natural mode of transmission—patients rarely give a history of contact with another case, either within or outside of the family.

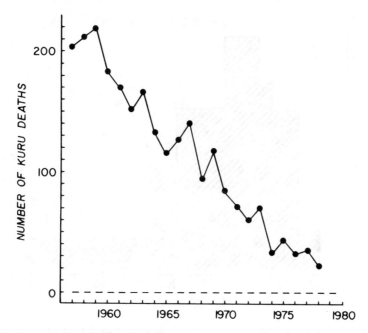

FIG. 11.2. Deaths from kuru, 1957–1975, among the Fore peoples of eastern New Guinea. Transmissions essentially ceased in 1957–1958 because of proscription of ritual endocannibalism. (After Alpers MP: Epidemiology and ecology of kuru. In SB Prusiner and WJ Hadlow (eds): *Slow Transmissable Diseases of the Nervous System*, New York pp 67–92 Academic Press, 1979.)

A few cases clearly are acquired by contact with infected tissues (corneal transplants) or instruments (neurosurgical electrodes). Three possibilities have been considered to explain transmission in most cases. First, some authorities have suggested consumption of scrapie sheep tissue as a course, but the disease is found in regions of the world (e.g., Australia) where scrapie has been eradicated from sheep. Second, the possibility of contact with subclinical human infections has been suggested, but such subclinical infections have never been documented in experimental animals. Finally, by exclusion of other possibilities, activation of a latent genome is a possibility.

Properties of the Spongiform Agents

The remarkable transmission experiments of Gibbs and Gajdusek established the resemblance of kuru and Creutzfeldt-Jacob disease agents to the scrapie agent. All spongioform agents show certain biologic properties similar to conventional viruses: (a) their infectivity can be reproducibly titered to endpoints of 10^8LD; (b) the incubation period increases inversely with the dose; (c) there is an apparently quantal dose-

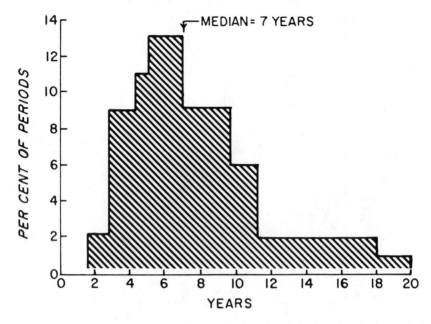

FIG. 11.3. Hypothetical estimate of the varying lengths of the kuru incubation period, based on the data in Fig. 11.2; assuming a log-normal distribution.

response relationship at the endpoint; (d) the incubation period shortens during adaptation of the agent to new hosts; (e) the agents replicate to high titers before either morphologic changes or clinical signs develop.

Nonetheless, the spongiform agents show some extraordinary properties that separate them from all animal viruses: (a) the incubation period is extremely long, varying from months to years, but is regular and predictable; (b) no antigen has been associated with these agents, and they have never been shown to induce any immune responses; (c) although filtration studies indicate that infectivity is associated with particles the size of small viruses, no structure has ever been observed on electron microscopic examination of infectious suspensions; (d) the agents are remarkably resistant to certain conventional modes of sterilization, such as formaldehyde, ethanol, boiling, and ultraviolet radiation (Table 11.2).

The composition of the spongiform agents has eluded definition to the present, although a variety of hypotheses have been advanced by many workers. No protocol has yet been devised to successfully purify and concentrate these agents; until this critical technical problem has been solved, their chemical nature will remain vague. It is even unclear whether the spongiform agents have a nucleic acid genome, and inactivation by x-radiation suggests an extremely small target size (less than

TABLE 11.2 Decontamination Procedures for Spongiform Encephalopathies

Procedure	Efficacy
Autoclaving (121C, 15 psi)	Effective
Sodium hypochlorite, 5%	Effective
Phenol	Effective
Alcoholic iodine	Effective
Ionizing radiation	Moderately effective
UV radiation	Ineffective
Formalin (3–10%)	Ineffective
Ethanol (70%)	Ineffective

Adapted from Chatigny and Prusiner. In Prusiner SB and Hadlow WJ (eds): *Slow Transmissible Diseases of the Nervous System*, pp 491–514. New York, Academic Press, 1979.

a single gene of conventional dimensions). Recently, Prusiner has suggested that the most highly purified preparations contain protein but lack nucleic acid; he therefore proposed the name *prions*. Until better data are available, it would probably be premature to accept any hypothesis about the nature of these highly unconventional agents.

Viroids

Several diseases of plants (e.g., potato spindle tuber disease, Cadang-Cadang disease of coconuts, and chrysanthemum stunt disease) have been shown to be caused by infectious nucleic acid particles. These particles, termed viroids (viruslike), are single-stranded circular RNA molecules, usually smaller than 100,000 daltons. They are capable of coding for (at most) a single, moderately sized peptide and are not associated with any structural proteins. Their discovery led to an intense search for similar agents in the spongiform encephalopathies, but, as yet, no viroids have been identified in any of these conditions, nor indeed in any other human or animal disease. Nevertheless, viroids are an excellent example of a single macromolecule that can disrupt the cellular apparatus of a higher organism. An exciting proposal for the mechanism of pathogenicity of viroids is that they may interfere with RNA splicing. The idea stems from the similarity between sequences of the RNA of potato spindle tuber viroid and of U_1 RNA—one of the small RNAs implicated in the processing of split genes. By forming stable complexes with unspliced messenger RNAs, viroids may prevent further processing and translation of the mRNAs into functioning proteins. Although the search for human viroids continues, at this point they are important only to molecular biologists and to those concerned with plant diseases.

CHRONIC CNS DISEASE DUE TO CONVENTIONAL VIRUSES

Progressive, Multifocal Leukoencephalopathy (PML)

In 1958 Astrom and colleagues described three cases of a progressive neurologic disease occurring in patients with lymphoproliferative disorders. The clinical picture was consistent with metastatic disease, but the patient's conditions did not improve with radiation therapy. At autopsy, brain lesions were found in the white matter that consisted of perivascular destruction of myelin sheaths with sparing of axones and proliferation of bizarre, gigantic astrocytes. The microglia surrounding the lesions had strongly basophilic nuclei. Some of these microglial cells were pleomorphic, and others also had intracellular inclusions. Astrom and associates termed the entity progressive, multifocal leukoencephalopathy. Although an infectious, possibly viral, cause was suspected soon after the initial report, not until 1965 did Zu Rhein and Chou report visualizing intranuclear papovaviruslike particles in patients with this disease. Their report was followed by isolation of a papovavirus from a patient with progressive, multifocal leukoencephalopathy. Isolation of the same virus—termed JC virus—in other laboratories confirmed the association.

Progressive multifocal leukoencephalopathy has been noted in previously normal persons, but, in general, the disease occurs in those having a background of immunosuppression, usually involving either lymphoma, leukemia, or acquired immunodeficiency syndrome (AIDS). Renal transplantation is associated with urinary excretion of JC and BK virus, a related papovavirus, and occasionally with clinical evidence of the disease. The symptoms all result from CNS involvement and, as in syndromes that affect the white matter, consist of paralysis, blindness, and dementia. Cerebral biopsy is the only definitive way to diagnose this condition.

Extensive pathologic studies have confirmed that progressive multifocal leukoencephalopathy is mainly an infection of oligodendrocytes and that demyelination follows cellular destruction. A recent immunochemical study showed that histologically normal areas in the affected brain may contain JC antigen (Fig. 11.4).

JC virus, the original isolate from a case of progressive, multifocal encephalopathy, is a polyoma virus, of the family Papovaviridae. Infection with JC virus occurs in childhood, and sera from 70 to 80% of adults have hemagglutination-inhibiting antibody. Until recently, only primary human fetal glial (HPFG) cells with a high proportion of spongioblasts were known to support viral replication. Although some strains have now been adapted to human amnion cells and adult brain cells, HPFG cells remain the only cell line acceptable for primary isolation,

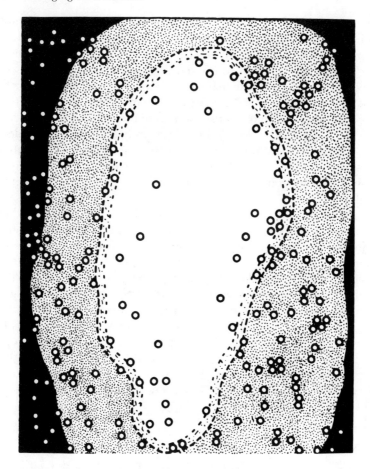

FIG. 11.4. Progressive multifocal leukoencephalopathy, showing the evolution of an acute lesion. In the outer zone, shown in black, white matter stains normally with antisera against myelin-associated glycoprotein (MAG) and myelin-basic protein (MBP). In the next zone, shown with stippling, myelin sheaths stain normally for MBP, but staining of MAG is decreased. This zone contains the highest density of infected oligodendrocytes, shown as small circles. The next zone, shown between dashed lines, includes the area of active myelin breakdown; MAG staining is decreased, MBP staining is abnormal, and there are fewer virus-infected cells. The central zone, shown in white, is demyelinated; MAG and MBP staining are deceased or absent, and there are the fewest virus-containing cells. (After Itoyama Y, Webster H deF, Stembeyer NH et al; Distribution of papovavirus myelin-associated glycoprotein, and myelin basic protein in progressive multifocal leukoencephalopathy lesions. *Ann Neurol* 11:398–407, 1982.)

despite the high proportion of the population with serologic evidence of infection. PML is not the only disease associated with JC virus. Brain tumors, primarily those of neuroectodermal origin, can be produced in hamsters and owl monkeys by intracerebral inoculation of JC virus. The histopathology of multifocal leukoencephalopathy has features suggesting neoplasia. This fact along with its ability to cause animal brain tumors has led to speculation regarding the oncogenic potential of JC virus. Epidemiologic studies in which the frequency of high-titered sera among cancer patients was compared with that among controls, as well as direct examination of tumor tissue for viral antigens, have so far failed to substantiate any role of JC virus in human neoplastic disease.

Like the other polyoma viruses, the genome of the JC virus consists of circular, covalently closed, double-stranded DNA with molecular weight 2.93×10^6 daltons. Restriction maps (see below) and antigenic studies of JCV, BKV, and SV40 have established the individual identity of each of these agents and the relative homogeneity of different isolates of JC.

Given the information currently available, it is unknown whether progressive, multifocal leukoencephalopathy represents reinfection with a common virus, primary infection in an immunosuppressed person not previously exposed to the agent, or reactivation of a virus the genome of which has remained latent after a primary infection in childhood. Two observations render reactivation an attractive hypothesis: the propensity of other polyoma viruses to remain latent in the host and experimental evidence in cell culture systems that shows that JC virus DNA can be integrated into host-DNA. If progressive leukoencephalopathy represents reactivation of a latent genome, the regulation of expression then must involve factors other than immunosuppression, because the incidence of the disease is very low even in immunocompromised patients.

Subacute Sclerosing Panencephalitis (SSPE)

A chronic progressive encephalitis of childhood, SSPE is fatal within 6 to 24 months after onset. Measles virus can be incriminated in most cases, although an occasional case may be caused by rubella virus. The causal agent may be difficult to isolate in conventional cell cultures, but viral antigens can be readily detected in CNS tissue or in cultures from CNS infections. Patients with SSPE give a history of an uneventful primary case of measles, usually 4 to 11 (average 7) years previously. The only unusual aspect of the primary infection is that it occurred before age 2 (compared with age 6 for the general population). From epidemiologic studies, SSPE is estimated to occur at the rate of 1 case in 100,000 primary measles infections. For children who have been immunized with live, attenuated measles virus vaccine, the risk of SSPE following naturally acquired measles is estimated to be ten times lower than in unimmunized children. This constitutes an additional argument in favor of active immunization.

The early stages in the pathogenesis of SSPE are obscure, and early age of infection is the only well-established predisposing factor. In experimental animals, a similar infection can be induced by pretreatment

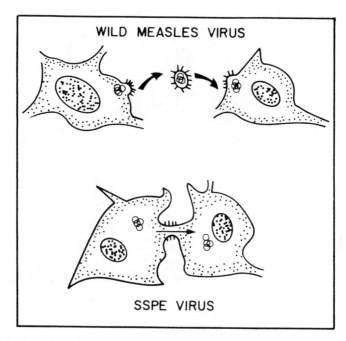

FIG. 11.5. Hypothetical mode of the cell-to-cell spread of subacute sclerosing panencephalitis virus compared with that of measles virus. Although competent measles virus may occasionally spread through intercellular bridges, infection is carried from infected to normal adults mainly by free virions, which are extracellular and therefore neutralizable by serum antibody.

with passive antibody before infection. The virus isolated from patients with SSPE often seems to be defective in that it is deficient in the matrix (M) protein. This protein is vital in the assembly of mature virions, and its deficiency would explain the difficulty in isolating free infectious virus. Apparently, SSPE virus can only spread slowly and inefficiently from cell to cell by means of intercellular bridges produced by the fusing function of a viral glycoprotein (Fig. 11.5). Neutralizing antibody, although present in supranormal titers in both serum and CSF of affected patients, cannot clear intracellular infection or prevent indolent spread of subacute sclerosing panencephalitis virus between adjacent CNS cells. As a result, there is a gradual but inexorable progression of the cytocidal infection through the CNS.

CHRONIC DISEASES THAT MAY BE CAUSED BY VIRUSES

Proof that disorders such as PML, SSPE, and C-J disease are caused by transmissible agents has given credence to the hypothesis that other

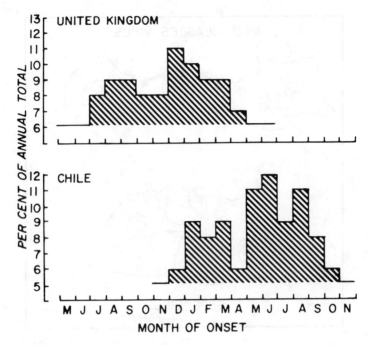

FIG. 11.6. Seasonal distribution of insulin-dependent diabetes, according to month of onset: United Kingdom, 903 cases aged 0–15 years; Chile, 525 cases, aged 0–30 years. (After Gamble DR: The epidemiology of insulin-dependent diabetes, with particular reference to the relationship of virus infection to its etiology. *Epidemiol Rev* 2:49–70, 1980.)

chronic diseases are triggered by virus infection. In this connection, much effort has been invested in the search for viral etiologies for diabetes mellitus and multiple sclerosis. Both of these disorders serve to illustrate the problems inherent in defining elusive relationships between chronic illness and preceding viral infections.

Diabetes Mellitus

Viruses and diabetes are linked by virtue of animal models and epidemiologic associations. Apparently several kinds of diabetes mellitus exist, but associations apply only to juvenile-onset, insulin-dependent diabetes.

There are several epidemiologic clues. First, only one of a pair of monozygotic twins may develop insulin-dependent diabetes, the second twin remaining unaffected and apparently normoglycemic in about half of the pairs. This finding strongly suggests that an exogenous triggering event is required to initiate disease. Second, new cases of insulin-dependent diabetes occur in a seasonal cycle, with rates in midwinter twice those in midsummer (Fig. 11.6). Third, although the disease is

TABLE 11.3 Influence of Host Genotype on Frequency of Elevated Blood Glucose (over 225 mg/dl) after Infection of Mice with the M Variant of Encephalomyocarditis Picornavirus

Strain or Designation	Genotype	No. of Mice	Percent Elevated	Percent Expected If Resistance Dominant
SWR/J	Susceptible	33	88	100
C56BL/6	Resistant	33	0	0
F_1	Susceptible × resistant	39	5	0
Backcross	F_1 × resistant	66	5	0
Backcross	F_1 × susceptible	93	46	50

After Onodera T, Yoon JW, Brown KS et al: Evidence for a single locus controlling susceptibility to virus-induced diabetes mellitus. *Nature* 274:693–696, 1978.

not rare in infancy, few cases begin before age 9 months, a pattern similar to that seen for most viral infections in which maternal antibody provides passive protection. Finally, the onset of the diabetes is often abrupt, compatible with an acute pancreatic insult. Such epidemiologic evidence is only circumstantial, however. Furthermore, individual case studies have variously implicated different viruses (Coxsackie B, mumps, rubella). If, in fact, several different viruses can occasionally cause insulin-dependent diabetes, one should not be surprised when the epidemiologic patterns tend to be blurred and fail to incriminate a single agent.

Several viruses can cause diabetes in mice, including picornaviruses (EMC and Coxsackie) and reovirus. These agents replicate in the islets of Langerhans and destroy beta cells during the acute infection; mice become hyperglycemic shortly thereafter. Manipulation of the animal model has shown that several variables are important. First, different inbred strains of mice vary considerably in their susceptibility, and resistance is dominant in crosses of susceptible and resistant strains; the age and sex of the animals also influence the outcome. Second, different clones of the same virus vary noticeably in diabetogenic potential. Finally, even with a high-risk combination of mouse genotype and virus strain, individual animals show striking differences in response, varying from hyperglycemia to normoglycemia (Table 11.3). The animal studies lend credibility to the possibility that viruses cause at least some cases of insulin-dependent diabetes; they also suggest why it might be difficult to prove virus association in epidemiologic studies.

Multiple Sclerosis

In multiple sclerosis (MS), a chronic, progressive neurologic disease, the underlying lesions are focal areas of primary demyelination ("plaques")

TABLE 11.4   Relative Incidence of MS in U.S. Military Veterans According to Place of Birth

Tier of States	MS Cases	Controls	MS Ratio (MS/Controls)
Northern	154	106	1.45
Middle	152	177	0.86
Southern	67	89	0.75

After Beebe G, Kurtzke JF, Kurland LT et al: Studies on the natural history of multiple sclerosis: 3. Epidemiologic analysis of the Army experience in World War II. *Neurology* 17:1–17, 1967.

disseminated throughout the CNS. The two lines of evidence suggesting a viral cause, as in insulin-dependent diabetes, arise from epidemiologic studies and investigation of animal models.

Twin studies of MS indicate that concordance, even among monozygotic pairs, is no greater than 30%, implying that a critical exogenous event triggers the disease. The correlation of MS incidence with increasing distance from the equator is well established; for example, rates in the northern United States are twice those in the southern United States (Table 11.4). In addition, migration between areas of different risk appears to influence incidence; after the move, the next generation exhibits the risk associated with the new environment. These observations parallel those of earlier studies of classic viral diseases such as poliomyelitis.

Animal models provide the most persuasive evidence for the proposition that MS could be a virus-initiated disease (Table 11.5). Several naturally occurring viral infections cause persistent CNS infections, as-

TABLE 11.5   Animal Models of Chronic Demyelination Associated with Persistent Virus Infection

Virus and Disease	Virus Group	Natural Host	Probable Mechanisms of Demyelination	Incubation from Infection to Disease
Theiler's murine encephalomyelitis	Picorna	Mouse	Immune	1–2 months
Mouse hepatitis	Corona	Mouse	Cytolytic	1–2 months
Visna	Retro	Sheep	Immune	0.5–8 years
Canine distemper	Paramyxo	Dog	Immune	1–2 months

After Martin JR, Nathanson N: Experimental models of virus-induced demyelination. In Zimmerman HM (ed): *Progress in Neuropathology*, pp 27–50. Vol 4, New York, Raven Press, 1979.

sociated with focal areas of primary demyelination. Myelin destruction may be started by direct viral cytolysis of oliogodendroglia or be secondary to an antiviral immune response, depending on the model.

Much fruitless effort has been expended in attempting to identify a specific virus (or viruses) in CNS tissue of patients with MS. Numerous preliminary reports of virus isolations have not been confirmed. Elevated levels of CNS immunoglobulin have long been recognized as diagnostic of MS in the absence of CNS infection. It is now known that these CNS immunoglobulins represent antibodies against measles, herpes simplex, and a variety of other viruses or nonviral infectious agents, according to the antecedent history of infectious diseases in the individual case. Such findings imply persistence of certain viruses in the CNS; but whether any of them, or the assorted immune responses, are related to the pathogenesis of MS remains unclear.

Acquired Immune Deficiency Syndrome (AIDS)

Beginning in 1980, cases of what appeared to be a new disease were first observed and reported in the United States. This syndrome is generally called AIDS, and is characterized by a variety of opportunistic infections and opportunistic neoplasms, such as Kaposi's sarcoma. Most cases progress over a few months or years to a fatal outcome in spite of supportive therapy. The disease appears to be transmitted by homosexual contact or blood and blood products. Hematologically, there is reduction or loss of certain subsets of thymus-derived (T) cells, particularly those involved in effector functions.

Recently, viruses that almost certainly represent the causal agent have been isolated and named HTLV-III (human T-cell leukemia virus) or LAV (lymphocyte-associated virus). HTLV appears to be a new group of human retroviruses that contains an additional gene of 1.5 to 2 kb, located at the 3' end of the retrovirus genome (between the env gene and the 3' LTR). This gene encodes a protein X (pX) that probably functions as a trans-acting element that may either stimulate cellular proliferation (HTLV-1 and HTLV-II) or interfere with replication leading to cell death (HTLV-III).

The state of knowledge regarding HTLV-III and AIDS is evolving rapidly, and studies have only recently begun to show promising results.

Delta Hepatitis

A new agent associated with hepatitis was discovered about 1978 by Rizzetto and colleagues. The delta agent, which is endemic in certain regions of Italy, is invariably associated with Hb_sAg (hepatitis B surface antigen) and is probably responsible for a significant proportion of all cases of chronic active hepatitis. Apparently it is incapable of replica-

TABLE 11.6   Viral Hemorrhagic Fevers of Humans: A Selected List

Virus Group	Specific Virus	Disease	Arthropod Vector	Animal Reservoir	Geographic Distribution
Bunya	KHF	Korean HF[a]	−	+	Northern Eurasia
Arena	Lassa	Lassa F	−	+	Africa
	Machupo	Bolivian HF	−	+	South America
	Junin	Argentine HF	−	+	South America
Rhabdo	Marburg	Marburg F	−	+	Africa
	Ebola	Ebola	−	+	Africa
Flavi	Dengue	Dengue HF	+	−	Pacific and Caribbean Coasts

[a] Also called hemorrhagic fever renal syndrome (HFRS).

tion on its own, and infection therefore has to occur either concurrently with hepatitis B infection or on a background of HB$_s$Ag carrier state. In infected persons its presence is detected through immunofluorescence of liver biopsies and by the detection of δ-antigen and anti-δ antibodies in ELISA.

The delta agent is 20–30 nm in diameter, and it has a small (1.7kb) RNA genome. Although it contains a core (δ antigen), it apparently requires the hepatitis B coat.

Recent outbreaks of δ hepatitis have been reported among groups known to be at high risk for HB infection (parenteral drug abusers, for example) in the United States. Fortunately, the absolute requirement for infection with the more conventional hepatitis agent makes control of the δ agent dependent on the vaccination of high-risk groups against hepatitis B.

HEMORRHAGIC FEVERS

In recent years a number of clinically similar illnesses, in which hemorrhage, shock, and fever are prominent symptoms, have occurred in epidemic form in different areas of the world (Table 11.6). Each of these virus-induced hemorrhagic fevers has a limited geographic distribution, due to localization of either the animal reservoir or the arthropod vector. Although the numbers of cases are small, the high mortality (often about 50%) has given dramatic prominence to these agents, which are considered Class IV (highest risk). The occasional transmission of infection to hospital contacts and the requirement for P4 containment enhance their notoriety.

The causative viruses belong to several different groups, and the mechanisms of shock and hemorrhage are undoubtedly diverse. Preventive vaccines and effective immunotherapy are under development for several of these agents.

FUTURE PROSPECTS USING NEW MOLECULAR AND IMMUNOLOGIC TECHNIQUES

In the past, transmissible agents were detected by morphologic studies, or by inoculation of cell cultures or animals, supplemented by explantation of viable tissues. Examples are the recovery of herpes simplex virus from the trigeminal ganglion, or Coxsackie B4 virus from acute-onset diabetes, and the identification of rotaviruses in outbreaks of diarrhea. However, these methods have failed to elucidate the cause of certain diseases in which a virus is implicated. Several powerful new technologies have been developed in the past 10 years, offering additional approaches from which there is evidence suggestive of a transmissible agent.

These new approaches can be categorized into two groups: (1) recombinant DNA technology, which enables one to search directly for homologous nucleic acid sequences by hybridization with DNA probes, and (2) monoclonal antibodies, which detect individual antigenic determinants and provide highly sensitive reagents. Both techniques are also powerful methods for comparing the genomes or proteins of different viruses to determine their molecular similarities or differences.

Nucleic Acid Probes

A segment of double-stranded DNA is radiolabeled (probe); the two strands are dissociated (usually by heating), and the individual strands are then permitted to reanneal with an unknown DNA or RNA sample, immobilized on paper or a glass slide. Temperature and pH are then altered to encourage reannealing (or hybridization). The probe hybridizes to homologous nucleic acid sequences in the sample. Binding is then detected by radioautography at gross or microscopic levels.

Specificity and sensitivity can be adjusted by altering the size (number of base pairs) of the probe or the conditions of hybridization. When applied to cells on a glass slide (in situ hybridization), the cellular localization of viral genomes can be determined. If the target is DNA that was extracted before testing, pretreatment with restriction endonucleases (each of which cuts DNA at specific sites) can be used to determine whether viral sequences are free or integrated into the host genome.

The same probe can be used to detect either viral DNA or viral RNA, by pretreatment of the target with DNase or RNase.

For DNA viruses, probes can be prepared directly from viral genomes, which are extracted and labeled (by nick translation or with ^{125}I). RNA virus genomes are unstable (due to ubiquitous RNases), and are therefore used as templates on which to construct complementary DNA (cDNA) probes, catalyzed by commercially available reverse transcriptase prepared from a retrovirus. Furthermore, if the viral genome sequence is known, at least in part, oligonucleotides synthesized up to about 20 base pairs can be synthesized chemically. Although currently expensive, commercial custom-made oligonucleotides can be bought and delivered in a short time. Recently (1982), "do-it-yourself" oligonucleotide synthesizers have also been marketed.

These methods have been used successfully to detect and analyze viral sequences in several animal models of persistent viral infection. Visna virus DNA and RNA have been shown to be present in nonvirogenic cells, and Theiler's viral RNA has been detected in the oligodendroglia of chronic, demyelinated lesions of the mouse spinal cord. In human spinal ganglia, varicella-zoster virus has been demonstrated for the first time by hybridization, although the virus has not yet been grown from explanted ganglia.

Enzymatic Digestion

Once a virus has been isolated, its genome can be extracted and characterized by enzymatic digestion, even if it cannot be grown in cell culture.

For RNA viruses, genomes can be cleaved with one of several endoribonucleases, such as T_1, which cut between specific bases, producing a series of oligonucleotides. When this T_1 digest is subjected to two-dimensional polyacrylamide gel electrophoresis, the oligonucleotides separate, producing a characteristic RNA "fingerprint." Even minor changes in the genome (a few bases) may result in alterations of a few oligonucleotide spots on the autoradiograph. Use of computer analysis for comparing the RNA fingerprints of related viruses has proved helpful in detecting wild-type contamination of attenuated viral vaccines and for delineating the taxonomy of a family of viruses.

For DNA viruses, genomic analysis depends on bacterial-restriction endonucleases, which cleave double-stranded DNA at specific sites. More than 350 enzymes have been identified, each with its own recognition and cutting sites, usually a defined sequence of 4 to 7 base pairs. After enzymatic digestion, the DNA fragments are sized on a polyacrylamide gel, where they are detected directly by radioautography or indirectly with a virus-specific radiolabeled DNA probe. Although a single enzyme will not, in most instances, be sufficient for genotyping different strains of a virus, a combination of enzymes can be used successfully.

Monoclonal Antibodies

Monoclonal antibodies are produced by cellular hybrids, resulting from the fusion of two different cells, usually of the same species: a replicating, nonsecreting myeloma cell and a single B cell, or clone taken from the spleen of an animal immunized with the antigen to be studied. After a bulk fusion of myeloma cells with disaggregated spleen cells, individual hybrid cells are isolated and grown out; those producing the desired antibody are selected and propagated. If the hybridoma is of murine origin, the antibody-producing hybridoma cells can be propagated as an ascites tumor in mice, the copious ascitic fluid serving as a plentiful source of monoclonal antibody. Monoclonal antibodies have now been prepared against a number of viruses in order to analyze individual viral proteins.

Work to date indicates that each viral protein has a limited number of antigenic sites (usually no more than 6); individual monoclonal antibodies will bind to somewhat different configurations (epitopes) at a given site. Hybridomas can be induced to secrete sharply specific antibodies in large amounts. Highly sensitive probes for single antigenic determinants in viral proteins can then be selected.

Monoclonal antibodies are being used to detect viral proteins in tissues chronically infected with measles and related viruses, in the search for antigens in chronic neurologic diseases; and to determine the degree to which different messenger RNAs of an established virus have been translated. Such information may shed light on the pathogenesis of diseases such as subacute sclerosing panencephalitis.

Comparison of Agents

A major question about newly discovered agents that are candidates for association with specific chronic diseases is their relationship to established viruses. Relationships may now be characterized at the genomic or protein levels. Once the DNA representing a viral genome (or segments of that genome) has been cloned, estimating its degree of homology to corresponding segments of newly identified agents will be possible by quantitative hybridization studies. Even more detailed information can be obtained when the segments of two related genomes have been sequenced.

Another type of data can be generated by the use of monoclonal antibodies. Information now available on established virus groups indicates that some antigenic sites on viral proteins are conserved during viral evolution and are shared by related viruses, other antigenic sites being subject to frequent mutation. Comparison of genomic sequences with patterns of monoclonal antibody may indicate the site at which mutation of a single base has occurred to produce the corresponding antigenic variants.

CONCLUSION

These new molecular techniques enable investigators to search for viruses in diseased tissues, even in the absence of overt viral replication, and to effect further refinements in analysis and comparison of subsequently isolated agents. Hepatitis B and rotaviruses are good examples of viruses that, although originally not propagable in any kind of cell culture, have nevertheless been characterized at the molecular level.

As with virus isolation, identification of viral genomes in somatic cells does not signify that the virus in question is the cause of a particular disease. Even though Koch's postulates cannot be strictly fulfilled in connection with agents that are neither propagable nor transmissible, there are other acceptable ways of showing causation. For example, the role of Epstein-Barr virus in Burkitt's lymphoma and that of hepatitis B virus in primary hepatocellular carcinoma have been elucidated entirely on the basis of molecular and epidemiologic studies.

Recommended Readings

GENERAL

Kimberlin RH (ed): *Slow Virus Diseases of Animals and Man.* Amsterdam, North-Holland, 1976.
Stevens JG, Todaro GJ, Fox CF (eds): *Persistent Viruses.* New York, Academic Press, 1978.
Wolinsky JS, Johnson RT: Role of viruses in chronic neurological diseases. *Comprehensive Virol* 16:257–296, 1980.
Youngner JS: Persistent viral infections. In Youngner JS (ed): *Microbiology—1977,* pp 433–540, Washington, DC, American Society for Microbiology, 1977.

SPONGIFORM ENCEPHALOPATHIES

Gajdusek DC: Unconventional viruses and the origin and disappearance of kuru. *Science* 197:943–960, 1977.
Gajdusek DC, Gibbs CJ, Alpers M: Experimental transmission of a kuru-like syndrome to chimpanzees. *Nature* 209:794–796, 1966.
Gibbs CJ, Gajdusek DC, Aber DM et al: Creutzfeldt-Jacob disease (spongiform encephalopathy): Transmission to the chimpanzee. *Science* 161:388–389, 1968.
Hadlow WJ: Scrapie and kuru. *Lancet* 11:289, 1959.
Oesch B, Westaway D, Walchli M, et al: A cellular gene encodes scrapie PrP 27–30 protein. *Cell* 40:(April), 1985.
Prusiner SB: Novel proteinaceous infectious particles cause scrapie. *Science* 216:136–144, 1982.
Prusiner SB, Hadlow WJ (eds): *Slow Transmissible Diseases of the Nervous System.* New York, Academic Press, 1979.

PROGRESSIVE MULTIFOCAL LEUKOENCEPHALOPATHY

Padgett BL, Walker DL: New human papovaviruses. *Prog Med Virol* 22:1–35, 1976.
Walker DL: Progressive multifocal leukoencephalopathy: An opportunistic viral infection of the central nervous system. In PJ Viken, GW Bruyn, HL Klawans (eds): *Handbook of Clinical Neurology* 34:307–330, 1978.

SUBACUTE SCLEROSING PANENCEPHALITIS

Johnson KP: Experimental subacute sclerosing panencephalitis in the hamster. *Adv Exp Med Biol* 134:303–309, 1981

Zeman W: Subacute sclerosing panencephalitis and paramyxovirus infections. In Viken PJ, Bruyn GW, Klawans HL (eds): *Handbook of Clinical Neurology* 34:343–368, 1978.

DIABETES MELLITUS

Brinton M, Nathanson N: Genetic determinants of virus susceptibility: Epidemiologic implications of murine models. *Epidemiol Rev* 3:115–139, 1981.

Craighead JE: The role of viruses in the pathogenesis of disease and diabetes mellitus. *Prog Med Virol* 19:161–214, 1975.

Gamble DR: The epidemiology of insulin-dependent diabetes, with particular reference to the relationship of virus infection to its etiology. *Epidemiol Rev* 2:49–70, 1980.

Notkins AL: Virus-induced diabetes mellitus. *Arch Virol* 54:1–17, 1977.

ACQUIRED IMMUNE DEFICIENCY SYNDROME

Barre-Sinoussi F, Chermann JC, Rey F, et al: Isolation of a T-lymphotropic retrovirus from a patient at risk for acquired immune deficiency syndrome (AIDS). *Science* 220:868–871, 1983.

Gallo RC, Salahuddin SZ, Popovic M, et al: Frequent detection and isolation of cytopathic retroviruses (HTLV-III) from patients with AIDS and at risk for AIDS. *Science* 224:500–503, 1984.

Ratner L, Haseltine W, Patarca R, et al: Complete nucleotide sequence of the AIDS virus, HTLV-III. *Nature* 313:(24 Jan), 1985.

MULTIPLE SCLEROSIS

Acheson, ED: Epidemiology of multiple sclerosis. *Br Med Bull* 33:34–39, 1977.

Cook SD, Dowling PC: Multiple sclerosis and viruses: An overview. *Neurology* 30:80–91, 1980.

Fraser KB: Multiple sclerosis: A virus disease? *Br Med Bull* 33:34–39, 1977.

Johnson RT: The possible viral etiology of multiple sclerosis. *Adv Neurol* 13:1–46, 1975.

Martin JR, Nathanson N: Experimental models of virus-induced demyelination. In Zimmerman HM (ed): *Progress in Neuropathology*, Vol 4, pp 27–50. New York, Raven Press, 1979.

Nathanson N, Miller A: Epidemiology of multiple sclerosis: Critique of the evidence for a viral etiology. *Am J Epidemiol* 107:451–461, 1978.

Norrby E: Viral antibodies in multiple sclerosis. *Prog Med Virol* 24:1–39, 1978.

HEMORRHAGIC FEVERS

Braude AI, Leelarasamee A: Dengue and other hemorrhagic fevers. In Braude AI, Davis CE, Fierer J (eds): *Medical Microbiology and Infectious Diseases*, pp 1534–1544. Philadelphia, Saunders, 1981, Chapter 202.

NEW METHODS

Abelson J, Butz E: Recombinant DNA. *Science* 209:1317–1438, 1980.

Evans AS: Causation and disease: The Henle-Koch postulates revisited. *Yale J Biol Med* 49:179–195, 1976.

Kennett RH, McKearn TJ, Bechtol KB: *Monoclonal Antibodies*. New York, Plenum Press, 1980.

Nathanson N: Persistent viral infections of man: An overview. *Microbiology, Am Soc Microbiol* 533–538, Washington DC, 1977.

Yelton DE: Monoclonal antibodies: A powerful new tool in biology and medicine. *Ann Rev Biochem* 50:657–580, 1981.

12

Viruses and Cancer

HENRY ROTHSCHILD and J. CRAIG COHEN

As documented in the preceding chapters, the role of viruses in acute, subacute, and chronic infectious diseases is well defined in humans. In most cases, the virus-host cell interaction observed in these disease states involves the destruction of the cell during virus replication. Early in the twentieth century, as the science of virology developed, investigators noted that some viruses, rather than destroying cells, were found to cause changes in the physiology of the cell, a process called transformation. Viral transformation refers to the changes in the biologic function, antigenic specificity, and morphology of a cell resulting from the infection of the cell and virus replication. These changes confer properties on the infected cell characteristic of neoplasia (Table 12.1). The study of virus-induced transformation is a medically relevant area, because it deals with the changes in morphology and metabolism of cells. Cells grown in culture represent the experimental counterpart of and generally mimic the growth and regulation of cancer cells in human and animal hosts.

Important to our discussion of virus-induced tumorigenesis is the pleiotropism exhibited by the oncogenic virus. After the virus enters the cell, expression of a single gene may cause the cell to undergo considerable changes, many of which are characteristic of a less-differentiated state. Because available data suggest that all of these changes are not encoded by a virus-specific gene, the morphogenesis of the normal cell to the malignant state must rely on aberrant expression of host genes. Consequently, one would be led to conclude that an oncogenic agent acts in some indirect manner to alter the normal gene-regulatory function.

In the early 1900s, shortly after the pathologist Peyton Rous began studying cancer, a chicken breeder brought to his laboratory a hen with a large lump on one leg. The mass turned out to be a spindle-cell sarcoma, a type of tumor that is often malignant in mammals, including humans. When Rous inoculated healthy chickens with this tissue, sarcomas developed. He then inoculated other chickens with extracts of

TABLE 12.1 Main Characteristics of a Viral Transformed Cell

Property	Nontransformed	Transformed
Multilayered cell growth	No	Yes
Random cell organization	No	Yes
Serum requirement	High	Low
Soft agar cloning	No	Yes
Passages in vitro	Low	High
Chromosomal changes	No	Yes
Altered cell surface	No	Yes
Anerobic glycolysis	No	Yes
Lectin agglutinability	Low	High

the cancerous tissue from which all the cells, including the cancer cells, had been removed by filtration. Tumors also developed in those chickens inoculated with this bacteria-free filtration. Rous concluded that a "filterable virus," too small to be seen with the microscopes of the day, was the cause of the tumor. At that time, however, the idea that cancer could be caused by an infectious agent was received with considerable skepticism. Not until a half-century later was Rous awarded the Nobel Prize for his discovery of tumor-inducing viruses.

Since the seminal observation made by Rous, many types of malignant growth have been shown to be inducible by viruses in animals. Although tumor virology today is a rapidly developing area of investigation, some of the most important advances took many years to unfold. An important early observation, made by Shope in 1932, was the isolation of a virus that caused benign papillomas in cottontail rabbits. These tumors could be induced in wild, but not in domestic rabbits, thereby showing the type of highly restricted host range that is characteristic of many oncogenic viruses. Bittner, in 1936, demonstrated that the milk of certain inbred strains of mice contained a virus (mammary tumor virus) that acted as an extrachromosomal agent to cause an increased incidence of breast cancer in successive generations of the same strain of mice; only animals that had been allowed to suckle developed cancer. It was not until 1950 that Ludwig Gross transmitted murine leukemia by inoculating newborn mice with filtrates of leukemic cell extracts; Ellerman and Bang in 1908 had demonstrated analogous findings in chickens.

Despite the abundant evidence now at hand for viral oncogenesis in animals, until recently, attempts to implicate particular viruses in human cancers have been, at best, circumstantial and therefore inconclusive. Early investigators, in trying to prove a viral causation of cancer, were guided by Koch's postulates, which may be summarized in the following four steps.

- The agent is always found in diseased, but not in healthy, animals.
- The agent, isolated from diseased animals, can be grown in culture.
- When the isolated agent is reinoculated into susceptible animals, the disease is produced.
- The agent can be reisolated from the experimentally infected animals.

These criteria are limited by the constraints against deliberately producing disease in a human host and the lack of an appropriate animal model. To overcome these difficulties, A. S. Evans proposed a new set of postulates using epidemiologic, seroepidemiologic, and virologic evidence that, because of new techniques (e.g., restriction endonuclease mapping), enable one to identify specific viral agents. The main features of Evans's postulates are as follows:

- Exposure to putative cause is more common in cancer patients than in controls.
- Immunologic response specific for the virus is observed.
- Temporally, the disease follows exposure to the putative agent.
- Virus or viral genetic material, or both, is present in affected tissues.
- Expression of viral information is detected in affected tissues.
- The virus or its genetic information, or both, is shown to induce neoplastic transformation in vitro.
- Elimination or modification of the putative cause decreases incidence of the disease.

From these criteria, strong evidence on possible human tumor viruses relates hepatitis B virus to cancer of the liver. Epstein-Barr virus clearly relates epidemiologically to African Burkitt's lymphoma and to nasopharyngeal carcinoma, whereas another herpesvirus, herpes simplex virus type 2, is epidemiologically linked with carcinoma of the cervix. Most recently, a retrovirus has been isolated from persons with T-cell leukemias.

VIRUSES KNOWN TO TRANSFORM CELLS

Many viruses have been shown to have oncogenic potential. As summarized in Table 12.2, all DNA viruses, with the exception of the single-stranded, DNA-containing parvoviruses, are known to transform cells in either culture or animal models, or in both. Among the RNA viruses, only retroviruses transform cells, and they are the only RNA viruses that replicate through a DNA intermediate. In the following sections, the viral genes and molecular biology involved in phenotypic transformation are considered in relation to the major groups of oncogenic viruses.

TABLE 12.2 Major Oncogenic Viruses

Virus Family	Subtype	Natural Host	TRANSFORMATION In vitro	TRANSFORMATION In vivo (Disease)
Papovavirus	Papillomavirus	Human	No	Human benign warts, alimentary carcinomas
	SV40	Rhesus monkey	Yes	Hamster, rat, mouse
	BK	Human	Yes	Hamster tumors
	JC	Human	Yes	Hamster, owl monkey tumor
	Polyoma	Mouse	Yes	Hamster, rat tumors
Herpesvirus	Herpes simplex			
	Type 1	Human	Yes	?
	Type 2	Human	Yes	? Human cervical cancer
	Varicella-zoster virus	Human		?
	Epstein-Barr	Human	Yes	Cottontail marmoset tumors Owl monkey tumors Human Burkitt's lymphoma; nasopharyngeal carcinoma
	Cytomegalovirus	Human		?
Adenovirus	Human adenovirus, Types 12, 18, 31	Human	Yes	Hamster, rat
Hepatic B virus	Human HBV	Human	No	Hepatocellular carcinoma
	Woodchuck HBV	Woodchuck	No	Woodchuck hepatocellular carcinoma
Retrovirus	Sarcoma virus	Chicken, mouse, cat	Yes	Sarcoma
	Leukemia virus	Chicken, mouse, primate	No	Leukemia
	Mammary tumor virus	Mouse	No	Mammary adenocarcinoma
	MC29	Chicken	Yes	Carcinoma
	Human T-cell leukemia viruses	Human	No	? T-cell leukemia ? Lymphoma

Papovaviruses

These DNA viruses characteristically have a covalently closed, double-stranded, superhelical genome of about 5000 base pairs. The group includes the common human wart viruses (papillomaviruses), the much-studied simian vacuolating virus 40 (SV40), and polyomavirus of primate and murine origin, respectively.

The human papillomaviruses (HPV) are the only viruses in which definitive evidence links infection and a human tumor (although usually nonmalignant). Because laboratory culture procedures suitable for this virus are lacking, however, little is known about its replication. Recent studies have demonstrated a remarkable number of human papillomaviruses. At least 16 distinct viruses have been isolated. HPV1 and 4 are preferentially associated with plantar warts.

The common wart (verruca vulgaris) is caused by HPV2. HPV3 is usually associated with flat warts and epidermodysplasia verruciformis, as is HPV5. The venereal wart (condyloma accuminata), a rapidly growing, benign hyperplasia that occurs at mucocutaneous junctions such as the vulva, is caused by HPV6. Subtypes of most of these viruses have been identified by restriction endonuclease digestion.

SV40 and polyomavirus infection of permissive cells (primate and murine, respectively) results in cell lysis and virus production. After in vitro infection of nonpermissive cells, including human cells, however, virus replication does not occur but cell transformation is observed. During papovavirus-induced transformation, limited expression of viral genes occurs and viral DNA randomly intergrates into the host genome. Infectious virus can be rescued from the integrated state by fusion of transformed cells to permissive cells and by use of certain physical and chemical agents.

Characterization of papovavirus functions has revealed that a single viral gene is necessary for oncogenic transformation. Treatment with mutagenic agents (e.g., ultraviolet irradiation) at doses that induce a single lesion per viral genome may produce temperature-sensitive (ts) mutants that transform cells at the nonconditional temperature (e.g., 35°C) but not at elevated, conditional temperatures (e.g., 41°C). Furthermore, if these ts mutants transform cells at the nonconditional temperature and subsequently the cells are raised to the conditional temperature, these transformed cells revert to the normal phenotype. These experiments indicate that a single, virus-specific protein is required for transformation of cells by papovaviruses and that the presence of the active protein is essential for maintenance of the malignant-cell phenotype.

Mapping of the viral gene responsible for the oncogenic potential of the papovaviruses has shown that this gene occurs in that region of the viral genome coding for functions required early in the replicative cycle. During lytic infection, two subsets (early and late) of viral genes, transcribed from separate strands of the viral genome, are expressed (Fig. 12.1). In nonpermissive cells transformed by these viruses, however, only the early viral gene subset is expressed. In SV40 and polyomavirus, the gene encoded in this region is called "T antigen," which in SV40 includes two separate but related proteins, small and large T antigens, synthesized from overlapping regions of the viral genome.

The exact function of T antigens in the induction and maintenance of the transformed state is not known. These proteins have demonstrable regulatory effect on transcription and thus might alter the expres-

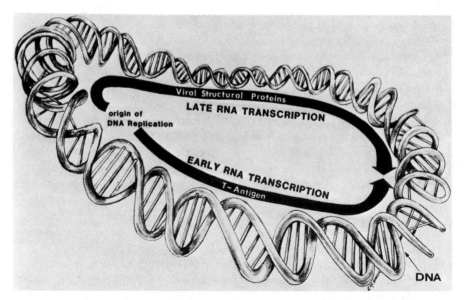

FIG. 12.1. The genetic structure of papovavirus as represented by SV40. The arrow indicates the direction of RNA.

sion of host genes. Alternatively, in lytic infection, early genes are involved in the induction of host DNA replication, which might be essential to transformation.

Papovaviruses are not associated with any spontaneous, malignant neoplasm in animals. The inadvertent inoculation of humans with SV40-contaminated, live poliovirus vaccine in the early 1960s was shown to have no effect on the occurrence of malignancies, although antibody to SV40 was found in immunized persons. Two human papovaviruses, BK and JC viruses, were isolated from an immunosuppressed transplant recipient and a patient with progressive, multifocal leukoencephalopathy, respectively. These two agents have been implicated in the pathogenesis of certain neurologic diseases.

Adenoviruses

Several adenoviruses, common pathogens of the human upper respiratory tract, induce tumors when injected into newborn hamsters, in which these viruses vary from highly oncogenic to nononcogenic. Transformation by the adenoviruses requires either infection at a high multiplicity in a nonpermissive host or inactivation with either physical or chemical agents of those viral genes required for lytic infection. In adenovirus-transformed cells, the entire viral genome need not be present; rather, subgenomic segments seem to be inserted randomly into the host DNA. The only viral DNA consistently observed is that derived from a small region of the viral genome. Virus particles are not observed and infectious virus cannot be rescued from transformed cells.

Characterization of viral gene products present in adenovirus-transformed cells has shown that several proteins are present, collectively referred to as T antigen. These proteins apparently are derived from genes normally transcribed during the early phase of lytic infection. Functional analysis of viral genes required for tumorigenesis has indicated that, unlike papovaviruses, initiation and maintenance of the transformed phenotype may depend on separate viral genes. Recently, the E1A gene of adenovirus was shown to exhibit properties of an oncogene (see below). As with papovaviruses, adenoviruses have not been associated with human malignancy.

Herpesviruses

The herpesvirus group of large DNA viruses is one of the most common human pathogens. Most persons are infected with one or more herpesviruses at an early age, with consequences varying from no clinical symptoms to life-threatening disease. In addition, these viruses tend to establish latent infection, in which state they may persist for long periods. Most herpesviruses are known to transform cells in vitro, yet assessing their role in human malignancies has been difficult, given their ubiquity.

Marek's disease virus of chickens and *Herpesvirus samiri* of primates are the only herpesviruses shown to induce tumors spontaneously. The first generally used vaccine against a malignancy was developed from a herpesvirus of turkeys. When administered to chickens, this live, attenuated vaccine prevents lymphoma induced by Marek's disease virus.

As with other lytic viruses, herpesvirus' lytic functions must be inactivated for in vitro transformation. This may be accomplished by partial inactivation of viral DNA by means of ultraviolet irradiation. Unlike either adenoviruses or papovaviruses, the herpesvirus DNA does not persist in transformed cells, and no virus-specific protein has been implicated in either initiation or maintenance of the transformed-cell phenotype. However, as discussed later, much indirect evidence has accumulated to support the role of herpesviruses in human malignant diseases.

Hepatitis B Virus

Found in the blood and liver of patients with serum hepatitis, hepatitis B virus (HBV) is poorly characterized because it cannot be grown in vitro. In recent years, however, the advent of DNA cloning has resulted in the production and subsequent study of the viral genome without use of infectious human materials. Furthermore, the development of animal model systems has contributed to a greater understanding of the structure and replication of this virus.

The virus replicates in the liver of infected hosts (human, woodchuck, and duck) and is released into the blood as a 42-nm particle. Within the virion the genome is a partial double-stranded DNA molecule that is 3.2 kilobases long. Recent data suggest that HBV may replicate by means of an RNA intermediate. This replication scheme is similar to that of the retroviruses, which make DNA from a virus-specific RNA template. HBV incorporates DNA into the virion, whereas the retroviruses incorporate RNA. In addition, like retroviruses, HBV sequences are found to be integrated into the host DNA sequences in hepatomas. As with that of the retroviruses, HBV DNA integration may be essential for the oncogenic potential of this virus.

Retroviruses

Retroviruses are the only RNA viruses that transform cells both in vivo and in vitro. Morphologically, these viruses are divided into three main groups. Type C virus, the most common, is seen as an enveloped capsid with a centrally located core. Type B virus, seen only with mouse mammary tumor virus, characteristically has an eccentrically located core. The A particle, which lacks a core, is an intracellular precursor to the B particle.

Retroviruses are unique in that the RNA genome is copied (transcribed) into a double-stranded DNA intermediate. DNA is synthesized from the RNA by the enzyme reverse transcriptase, a component of the mature virion. The virus-specific DNA undergoes genetic recombination (integration) with host DNA. Virus-specific RNA is transcribed from the integrated viral DNA molecule for both production of viral proteins (translation) and incorporation into virions.

The structure of the retrovirus DNA molecule bears consideration because of its functional and structural resemblance to a genetic element found in prokaryotes, the transposon. A DNA molecule involved in bacterial genetic variation, the transposon is able to migrate among different sites on the host genome and among individual organisms, as well as to effect gene transfer. Antibiotic-resistance genes are often located on transposons. The structure of the DNA within the transposon characteristically includes a region of a unique sequence bound at each end by terminal repeated sequences, a structure homologous to that of retrovirus DNA. Therefore, the retrovirus may be considered a transposon of eukaryotic cells because it is mobile, and, as discussed later, these viruses are capable of gene transfer within and among eukaryotes.

The retroviruses may also be classified functionally into two groups, based on characteristics of the cell transformation they effect. The first

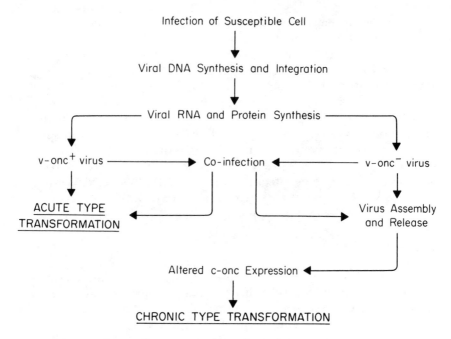

Infection of Susceptible Cell

Viral DNA Synthesis and Integration

Viral RNA and Protein Synthesis

v-onc$^+$ virus ⟶ Co-infection ⟵ v-onc$^-$ virus

ACUTE TYPE TRANSFORMATION

Virus Assembly and Release

Altered c-onc Expression

CHRONIC TYPE TRANSFORMATION

FIG. 12.2. Retrovirus replication and cell transformation.

group produces tumors or foci of transformed cells either in vitro or in vivo after acute (weeks) infection. Genetic studies show that a single viral gene is required in both induction and maintenance of the transformed cell phenotype. The second group of retroviruses transforms cells in vivo only after chronic (months) infection. A viral gene has not been implicated in the transformation process by the latter group of viruses. The many sarcoma-inducing viruses are found in the group of viruses that transform after acute infection, whereas the leukemia and mammary tumor viruses are in the group that transform after chronic infection.

Retroviruses that transform cells during acute infection are, in most cases, defective in one or more functions required for viral replication. Propagation of these defective viruses depends on the presence of a helper virus that provides the necessary gene products (Fig. 12.2). Transforming activity can be demonstrated in the helper virus, but it belongs to the group that induces tumors only after chronic in vivo infection. Treatment of the acutely transforming viruses with mutagenic agents results in mutants that have lost all transforming activity. Therefore, the oncogenic potential of these viruses depends on a virus-specific gene, called the viral oncogene (v-onc).

From the acutely transforming retroviruses, many different oncogenes have been identified and characterized (Table 12.3). Several of these have an enzymatic protein kinase activity, that is, phosphorylation of proteins. The oncogene is not involved in viral replication and usually, in fact, replaces genes required for such replication, resulting in a defective virus, as noted (Fig. 12.2).

Perhaps the most significant aspect of the oncogene is the occurrence of copies of all oncogenes in the uninfected host. These homologous sequences, found in all eukaryotic cells, are called protooncogenes or cellular oncogenes (c-onc). These genes are evolutionarily conserved, as shown by the substantial DNA and amino acid sequence homology that exists between genes of such divergent species as chicken and human. The acutely transforming retroviruses during their evolution are thought to have acquired their oncogenes from the host cells. Thus, this group of retroviruses transfers a new copy of an oncogene into the eukaryotic cell, a phenomenon parallel to that observed with the transposons in prokaryotes. In the normal, uninfected cell, the protooncogenes are not expressed. In the infected cell, however, the viral oncogene is integrated into the host genome as part of the virus replicative cycle and is expressed. This uncontrolled expression of the viral oncogene is the essential difference between a normal and an acutely transformed cell.

The other group of retroviruses routinely induce malignancies in experimental animals only during chronic infection. These viruses do not transfer an oncogene, and no viral gene is required for tumorigenesis. Experimental evidence suggests that these viruses act as biologic mutagens of protooncogenes. Rather than insert a new copy of an oncogene into the infected cell, these viruses transform cells by the integration of the viral DNA either immediately adjacent to, or in the area of, a protooncogene. Insertion of the viral DNA molecule results in aberrant expression of the protooncogene. Thus, both groups of retroviruses that cause transformation during either acute or chronic infection rely on the expression of an oncogene, either viral or protooncogene.

The scenario for tumorigenesis during chronic retrovirus infection correlates well with both in vivo and in vitro characteristics of replication by this group of viruses. Because these viruses integrate at many, perhaps random, sites in the host genome, integration adjacent to any of the 17 or more protooncogenes (Table 12.3) would be a statistically rare event, in view of the more than a million genes in the genome of higher eukaryotes. Therefore, many individual infection events must take place before transformation occurs. After a prolonged infection, the probability of transformation in the animal is significant, and tumors would be expected. Because cells in culture must be passaged regularly and all cells in the population cannot be maintained continuously over many months, the chance occurrence of the transforming event is too infrequent to be observed in vitro.

The dependency of transformation on oncogenes is applicable to chemical carcinogenesis as well. If viruses can mutagenize a protooncogene, then these genes may also be mutated by chemical carcinogens.

TABLE 12.3 Retroviral Oncogenes

Acronym	Origin	Species of Isolation	Malignancy	Human Chromosome	Biochemical Activity
src	rous sarcoma	chicken	sarcoma	20	tyrosine kinase
fps = fes	Fujinami (ST Feline) sarcoma virus	chicken (cat)	sarcoma	15	tyrosine kinase
yes	Y 73 sarcoma virus	chicken	sarcoma	—	tyrosine kinase
abl	abselson murine leukemia virus	mouse	B-cell lymphoma	9	tyrosine kinase
ros	UR II avian sarcoma virus	chicken	sarcoma	—	tyrosine kinase
Fgr	Gardner-Rasheed Feline sarcoma virus	cat	—	—	tyrosine kinase
erb B	Avian erythroblastosis virus	chicken	erythroleukemia & sarcoma	—	—
fms	McDonough Feline sarcoma virus	cat	sarcoma	5	—
mos	Moloney murine sarcoma virus	mouse	sarcoma	8	—
raf	3611 murine sarcoma virus	mouse	sarcoma & erythro-leukemia	3	protein kinase
Ha-ras 1	Harvey murine sarcoma virus	rat	—	11	Guanosine diphosphate or guanosine triphosphate binding
Ki-ras 2	Kirsten marine sarcoma virus	rat	—	12	—

Acronym	Origin	Species of Isolation	Malignancy	Human Chromosome	Biochemical Activity
myc	Avian MC 29 myelocyto-matosis virus	chicken	carcinoma, sarcoma, & leukemia	8	DNA binding
myb	Avian myeblastosis virus	chicken	myeloblastic leukemia	6	—
fos	FBJ osteosarcoma virus	mouse	sarcoma	2	—
ski	Avian SK V770 virus	chicken	sarcoma	1	protein kinase
rel	reticuloendotheliosis virus	turkey	lymphatic leukemia	—	—
sis	Simian sarcoma virus	woolly monkey	sarcoma	22	related to platelet growth factor
bas			sarcoma	?	?

Related oncogenes known from sequence hybridization or transfection

Acronym	Origin	Species of Isolation	Malignancy	Human Chromosome	Biochemical Activity
N-myc	Neuroblastoma	human	—	2	—
N-ras	Neuroblastoma leukemia, sarcoma	human	—	1	—

Unrelated oncogenes known from sequence hybridization or transfection

Acronym	Origin	Species of Isolation	Malignancy	Human Chromosome	Biochemical Activity
Blym	Bursal lymphomas	chicken	—	1	—
mam	Mammary carcinomas	mice, human	—	—	—
neu	neuro-, glioblastomas	rat	—	—	—

Recent experimental data indicate that this alteration does indeed happen, because point mutations known to result from the interaction of chemical carcinogens with cellular DNA have been shown to occur in protooncogenes isolated from several malignancies in humans. Thus, oncogenes provide a unified concept for the induction of malignancies by both acute and chronic transforming viruses as well as by chemical carcinogens.

VIRUSES AND HUMAN CANCER

As stated earlier, definitive evidence for the involvement of viruses in human cancers, as defined by Koch's postulates, is not available. For many human malignancies, however, there is epidemiologic, genetic, and biochemical evidence for virus-induced or virus-associated neoplastic disease. Based on Evans's revisions of Koch's postulates, those cancers for which the strongest evidence of viral etiology exists are considered here.

Cervical Carcinoma

Squamous-cell carcinoma of the cervix, which was noted to be distinct from cancer of the uterine body as early as the nineteenth century, is almost always preceded by a preinvasive phase (dysplasia and carcinoma in situ). An Italian physician, Rigoni-Stern, in 1842, noted that this cancer was rare in nuns but frequent in married women. Epidemiologic studies in this century confirmed this observation and demonstrated an increased risk associated with initiation of sexual intercourse at an early age.

The possible role of herpes simplex virus (HSV) in cervical cancer was suspected when women with HSV2 infections were found to be at higher risk of developing cervical carcinoma than uninfected women in prospective studies. In addition, frequency of HSV2 antibodies was higher in affected women than in controls. A similar response to other viruses or in patients with other malignancies is not seen. The finding of HSV-specific DNA and RNA in cells has not been consistent, suggesting that if this virus were involved, it would be required for initiation and not for maintenance of the transformed state.

Burkitt's Lymphoma

Burkitt described a malignant tumor that primarily affects the jaw, usually in children between ages 1 and 14 years, with the highest incidence between ages 6 and 8 years. Burkitt's lymphoma is limited in geographic distribution to the hot, wet lowlands of equatorial Africa. The disease is endemic in several regions of both East and West Africa and in New Guinea, and its occurrence is sporadic and infrequent outside these endemic areas. African Burkitt's lymphoma accounts for as much as 60% of all childhood cancers in Nigeria.

Extensive prospective epidemiologic studies have indicated a relationship between Burkitt's lymphoma and infection with Epstein-Barr virus (EBV). Practically all children in the endemic areas have antibodies to, and contain the DNA of, EBV by age 3 years. As in the endemic cases, a small but significant number of cases outside the endemic areas, including those found in the United States, are found to have a wide range of serum antibodies to EBV-specific antigens and their tumor cells contain EBV DNA sequences.

However, cells from most cases in nonendemic areas lack EBV-specific DNA. Burkitt's lymphoma may therefore be another example of disease heterogeneity, that is, two diseases with similar manifestations but different causes. Cases from nonendemic areas also differ from African Burkitt's lymphoma with regard to age distribution and primary tumor sites. Otherwise endemic and nonendemic Burkitt's lymphoma seem to be clinically indistinguishable: both originate from B lymphocytes, both have similar histology, and both have a unique cytogenetic abnormality, a reciprocal translocation between chromosomes 8 and 14. Recent data suggest that this chromosomal translocation may involve a cellular protooncogene whose expression is altered by the translocation. A similiar translocation and chromosome rearrangement involving altered oncogenes has been observed in retinoblastoma and Wilms' tumor.

Nasopharyngeal Carcinoma

Nasopharyngeal carcinoma (NPC), a squamous-cell carcinoma with lymphocytic infiltration of the epithelial lining, affects mostly adults, with a peak incidence at age 40. Its highest incidence is in southern China, especially in Guandong Province, where the overall frequency is as high as 39.9 per 100,000 persons annually—about 40 times that of whites living in California. The incidence of the tumor is also high among Chinese living in other areas. In Singapore the age-standardized incidence per 100,000 is 18.7 for Chinese males and 7.1 for Chinese females, with variation among the major Chinese dialectic groups. For other peoples of Southeast Asia, for example, Thais, the rates appear to be intermediate. The tumor is frequently found among natives in Kenya, especially the Nandi tribe, and is rare in northern China, Japan, Western Europe, the United States, and elsewhere in the world. EBV-specific DNA sequences are found in nasopharyngeal carcinomas.

The explanation for its ethnic distribution seems to be neither wholly genetic nor wholly geographic. Among first-generation immigrants from South China to the United States, the high frequency of nasopharyngeal carcinoma declines somewhat, but it remains considerably above that of the white population. The incidence of this tumor among emigrants of this ethnic group, whether to other parts of China or to the United States, remains higher than among other ethnic groups. For offspring of mixed marriages between South Chinese and other ethnic groups, the tumor frequency is intermediate.

As with Burkitt's lymphoma, EBV infection occurs at an early age in areas where NPC is endemic, similar to that found in the lower socioeconomic population in Western societies. The antibody titers in patients with these tumors, however, are 10- to 15-fold higher than in the respective control population in each locale. Thus, unlike protection against infectious mononucleosis, in these ethnic groups or locales, early antibody response seems to be associated with tumor formation.

Some investigators have suggested that the protean manifestations and the diverse ethnic response to the virus could be due to the geographic distribution of different strains of EBV. That explanation does not seem adequate, however, because all strains of EBV, from whatever source, have remarkably uniform structure and biologic behavior. The one variant of EBV that has been isolated was probably the result of laboratory manipulations, because the parent virus from which the variant virus was derived was antigenically similar to other EB viruses.

Hepatocellular Carcinoma

The correlation between primary hepatocellular carcinoma and infectious hepatitis B virus was noted in epidemiologic studies. In areas of the world in which HBV is endemic, as in parts of Africa and Asia where up to 10% of the population are infected, the incidences of liver cirrhosis and cancer are correspondingly high. The incidence of these tumors varies in different parts of the world. In the United States this disease accounts for 1 to 2% of all malignancies and is found in 0.1 to 0.7% of all autopsies. A striking 3:1 male preponderance of cases occurs with this disease.

Serologic evaluation, clinical correlation, and biochemical analysis for HBV were done in both retrospective and prospective studies of primary hepatocellular carcinoma. In more than 80% of affected persons, HBV surface antigen or antibodies to it were found in case-control studies. Furthermore, cirrhosis or chronic hepatitis, or both, were observed in an equivalent number of cases. Prospective epidemiologic studies revealed a 200-fold increase in liver cancer risk in carriers over noncarriers. Finally, HBV-specific DNA was detected in most of these cancer tissues.

In addition to the substantial evidence derived from HBV in humans, an animal model for this disease was found. A virus similar to human HBV was isolated from woodchucks in a colony maintained in captivity. In these animals, as in humans, the naturally occurring infection resulted in a chronic carrier state and was associated with a higher incidence of primary liver cancer than in infected animals. Furthermore, HBV has been found to induce hepatitis and cancer in the duck.

The combined epidemiologic and experimental data strongly substantiate a cause-effect relationship of primary hepatocellular carcinoma and chronic infection with hepatitis B virus. This positive correlation may be further strengthened should the introduction and use of human vaccines against HBV lower the incidence of this neoplasm.

T-Cell Leukemia

Although retroviruses have been implicated in malignancies of many animals since the time of Rous, the search for a human retrovirus was largely unsuccessful until a human type C retrovirus, the human T-cell leukemia viruses (HTLV), were found. HTLV1 was isolated from both T-lymphoblastoid cell lines and individual cases of T-cell leukemia.

T-cell malignancies that occurred in a cluster in southwest Japan were shown to correlate with the presence of HTLV. Subsequently, a similar virus was isolated from other patients in the United States and the West Indies. Clinically, these malignancies were characterized by leukocytosis, skin lesions, hepatosplenomegaly, lymphadenopathy, and hypercalcemia and by a rapidly deteriorating clinical course. Patients had high titer of antibodies to the internal protein of HTLV, and the virus has been isolated from peripheral blood cells obtained from patients.

Biochemical analysis of DNA sequences in T-cell leukemic cells from adults indicated the presence of HTLV-specific sequences. In addition, virus-specific DNA was detected in other T-cell malignancies (e.g., cutaneous T-cell lymphoma, T-lymphosarcoma, leukemia, and diffuse mixed lymphoma). Furthermore, HTLV sequences were found in DNA from healthy family members of a patient with persistent leukemia. These studies also indicated that these T-cell leukemias were clonal, suggesting that HTLV is similar to chronic transforming retroviruses of other species. Therefore, HTLV-induced oncogenesis may occur by viral insertion near a protooncogene, as discussed earlier. Recent data suggest the involvement of a viral gene in HTLV-1 induced tumor transformation in combination with other oncogenes.

CONCLUSION

After 70 years, the great questions of how viruses cause cancer and whether human tumor viruses exist have been resolved. As with most great leaps in human understanding, finding answers to these two questions has only created new ones. Now, however, investigators can begin to explore areas of cell growth and metabolism that deal not only with the malignant process but also with normal development and differentiation. Furthermore, we can begin to explore areas that are directly applicable to cancer therapy. If we can identify the gene products required for maintenance of the oncogenic state, then we can design specific inhibitors to revert the cell to the normal phenotype. This procedure would be analogous to that used to develop chemotherapeutic agents effective against herpesviruses.

Because the herpesvirus encodes for a specific thymidine kinase, agents that will inhibit this enzyme but not that found in the normal, uninfected host cell will kill only the virus-infected cell. Thus, in the future, the specific oncogene or oncogenes responsible for the malignancy will be diagnosed by a molecular biologist, and agents (chemical or immunologic) with a specific activity against the oncogene will be administered by the oncologist. Thereby only the tumor cell, which requires elevated levels of the oncogene product for growth, will be affected.

Is there an anticancer vaccine in the future? The effectiveness of an antihuman T-cell leukemia virus is questionable for the general population, although it could be useful in reducing the occurrence of T-cell leukemias in populations where HTLV is endemic. It seems unlikely, however, that a single related group of viruses transmitted among humans is responsible for a significant proportion of human malignancies. Thus, developing an adequate number of vaccines would seem impossible. Furthermore, because these viruses are transmitted genetically, extracellular expression might not be required, which renders a vaccine useless. Nonetheless, viruses have been, and will continue to be, our guide to unlocking the secrets of the tumor cell.

Recommended Readings

Bishop JM: *Ann Rev Biochem* 52:301, 1983.

Cline, MJ: *Annals of Internal Medicine* 101:223–233, 1984.

Levine AJ, VandeWoude GF, Topp WC, Watson JD: *Cancer Cells: The Transformed Phenotype.* Cold Spring Harbor, NY, Cold Spring Harbor Laboratory, 1984.

Slamon, DJ, deKernion, JB, Verma, IM, and Cline, MJ: *Science* 224:256–262, 1984.

Tooze J (ed): *The Molecular Biology of Tumor Viruses, DNA Tumor Viruses.* Cold Spring Harbor, NY, Cold Spring Harbor Laboratory, 1980.

VandeWoude GF, Levine AJ, Topp WC, Watson JD: *Cancer Cells: Oncogenes and Vital Genes.* Cold Spring Harbor, NY, Cold Spring Harbor Laboratory, 1984.

Varmus HE: *Cancer Surveys,* p 309. Oxford, Oxford University Press, 1982.

Weiss R, Teich N, Varmus H, Coffin J (eds): *The Molecular Biology of Tumor Viruses, RNA Tumor Viruses.* Cold Spring Harbor, NY, Cold Spring Harbor Laboratory, 1982.

13

Diagnostic Virology*

EVELYN W. KELLER and MARTIN S. HIRSCH

Diagnostic virology has come of age. Once thought to be a luxury of little value to patient management, the clinical virology laboratory has now become a useful resource for optimal clinical care. Development of rapid diagnostic techniques, together with the increasing spectrum of antiviral chemotherapy, will further expand diagnostic virology facilities during the next decade.

Most laboratories remain associated with universities or governmental institutions. Nonetheless, private laboratories have recently begun to offer limited virologic and serologic testing and will doubtless have to expand these activities to meet increasing demand. Establishing regional diagnostic virology services with satellite laboratories for rapid specimen-processing (one laboratory per 1 million population) has been suggested for maximal efficiency at reasonable cost.

Increased availability of services has generated a desire for better diagnostic reagents at reasonable prices. Using monoclonal antibody technology, a number of commercial enterprises are striving to provide such reagents. These are increasing the specificity and sensitivity of diagnosis. The application of this technology will further allow the discipline of clinical virology to offer services to the physician that heretofore were unavailable.

Undue delay in establishing a specific virologic diagnosis diminishes the potential effectiveness of any antiviral agents that may be indicated, such as acyclovir, vidarabine, and interferon. Advances in rapid diagnosis will allow prompt institution of effective therapy and, of almost equal importance, the cessation of inappropriate antibiotic therapy. Moreover, quick viral diagnosis may, in many instances, prevent unnecessary laboratory testing, and thereby shorten hospitalization and lower resultant medical costs. Finally, early diagnosis of potentially epidemic organisms, such as influenza virus or arbovirus strains, may permit more

*The writing of this chapter was supported in part by the Mashud A. Mezerhane B. Fund.

adequate public health efforts to control the spread of infection, in both the hospital and the community.

The following case reports illustrate some of the benefits and limitations of today's capabilities in diagnostic virology:

CASE 1: A 63-year-old obstetrician was admitted to the hospital with fever (40°C [104°F]) and chills 1 month after an uncomplicated coronary artery bypass graft. His fever was accompanied by malaise, profound fatigue, nausea, and vomiting. Liver enzyme levels were slightly above normal limits, and his peripheral blood leukocyte count was 10,000/mm^3, with 60% lymphocytes and 10% atypical lymphocytes. Heterophile antibody tests were negative, but a cytomegalovirus (CMV) antibody enzyme-linked immunosorbent assay (ELISA) showed a significant rise between his admission and a second sample taken 1 week later. Culture of buffy-coat leukocytes from admission grew CMV, first detectable 20 days after admission, whereas admission urine culture was negative for CMV. Repeated cultures of urine and throatwash specimens taken 14 days after admission both yielded CMV. The patient remained febrile for 3 weeks but gradually improved and was discharged to convalesce from his CMV post-transfusion mononucleosis.

This case demonstrates both the utility and the frustration of diagnostic virology. The diagnosis of CMV infection was of financial and emotional benefit to the patient. Painful and potentially dangerous procedures were avoided, and expensive antibiotics were withheld. Moreover, reassurance could be given the patient that he would recover completely. In addition, the diagnosis was of public health importance for the protection of any pregnant hospital staff members. The obstetrician was also advised to have repeated cultures for CMV before he resumed his practice, to avoid excessive risk to his own patients.

However, the foregoing case also illustrates the current limitations of diagnostic virology. What was needed was a rapid assay to detect virus, viral antigen, or early antibody. These techniques, as will be discussed later, are either unavailable or still in experimental stages. A specific diagnosis made early in the course of this disease might have prevented hospitalization altogether.

CASE 2: A 10-year-old boy was admitted with a 3-day history of headache, fever, and inappropriate behavior. Cerebrospinal fluid contained 60 lymphocytes per cubic millimeter, and a computed tomographic (CT) scan demonstrated an area of low density in the left temporal lobe. A temporal-lobe biopsy specimen was positive for herpes simplex antigen by immunofluorescence within 3 hours of surgery. Viral cultures from the specimen grew herpes simplex virus type 1 (HSV1) within 24 hours. Vidarabine administration was begun immediately after operation and continued for 10 days. The patient made a good recovery and was left with only minor neurologic impairment.

This case shows how HSV encephalitis can be diagnosed rapidly if appropriate specimens are tested by currently available techniques.

However, it also shows how inadequate our diagnostic methods are in requiring a major neurosurgical procedure. What is needed is a rapid, sensitive, and specific assay for HSV antigens or viral products in cerebrospinal or other body fluids. Such a test is not currently available, but as explained below, may well be on the horizon. In what follows, we outline both techniques that are established and still useful for detection of viruses and promising avenues for future advances in diagnostic virology.

STANDARD TECHNIQUES

Virus Isolation

Early efforts at virus isolation involved attempts to induce disease experimentally in animals by inoculation of potentially infected tissues or body fluids. The first revolution in diagnostic virology occurred with the development of in vitro growth of viruses. In 1929, the first attempts at tissue culture were recorded, but not until the availability of antibiotics after World War II did isolation of viruses in vitro become practical. By the 1960s, basic techniques for the cultivation of viruses in the laboratory were well established. With further modifications, as more viruses have become recognized, isolation of viruses by growth in cell culture along with previously established techniques using embryonated eggs have become the "bread-and-butter" work of most diagnostic virology laboratories.

The choice of experimental host systems for recovery of viruses from clinical specimens depends on which agent or agents are suspected of being present. Inoculation of animals, embryonated eggs, or appropriate cell cultures is carried out after treatment of specimens with antibiotics to inhibit bacterial growth. In addition, some laboratories concentrate small amounts of virus present by any of several procedures, for example, extraction of fecal specimens or ultracentrifugation.

ANIMAL INOCULATION

Animal inoculation is usually done on the suckling or adult mouse, or occasionally the guinea pig, by intracerebral, intraperitoneal, or subcutaneous injection as indicated by the suspected agent. The animals are monitored daily for signs of illness or death. Moribund animals are killed and the tissues are examined for the presence of virus by passage to other animals or inoculation into cell cultures. Final identification of an agent rests in neutralization with antisera of known specificity, usually in the same host system as that used for primary isolation.

EGG INOCULATION

Generally this procedure is done on embryonated eggs of hens or, less frequently, ducks. The age of the egg and the choice of inoculation route vary according to the virus in question. The egg is first candled to determine fertility, location of the embryo, and position of the fetal eye. Inoculation is through a small puncture into the amniotic, allantoic, or yolk sac, or through a larger opening onto the chorioallantoic membrane. The opening is then sealed, and the eggs are incubated for several days, with candling done periodically to detect changes in the developing embryo, such as reduction or cessation of movement, which suggests illness or death.

After appropriate incubation, or when signs of infection in the embryo are noted, tissues or fluids are harvested and examined for the presence of virus. If the chorioallantoic membrane was inoculated, it is removed and examined macroscopically for the presence, size, and shape of lesions on the membrane. Amniotic or allantoic sac inoculation is later followed by withdrawal of the enclosed amniotic or allantoic fluid, which is tested with washed fowl or guinea pig erythrocyte suspensions for the presence of a hemagglutinating agent. In addition, the amniotic or allantoic membrane may be separated from the embryo and shell and prepared for testing against an appropriate antiserum.

CELL CULTURE

Many agents that may be harbored in clinical specimens can be isolated in appropriate cell cultures. Table 13.1 lists the types of cells that are most widely used in clinical laboratories. Primary cells have to be cultured from freshly obtained tissues and will not survive through more than a few passages. Cell lines, on the other hand, can be maintained in the laboratory indefinitely under appropriate conditions. When the appropriate cell systems have been selected according to the clinical information supplied with the specimen, cell monolayers are drained of maintenance medium and inoculated with previously prepared samples, usually containing antibiotics. Inoculated cultures are incubated at optimal temperature (usually 37°C) over several days or weeks and are periodically examined for cytopathology. Cultures that show no changes may have to undergo "blind passage" to fresh cultures before being called negative.

Virus Detection

Several methods can be used to detect virus replication. By light microscopy, cell culture monolayers are examined periodically for cytopathic effects (CPE), that is, visible cell alterations due to the presence of replicating virions. Of the viruses that are cultivated in vitro, most elicit a characteristic cytopathic effect that the experienced observer will recognize as representative of one virus or a group of viruses. Daily

TABLE 13.1 Frequently Used Host Systems for Major Human Viruses and Approximate Culture Times

Virus	Host System[a]	Method of Detection	Approximate Period from Inoculation to Detection
Adenoviridae	HEL, HEK, HEp-2	CPE	2 days to 3 weeks
Herpesviridae			
Herpes simplex	HEL, HEK	CPE	12–72 hours, occasionally longer
Cytomegalovirus	HEL	CPE	Infrequently within 24 hours, usually 5–21 days
Varicella-zoster	HEL, HEK	CPE	3–5 days minimal, may take 3 weeks
Epstein-Barr	B lymphocytes	Transformation	2–8 weeks
Papovaviridae			
BK	HEL, HEK, Vero	CPE	1–6 weeks or longer
JC	HFG, HAmn	CPE	Testing not available in most laboratories
Poxviridae			
Vaccinia	HEK, MK, HEp-2	CPE	24 hours to a few days
Hepatitis B (Hepadnavirus type 1)	Not cultured	RIA, EIA	
Orthomyxoviridae			
Influenza A, B	MK, eggs	HAd, CPE, HA	Influenza A—2–7 days; influenza B—4–7 days
Influenza C	Eggs	HA	3–5 days
Paramyxoviridae			
Parainfluenza	MK	HAd, CPE	4–7 days
Mumps	MK, eggs	HAd, CPE, (eggs-HA)	7–12 days in tissue culture; 4–5 days in eggs
Measles	HEK, MK	HAd, HA, CPE	5–10 days up to 4 weeks
Respiratory syncytial	HEp-2, HEL, MK	CPE	2–7 days, occasionally longer
Bunyaviridae	Mouse, HeLa, BHK$_{21}$	Signs of illness, CPE	2–3 days to 2 weeks
Togaviridae	Chick embryo tissue culture, mouse	CPE, signs of illness	Several days in tissue culture, 2–3 days to 2 weeks, occasionally longer in mice
Rubella	AGMK, RK$_{13}$	Interference, CPE	5–7 days
Picornaviridae			
Rhinovirus	HEL, some in MK	CPE	Several days to 2 weeks
Enterovirus			
Polio	MK, HEL	CPE	1–7 days
Coxsackie A	Mouse, some in MK	Signs of illness, CPE	3–5 days in mice, 1–7 days in tissue culture
Coxsackie B	MK, Mouse	CPE, signs of illness	1–7 days in tissue culture, 3–5 days in mice
Echovirus	HEL, MK	CPE	1–7 days

TABLE 13.1 Frequently Used Host Systems for Major Human Viruses and Approximate Culture Times

Virus	Host System[a]	Method of Detection	Approximate Period from Inoculation to Detection
Hepatitis A (Enterovirus 72)	Not cultured	EM, RIA, EIA	
Reoviridae			
Reovirus	MK, HEK	CPE	Several days to 3 weeks
Rotavirus	Not cultured	EM, ELISA, RIA	
Orbivirus	Mouse	Signs of illness	2–4 days
Coronaviridae	Tracheal organ cultures, HEL (few) HEK, HeLa	CPE	≥6 days
Arenaviridae			
Lymphocytic-choriomeningitis	Mouse, Mouse embryo fibroblasts, BHK$_{21}$	Signs of illness, immunofluorescence	6–10 days in mice, several days in tissue culture
Lassa virus Tacaribe complex viruses	Vero, Mouse	CPE, signs of illness	Suspected cases should be reported to a public health facility and laboratory testing should not be attempted at the local level
Rhabdoviridae			
Rabies	Mouse, immunofluorescence of clinical samples	Signs of illness; immunofluorescence	Suspected cases should be reported to a public health facility and laboratory testing should not be attempted at the local level
Retroviridae	T lymphocytes	Transformation or CPE Reverse transcriptase Immunofluorescence	Testing should not be attempted at the local level

[a] Tissue Culture Host Systems:
   Hel—diploid human embryonic lung
   HEK—primary human embryonic kidney
   MK—primary monkey kidney
   HEp-2—cell line derived from human carcinoma of the larynx
   Vero—cell line derived from African green monkey kidney
   HFG—primary human fetal glial cells
   HAmn—primary human amnion
   AGMK—primary African green monkey kidney
   RK$_{13}$—cell line derived from rabbit kidney
   HeLa—cell line derived from carcinoma of the cervix
   BHK$_{21}$—cell line derived from baby hamster kidney

readings for the first 7 to 10 days will offer the earliest recognition of viral infection, and many positive cultures can be detected within that period. Some viruses may require only 12 to 24 hours for earliest detection of cytopathic effects, and others as much as 4 to 6 weeks before a specimen is considered negative for infectious virus.

The orthomyxoviruses (influenza types) and paramyxoviruses (parainfluenza, measles, mumps) produce a hemagglutinin that can be detected on the host-cell membrane or in supernatant fluids with the application of a washed suspension of fresh erythrocytes. After incubation at a predefined temperature, infected monolayers are examined under light microscopy for the presence of red blood cells adsorbed onto the surface of infected cells. This hemadsorption technique is a more sensitive method for early detection of these groups because a CPE may not occur or, if it does occur, may not be easily recognized.

Another traditional virus detection method is worth mentioning, although its clinical application has been limited to detection of rubella virus. Viruses have been known to interfere with one another in culture. This interference phenomenon can be applied in the laboratory to those viruses that may not be detectable in tissue culture by CPE but that can inhibit growth of a second cytopathic or hemadsorbing agent. In practice, a clinical sample of suspected rubella virus is inoculated onto a suitable tissue culture and allowed to replicate over several days. At a later date, usually 5 to 7 days, a second challenge virus of known concentration and character is inoculated onto the same cultures. During the next few days, cultures and appropriate controls are examined for evidence of viral interference. Those cultures showing significant reduction in degree of expression of challenge virus are considered to be positive for an interfering agent, such as rubella virus.

Virus-Host Systems

Table 13.1 summarizes frequently used culture methods for the major human viruses and the approximate time from inoculation to detection of the virus. This time will vary somewhat, depending on the sensitivity of individual host systems and initial titer of virus.

Human embryonic lung cells are a semicontinuous, human diploid fibroblastic cell culture derived from fetal lung tissue. Several human embryonic lung cell strains are available. These cells are useful for cultivating cytomegalovirus, echoviruses, and rhinoviruses. Herpes simplex, varicella-zoster, respiratory syncytial, and BK viruses also grow readily in human embryonic lung cells. Other diploid fibroblasts, such as human foreskin, may be comparably sensitive for these agents. Primary cell cultures are valuable in the laboratory because of their great sensitivity. Those most often used are human embryonic kidney and monkey kidney. The former is extremely sensitive for measles, adenovirus, herpes simplex, varicella-zoster, vaccinia, BK virus, occasional enteroviruses, and reovirus. The Old World rhesus monkey or the cynomolgus monkey provides kidney tissue culture cells that are used for culture of influenza viruses, parainfluenza viruses, measles, mumps, respiratory syncytial virus, enterovirus, some rhinoviruses, vaccinia, and reovirus. Continuous epithelial cell lines such as HeLa and HEp-2 may be used for cultivating certain viruses, for example, some strains of HEp-2 cells are far superior to human embryonic lung cells for primary culture of respiratory syncytial virus.

Identification

VIRUS CHARACTERIZATION

Once its presence has been demonstrated, the virus should be identified. Identification may proceed quickly if the technician has a reasonable idea of which virus or group of viruses is under consideration. If the isolate's identity is not clear, the laboratory will select appropriate procedures to define the biochemical and biophysical properties of the agent. These procedures might include hemagglutination capability, determination of nucleic-acid type, size, lipid content, and temperature and pH stability of the isolate.

A number of viruses can agglutinate the erythrocytes of certain animal species at specific temperatures. The resulting pattern is that of a lacy, diffuse network of red blood cells coating the bottom of a U-bottom tube or well, and not the clumped pattern observed when hemagglutinins are not present. This "hemagglutination" phenomenon is shared by a number of viruses, and the choice of erythrocytes as well as the temperature at which agglutination optimally proceeds are geared to the suspected virus.

Determination of nucleic-acid type by use of acridine-orange staining or halogenated pyrimidines, and of a lipid envelope presence using solvents such as ether or chloroform, may be sufficient to narrow the range of possibilities for serologic identification. Occasionally, determination of pH stability, heat stability, or virion size may be necessary. Once the list of possibilities is small, final identification is accomplished by antigenic analysis.

IDENTIFICATION PROCEDURES

Various serologic methods are available, all of which involve treating a predetermined quantity of the unknown virus with a standardized quantity of specific antiserum. The identification procedures will vary, depending on the agent in question. Thus, a hemagglutinating agent will be identified by a procedure using hemagglutination-inhibition as an indicator system, whereas a cytopathic agent isolated in tissue culture may be identified by neutralization of the cytopathic effects. Other commonly used identification tests include hemadsorption-inhibition and immunofluorescence.

The neutralization test has wide application because it can be applied to virus groups in all host systems. The principle involved is that of elimination of virus infectivity by incubation of virus with hyperimmune serum prepared by immunizing animals with specific virus serotypes. A known quantity of virus in the form of a virus suspension is combined with a standard amount of antibody and incubated for a period at a predetermined temperature. The mixture is then inoculated into a suitable host system. Cell cultures, eggs, or animals are monitored daily for signs of infectivity, together with appropriate positive and negative controls. The unknown virus is identified by the virus/antiserum combination that results in loss of viral infectivity. Although this assay has specificity, with broad application to any virus that can be cultivated, it is usually expensive and time-consuming, requiring days to weeks for satisfactory results.

For those viruses that can hemagglutinate, the hemagglutination-inhibition test offers results in hours rather than days. It, like the neutralization test, begins with combining specific amounts of virus and antiserum, but, after incubation, a sensitive erythrocyte suspension is added and incubation continued; results are available as soon as the erythrocytes settle. The hemagglutination-inhibition method is specific for given viruses, and does not require a virus replication period; thus, identification by hemagglutination-inhibition can be completed the same day that virus hemagglutination is demonstrated. The main disadvantages are that not all viruses can hemagglutinate, and even those that do may not initially achieve sufficiently high titers to produce enough hemagglutinin for detection.

Those viruses that produce hemagglutinin but in insufficient quantities for early hemagglutination demonstration may be detected by hemadsorption and identified by means of the hemadsorption-inhibition technique. For that technique the unknown virus is inoculated onto a permissive tissue culture system, and cultures are incubated for several days until sufficient viral hemadsorption can be demonstrated. At this time, cultures are washed free of extracellular virus and appropriate antisera are overlaid on the monolayers. After an incubation period, a washed erythrocyte suspension is added and results are available after a further short incubation. Hemadsorption-inhibition is a sensitive and specific method but requires a virus replication period of several days, similar to the neutralization test, so results are not available with the speed of the hemagglutination-inhibition test.

The fluorescent antibody technique is an immunofluorescent procedure that uses a fluorescent dye, usually fluorescein isothiocyanate, to indicate a specific antigen-antibody reaction. The antigens are contained in virus-infected cells. In the direct fluorescent antibody method, specific antibody is conjugated with the fluorescent dye. Virus-infected cells are fixed onto glass slides, then overlaid with an appropriate fluorescent-tagged antibody. After an incubation period, slides are washed and examined under an ultraviolet microscope for a pattern of fluorescence characteristic of a particular virus. The procedure is rapid and straightforward, but it lacks the sensitivity of the indirect method.

The indirect fluorescent antibody test is a sandwich technique that involves reaction of a virus antigen with a known antiserum followed by a further reaction with a fluorescein-tagged anti-immunoglobulin. This last reagent is prepared by immunizing an animal against the globulin fraction (IgM, IgG, or total immunoglobulin) of serum from the same species used to produce the viral antibody. In practice, virus-infected cells are overlaid with antiserum and incubated, followed by washing to remove any free antibody not bound in an antigen-antibody complex on the slide. A fluorescein-conjugated antispecies immunoglobulin is then added. After incubation and final washing, the slides are examined under an ultraviolet microscope for specific viral immunofluorescence. This technique takes longer than the direct fluorescent antibody technique because of an additional step, but it has been estimated to be in the range of 4 to 12 times more sensitive.

Practical Considerations

The ability to diagnose an illness through virus isolation depends on several considerations: appropriate specimens, proper handling, selection of test systems, titer of virus in the specimen, carefully controlled culture conditions in the laboratory, and an experienced technician.

SPECIMEN SELECTION

The choice of specimen is essential for meaningful results. Specimens should be selected based on the type of illness and the implicated viruses, as shown in Table 13.2.

TABLE 13.2   Specimen Selection for Clinically Important Viruses

Type of Illness	Viruses	Specimens
Respiratory	Influenza Parainfluenza Respiratory synctial virus Rhinovirus	Throat or nasal washes, swabs, nasopharyngeal swab, aspirates of lower respiratory tract, lung biopsy/necropsy
	Enterovirus Adenovirus	Same as above, also stool
Gastrointestinal	Reovirus Adenovirus Enterovirus	Stool, throat secretions
	Rotavirus Norwalk agent Calicivirus Astrovirus	Stool, intestinal biopsy/necropsy (not currently detected by culturing)
	Hepatitis A Hepatitis B	Serum, stool Serum

Type of Illness	Viruses	Specimens
CNS	Mumps	CSF, saliva, urine
	HSV1, CMV, EBV	Brain, rarely CSF
	HSV2	CSF, genital swab
	Enterovirus	CSF, stool, throat secretions
	Papovavirus	Brain
	Togavirus	
	Bunyavirus	
	Lyphocytic choriomeningitis	CSF, anticoagulated blood
Genitourinary	HSV1 and 2	Genital secretions
	CMV	Urine, genital secretions
	Adenovirus	Urine
	Papovavirus	Urine
Exanthem		
	Vaccinia	
Vesicular	HSV	Vesicle aspirate, vesicle scrapings, also
	Varicella	stool for enterovirus
	Enterovirus	
Macular/	Enterovirus	Throat secretions, also stool for adenovi-
papular	Adenovirus	rus and enterovirus, urine for measles,
	Measles	rubella
	Rubella	
Hemorrhagic	Bunyaviruses	Blood (anticoagulated), throat secretions,
	Arenaviruses	urine, tissues
Congenital or	CMV	Throat/oral secretions, urine; depending
perinatal	Rubella	on clinical signs: vesicle material, ocular,
	HSV	throat secretions, CSF, brain urine
	Enteroviruses	Throat secretions, stool
Mononucleosis	CMV	Anticoagulated blood, oral secretions,
	EBV	urine, saliva

SPECIMEN COLLECTION AND HANDLING

Timing of specimen collection in relation to onset of infection is critical. With few exceptions, viruses are generally not recoverable in body fluids beyond the first week of illness. A good rule of thumb is to collect virus isolation material as soon as possible after onset, preferably within the first 3 days. Respiratory secretions, CSF, and vesicle material have highest yields if collected within 2 to 3 days of onset. Urine can remain virus positive for from several days to 2 weeks and for months in the case of cytomegalovirus. Some viruses found in stool (reoviruses, adenoviruses, enteroviruses) may occasionally be detectable for as long as 1 month. Immunocompromised persons will shed virus for longer periods, and viruses congenitally acquired may be present for months— and sometimes years.

Handling of any specimen should be brief, with immediate delivery to the laboratory. Dry swabs are not useful. Usually the laboratory will supply special transport medium that contains salts in an isotonic solution with protein in the form of gelatin, whole serum, or albumin, and antibiotics to control growth of other microorganisms. If nothing is supplied by the laboratory, physiologic saline, or preferably saline buffered to a pH of 7.2 to 7.4, may be used for washes; beef heart or veal infusion broth may also be used.

SPECIMEN TRANSPORT

Swabs and tissue are best transported in medium. Stools and body fluids should be sent in sterile containers and delivered by the fastest route possible, with specimens immersed in an ice slurry to maintain temperatures of approximately 4°C. Whereas some viruses, such as enteroviruses and adenoviruses, are stable at ambient temperature, many are not, and a few, such as varicella and respiratory syncytial virus, will not withstand prolonged transport at room temperature. The practice of bedside culture of specimens may increase the recovery of such temperature-labile agents in certain situations.

Because specimens will likely contain infectious virus, care should be taken in the handling and shipment of medical material. Sturdy containers that can be tightly sealed and cushioned should be chosen. Specimen containers should, in turn, be placed into a plastic bag or some impermeable container that will hold wet ice. Pertinent information regarding the specimen and the patient's history should be kept separate from the inner package to avoid wetting or contamination from the contents. If 24-hour shipment cannot be ensured, a dry ice/95% ethanol bath should be prepared and the specimens frozen with gentle agitation, and then stored below −60°C. Shipment should include packing in dry ice. Although some viruses will withstand this treatment, others will be lost, or the titers diminished, unless protein is added to the specimen for stabilization. Varicella-zoster and respiratory syncytial virus are two that require culturing from fresh unfrozen material.

Conventional Serologic Methods

In addition to virus isolation, conventional diagnostic virology services provide information concerning antibody status to particular viruses. As a general rule for serological diagnosis, two blood samples are necessary to provide useful information. One should be drawn during the early or acute stage of illness, the second after sufficient time has elapsed to demonstrate a fourfold or greater titer rise. Diagnosis based on a single serum titer is hazardous. Low antibody titers on a single specimen may reflect a previous infection or insufficient time for a significant response to a current one, whereas high titers may reflect a prolonged response to a previous infection.

The timing of blood sampling depends on the serologic test to be performed. Those most frequently used are complement fixation, indirect hemagglutination, hemagglutination inhibition, and immunofluorescence. Newer serologic techniques include immune adherence hemagglutination, serum radial hemolysis, and enzyme linked immunosorbent assay.

COMPLEMENT FIXATION

The complement fixation (CF) test is the classic method used by most laboratories for serologic diagnosis of a primary viral infection. The operative principle combines both the binding of complement in an antigen-antibody reaction and its capacity to lyse sensitized erythrocytes. The test lacks the sensitivity of some other serologic procedures, but it is particularly useful for diagnosis of a primary infection. It is less useful for immune status testing or demonstration of recurrent infections.

Viral antigen preparations are incubated with serial dilutions of the patient's heat-inactivated serum in the presence of standardized, fresh guinea pig complement. This is usually an overnight incubation at 4°C, followed by the addition of sheep erythrocytes that have been coated (sensitized) with rabbit antisheep erythrocyte serum. Complement that is bound in the antigen-antibody complex formed during a specific reaction is unavailable for lysis of the sensitized erythrocytes. Free, unbound complement will lyse available erythrocytes. The reciprocal of the highest serum dilution demonstrating a 50% reduction in hemolysis is the titer of antibody to a particular virus or virus group. Occasionally a serum will react with the sensitized sheep erythrocytes, binding complement in the process. This is known as anticomplementary activity and may prevent determination of accurate antibody titers by this method. Reagents for viral CF testing are widely available through commercial sources. Generally, the acute-phase serum should be drawn within the first 7 to 10 days after onset of illness to establish baseline titers. The follow-up serum should be collected 2 to 4 weeks into the illness. Occasionally a third, later convalescence serum is necessary if no seroconversion is yet noted. Peak antibody titers are generally seen by 1 month after onset of illness, with subsequent decline in titers over time.

HEMAGGLUTINATION INHIBITION

Serum antibody titers by hemagglutination inhibition may be performed against viruses that have hemagglutinin antigens. The antigens are incubated with serial dilutions of a patient's serum, followed by addition of a suitable erythrocyte suspension. After the red blood cells settle,

the test is read for the highest dilution of serum giving no hemagglutination. The antibody titer is the reciprocal of this dilution. Viral antibodies that are frequently detected by this method include influenza,
parainfluenza, measles, mumps, togaviruses, and adenoviruses. Antibody responses detected by the hemagglutination inhibition assay are
demonstrated sooner than with the complement fixation test in most
situations, and with earlier peak titers. This permits earlier diagnosis of
infection. The test is sensitive and specific but has the disadvantage of
limitation to only those viruses that produce hemagglutinins.

INDIRECT HEMAGGLUTINATION
A hemagglutination technique has been developed that permits detection of antibodies to viruses that cannot by themselves agglutinate
erythrocytes. This procedure is known as the indirect hemagglutination
or passive hemagglutination test. It incorporates erythrocytes that are
treated with a tanning agent to enhance the attachment of soluble viral
antigens onto the cell membrane. Introduction of a patient's serum
containing antibodies specific to the antigen will result in a positive agglutination pattern. This procedure is often used for herpes group antibody determinations.

INDIRECT FLUORESCENT ANTIBODY TEST
Indirect immunofluorescence, when properly controlled, is another
highly specific technique. Its sensitivity is well established, with earlier
detection of antibodies and higher titers compared with complement-
fixation results. For determination of antibody titers to viruses of the
herpes group, however, difficulty is encountered because of virus-induced
F_c receptors on the surface of infected cells. IgG binds nonspecifically
to these receptors, giving fluorescent staining that makes interpretation
difficult.

FLUORESCENT ANTIBODY TO MEMBRANE ANTIGEN
For varicella-zoster antibody determinations, the problem of nonspecific staining has been circumvented by development of an indirect assay known as fluorescent antibody to membrane antigen. The procedure is similar to other assays, except that membrane antigens are
detected because the varicella-zoster infected tissue culture cells are in
suspension rather than fixed onto slides. Gluteraldehyde-treated or live
infected cells are incubated with the patient's serum, followed by washing and incubation with fluorescein isothiocyanate-conjugated antihuman globulin. Cells are placed onto slides and examined for a halo or
ring of fluorescence around the cell membrane. The fluorescent antibody to membrane antigen assay seems to prevent F_c-receptor-mediated,
nonspecific staining and has become the most practical approach to
sensitive determination of varicella-zoster immune status.

IMMUNE ADHERENCE HEMAGGLUTINATION

Similar to the complement fixation test in specificity, but far more sensitive, is the immune adherence hemagglutination serologic test. Like the complement fixation test, it detects antibodies found in an immune complex—incorporating complement, but unlike its dependence on complement's lysing capacity, the immune adherence hemagglutination assay detects immune complexes that bind to the C'3b receptors on primate erythrocytes.

The immune adherence hemagglutination test can use the same viral antigen preparations used for the complement fixation test. Serial dilutions of a heat-inactivated serum are mixed with standard antigen preparations and then incubated with guinea pig complement. Later, a volume of dithiothreitol-ethylenediaminetetraacetic acid (EDTA) in veronal-buffered saline is added to stop the reaction, followed by addition of a human type O erythrocyte suspension. The test is left to permit erythrocytes to settle out, and endpoints of agglutination are determined. The results are available within a matter of hours. The immune adherence hemagglutination test is as specific as the complement fixation test but 4 to 20 times more sensitive, thus resulting in higher titer and earlier detection of a seroconversion.

RADIOIMMUNOASSAY/ENZYME-LINKED IMMUNOSORBENT ASSAY

The newer techniques of radioimmunoassay and enzyme-linked immunosorbent assay are similar in principle, performance, sensitivity, and specificity. They seem to be superior to the complement fixation test for sensitivity and specificity and are considerably easier to perform. Both tests rely on indicator systems to identify a specific antigen-antibody reaction. One binds an enzyme-labeled antibody into the reaction, which is detected by addition of a suitable substrate, and the other binds an isotope-labeled antibody into the complex, which is recognized by radioactive emissions. As with immunofluorescent techniques, both assays can be performed using direct and indirect methods. These procedures will be discussed further under rapid viral diagnostic techniques.

SERUM RADIAL HEMOLYSIS

The last serologic method for discussion here is known as serum radial hemolysis. In practice, it has been used largely for detection of rubella and influenza antibodies. The test involves viral antigen bound to erythrocytes in an agar medium. Patient serum is added to a well punched in agar and allowed to diffuse through the agar for 18 to 24 hours. In a specific reaction, antigen-antibody complexes are formed in the agar. Complement is added to the well, whereupon it diffuses through the agar and lyses any erythrocytes not coated with antigen-antibody complexes. The zone size of hemolysis reflects the amount of antibody present. The serum radial hemolysis test is a simple and reliable procedure considered to be comparable to the hemagglutination inhibition test in sensitivity, yet its cost is far less, making it a useful procedure to perform.

With the exception of complement fixation, all these serologic tests have a degree of sensitivity that permits single serum testing for immune status, as well as multiple sample testing for diagnosis of a current infection. However, for multiple sample testing, all sera must be assayed at the same time for meaningful comparisons. Fluctuations in test variables make comparison of data from different tests inadvisable. One characteristic of the more sensitive procedures is that early blood sampling is even more important than for complement fixation. Whereas antibodies by the latter test may not be evident for the first week of an illness, the other tests will show levels that peak at a time when levels of complement-fixing antibodies may be only moderately elevated. One must draw blood samples as early as possible, to maximize demonstration of significant antibody rises before titers peak. Sometimes, if results are inconclusive by the more sensitive tests, a complement-fixation assay may be called on to clarify interpretation.

Certain modifications in complement-fixation procedures from those usually used may enhance sensitivity. Reducing the amount of complement used may allow such assays to approach the sensitivity of other serologic tests.

RAPID DIAGNOSTIC TECHNIQUES

The discussion of diagnostic virology to this point has focused on demonstration of viral infectivity and rising antibody levels for diagnosis. Results with this approach are often too delayed to be of immediate clinical benefit. In addition, some viruses are not easily isolated or have not been successfully cultured in the laboratory. Many laboratories have therefore sought rapid viral diagnostic techniques that offer sensitive, specific results independent of infectivity and free of the "acute-convalescent" approach to serology.

Within the last decade, techniques have been developed that can provide results within 2 to 24 hours of specimen collection. These tests can be divided into two categories: viral antigen detection and early antibody demonstration. Many of the procedures do not necessitate virus isolation backup facilities and can easily be adapted to the smaller laboratory whose staff is inexperienced in diagnostic virology.

Antigen Detection

Three techniques are currently enjoying popularity for viral antigen detection—immunofluorescence, radioimmunoassay, and the solid-phase, enzyme-linked immunosorbent assay. Further, electron microscopy has been applied successfully to detection of viral particles in stools, tissues, and urines, and some laboratories are performing immunoperoxidase assays for rapid diagnosis.

IMMUNOFLUORESCENCE

The basic techniques of direct and indirect immunofluorescence have already been described. These procedures have been applied to rapid viral antigen detection in two ways. A number of laboratories have reported success in detection of virus antigen in tissue cultures inoculated with clinical samples. The greatest application seems to be with respiratory virus infections in which antiserum, either in pools or singly, can be applied in the indirect method as early as 16 to 24 hours after inoculation, with positive results. One study reported that diagnosis was made in 30% of specimens within 16 hours of inoculation.

Even more rapid results can be achieved by direct screening of clinical samples for virus-infected cells. Respiratory secretions containing ciliated epithelial cells, as well as touch preparations or tissue sections of biopsy/necropsy material are fixed on glass slides, or processed to remove potential interfering noncellular material and then fixed. Immunofluorescent staining is performed as previously described. Because these fixed viral antigens are more stable than infectivity, the usual concerns regarding loss of viability of a specimen do not apply, and slides can be stored at $-20°C$ for later analysis. Considerable success has been reported for diagnosis of respiratory infections using respiratory secretions, with greater than 95% correlation for certain viruses compared with culture findings. Success depends on the presence of sufficient cellular material in a specimen, sampling early in the course of illness, and good-quality reagents. Currently, the major difficulty with fluorescent antibody tests for viral detection is the limited supply of high-quality reagents. The commercial availability of monoclonal antibodies is helping to alleviate this problem.

Immunofluorescence is also useful in the rapid diagnosis of viral infections involving the skin or central nervous system. Reliability, that is, agreement with culture results, occurs in 60 to 90% of brain biopsy specimens from patients with herpes simplex encephalitis. Similarly, diagnosis of vesicular eruptions secondary to herpes simplex or varicella-zoster viruses can be confidently made by immunofluorescence. Sensitivity of antigen detection varies, depending on specimen collection time in relation to course of disease, quality of the specimen, and quality of reagents. Although it is possible to detect viral antigen by immunofluorescence for several days after infectivity is no longer present, a negative rapid test result does not rule out infection by the suspected virus, and isolation attempts may still be required.

IMMUNOENZYME ANALYSIS

Similar to immunofluorescence in sensitivity, immunoperoxidase methods require only light microscopy rather than a costly ultraviolet microscope. Clinical material, either processed to remove endogenous peroxidase activity or used unaltered, is fixed on slides with acid alcohol. Depending on whether a direct or indirect procedure is used, a peroxidase (e.g., horseradish peroxidase) is conjugated to specific viral antiserum (direct method) or to an antispecies immunoglobulin (indirect method). In the direct method, peroxidase-tagged viral antiserum is flooded over the clinical sample, followed by incubation, washing, and addition of a staining reagent (e.g., benzidine hydrochloride, aminoethylcarbazole), which will result in colored deposits at the sites of specific antigen-antibody peroxidase complexes. The indirect method uses viral antiserum followed by the peroxidase-conjugated antispecies immunoglobulin, and, finally, a staining reagent.

The specificity and sensitivity of immunoperoxidation are comparable to those of immunofluorescence, but here, too, availability of quality reagents is a limitation, unless the laboratory can prepare its own hyperimmune serum and accomplish the conjugation.

RADIOIMMUNOASSAY/ENZYME IMMUNOSORBENT ASSAY

Radioimmunoassays and enzyme immunosorbent assays are being used increasingly for detection of viral antigens. Those most widely available today are for detection of hepatitis B and rotavirus antigens. It was by electron microscopy that the hepatitis agents and those causing gastroenteritis in children were first detected, but without sensitive culture systems the viruses were not identified by most laboratories until the techniques of radioimmunoassay and enzyme immunosorbent assay were applied. These procedures are basically similar in principle and performance. Both rely on indicator systems to identify a specific antigen-antibody reaction and are known as solid-phase immunoassays. They provide sensitive, specific results using straightforward procedures. In some cases radioimmunoassay is more sensitive, but the enzyme immunosorbent assay offers a more stable system without the need for handling and disposal of gamma-radioactive wastes. As a consequence, the trend in rapid viral diagnosis has been away from radioimmunoassay and toward developing enzyme immunosorbent assay reagents for detection of viruses other than rotaviruses. The small laboratory without experienced virologists can now diagnose certain viral infections by detection of antigen independent of infectivity.

As a means of rapid antigen detection, specific viral antibody is coated onto a microtiter well, bead, or some solid-phase system. The clinical sample is added and incubated with this antibody for a prescribed period, followed by washing to remove unbound materials. Another viral antibody that has been tagged with a peroxidase (enzyme immunosorbent assay) or radioisotope (radioimmunoassay) is added for an incubation period, and then washed. Finally, in the enzyme immunosorbent method, a substrate is added that triggers a color change, permitting visual or spectrophotometric reading. These two procedures are known as the direct antigen assay and are similar to the direct fluorescent antibody test. Indirect radioimmunoassay and enzyme immunosorbent assay methods are also possible, similar to the indirect immunofluorescent procedure discussed elsewhere (Table 13.3). The indirect methods permit a broader use of the enzyme- or isotope-labeled immunoglobulins against more than one viral antigen system.

ELECTRON MICROSCOPY

In recent years electron microscopy has been applied with some success to diagnostic virology. Although it has the disadvantages of requiring expensive, elaborate equipment and highly skilled personnel, the technique can offer rapid results. Its greatest application has been in the diagnosis of gastrointestinal illness. Routinely cultivatable viruses from stool have been limited to adenoviruses, reoviruses, and enteroviruses. With the application of electron microscopy to examination of stool, astroviruses, caliciviruses, rotaviruses, Norwalklike agents, and noncultivatable adenoviruses have been detected.

A number of ultrastructural assays are possible, but the one most adaptable to examination of clinical samples is negative staining with phosphotungstic acid. Fluids can often be examined without concentration, although the level of sensitivity is lower. When high yields of viruses are present, however, direct examination provides diagnosis within hours. For greater yields, the specimen can be concentrated first by ultracentrifugation. The resulting pellet is resuspended in a small amount of distilled water and mixed with phosphotungstic acid. The final stained material is placed on a Formvar carbon-coated grid, air-dried, and examined by electron microscopy for the presence of a characteristic virus morphology suggestive of a particular group of viruses. The disadvantage of this technique is its inability to distinguish among morphologically similar viruses and the requirement for large amounts of virus (10^6 particles/milliliter). However, by introducing specific viral antibody to the specimen before fixation, sensitivity can be further increased by aggregation of virus particles within antigen-antibody complexes. The virus is also identified by the specific antibody used.

Enzyme-linked immunosorbent and immunofluorescent assays have become major approaches to rapid viral antigen detection. Radioimmunoassay produces radioactive waste that must be discarded safely, and electron microscopy remains of limited use because of the unavailability of special equipment.

Antibody Detection

VIRAL IgM

Viral serology on acute and convalescent specimens has been the historic approach to diagnosis of viral infections. Newer, more sensitive

TABLE 13.3 Virus Identification Procedures

CONVENTIONAL METHODS

Neutralization
Virus + antibody → + suitable host system → neutralization of infectivity

Hemadsorption-inhibition
Virus replication in tissue culture → + antibody → + erythrocytes → inhibition of hemadsorption

Hemagglutination-inhibition
Virus + antibody → + erythrocytes → inhibition of agglutination

RAPID METHODS

Immunofluorescence

Direct
Virus-infected cells + fluorescein conjugated viral antibody → specific viral fluorescence

Indirect
Virus-infected cells + antiviral antibody → + fluorescein antibody conjugate → specific viral fluorescence

ELISA

Direct
Antibody-coated solid phase + specimen → + enzyme-labeled antibody → substrate → color change

Indirect
Antibody-coated solid phase + specimen → + specific antibody → + enzyme-labeled antispecies globulin → + substrate → color change

RIA

Direct
Antibody-coated solid phase + specimen → radioisotope-labeled antibody → gamma emission

Indirect
Antibody-coated solid phase + specimen → + specific antibody → + radioisotope-labeled antispecies glob-ulin → gamma emission

serologic techniques have reduced the response time, yet in most cases two sera are still necessary. Techniques have been developed that now permit diagnosis based on a single serum. Antibodies of the IgM class are often present early in the course of a viral infection, and detection of specific IgM antibodies may be diagnostic of active infection. IgM antibodies usually peak within 1 to 3 weeks, and then decline rapidly, although in congenital infections they may remain elevated for several months. Virus-specific IgM is usually associated with a recent infection that may or may not be primary, since IgM can be produced following reinfection. The challenge has been to develop simple, efficient methods of IgM detection in serum that will be specific for viral IgM and will not be affected by the presence of rheumatoid factor or competition of IgG.

A major drawback to detection of IgM is the frequent presence of rheumatoid factor, an anti-IgG globulin that is usually of the IgM class. It may be present in connective-tissue diseases, certain infections (e.g., toxoplasmosis, rubella, cytomegalovirus, varicella-zoster, influenza, syphilis), and occasionally in other diseases or in some healthy persons. Removal of rheumatoid factor by absorption may have the unfortunate result of removing specific viral immunoglobulins as well, so the choice of absorption procedure should be made carefully. Removal of rheumatoid factor is recommended for any serum that is to be tested for IgM levels by competitive receptor-binding tests.

The presence of both IgM and IgG in a sample may result in IgG inhibiting the detection of IgM, especially if the latter is in low concentration. When IgM is in high concentration, such as is seen in an early acute infection, IgG inhibition is usually not a problem. Physical separation of IgM and IgG is possible by sucrose gradient centrifugation, by coagglutination with protein A from *Staphylococcus aureus*, or by chromatography and precipitation methods using 2-mercaptoethanol or gamma-chain-specific antihuman IgG. Whatever procedure is used, caution is necessary not to report false-positive results stemming from presence of rheumatoid factor or false negatives resulting from a masking of IgM activity by specific IgG.

Viral IgM can be detected by means of the competitive receptor-binding assays of immunofluorescence, enzyme-linked immunosorbent assay, and radioimmunoassay. In these procedures, antihuman IgM immunoglobulin conjugated to its appropriate indicator is used. IgM detecting enzyme-linked immunosorbent assay, and fluorescent antibody procedures have been used to diagnose congenital infections due to cytomegalovirus and rubella.

EARLY ANTIGENS

Finally, another approach to rapid immunodiagnosis is illustrated by Epstein-Barr virus (EBV) serologic testing. Antibody to viral capsid antigen is usually at or near its peak at the time serology is requested to confirm an EBV infection. However, antibodies to other antigens such as nuclear antigen (EBNA) and early antigen (EA) are produced at different stages of infection. Measurement of these antibodies by immunofluorescent techniques may distinguish acute from chronic EBV infections. Definition of early antigens and early antibodies against other viruses has not progressed to the state that it has in EBV serology, but it may well become useful in the future.

PRESENT REALITIES AND FUTURE EXPECTATIONS

Enzyme-linked immunosorbent assay and radioimmunoassay techniques have already proved useful in the diagnosis of hepatitis and rotavirus infections. They are being extended to other viruses, as antibodies with sufficient sensitivity and specificity are produced. The development of monoclonal antibody technology is facilitating the use of such solid-phase assays. Herpes simplex virus antigen screening assays are available; these assays may identify women at risk for transmission of virus at delivery, thereby serving as a guide for determining route of delivery. Similar adaptations of this technology to cytomegalovirus, respiratory viruses, and enteroviruses should prove useful in monitoring high-risk patients, including organ transplant recipients, resulting in prompt intervention measures (isolation, therapy) when appropriate.

Some virus infections are associated with little release of extracellular antigen. Because antibodies diffuse poorly into intact cells, techniques are needed to expose sufficient intracellular antigen for detection or to improve penetration of antibodies. Alternatively, indirect techniques are necessary to amplify detection sensitivity for minute quantities of extracellular antigen in solid-phase immunoassays. Several such amplifying systems using complement components (e.g., Clq), newer conjugation methods (e.g., avidin-biotin), newer solid phases (e.g., nylon, activated paper), or higher energy substrates (e.g., umbelliferone) are under study. Detection of certain viral products, such as HSV thymidine kinase or influenza neuraminidase, by enzyme-linked immunosorbent assay or radioimmuoassay, may soon become practicable. Viral product detection by gas liquid chromatography or mass spectrophotometry may also be feasible within the next few years.

Detection of viral nucleic acids by molecular hybridization is a useful research procedure that is now under study as a diagnostic technique. Radio-labeled probes of DNA can be used to demonstrate small amounts of viral DNA in clinical specimens. Although the technique is not yet widely available, preliminary studies suggest that CMV DNA can be demonstrated in patient urine by this technique and that other applications of similar procedures are likely.

Local production of specific antibodies may reflect viral replication at a particular site. Measurement of secretory IgA in respiratory secretions during infections with respiratory syncytial virus or other agents may be useful tests. Cerebrospinal fluid/serum antibody ratios have been studied in HSV, VZV, and measles-associated infections. In chronic infections, these tests are helpful, but in acute infections ratios may be falsely normal.

Finally, investigators are now considering "nonviral" tests that are associated with viral infections. Measurement of serum interferon levels often reflect ongoing viral replication, and development of rapid assays for measurement of interferon may obviate the need for cumbersome biologic procedures. Moreover, measurement of interferon-induced proteins, such as $2' - 5'$ oligoadenylate synthetase, may also rapidly reflect ongoing viral infections. However, the specificity of such procedures is doubtful; conditions such as systemic lupus erythematosus, rheumatoid arthritis, and Kaposi's sarcoma may give false-positive results.

Measurement of peripheral blood T lymphocyte subsets using monoclonal antibodies and immunofluorescence or flow cytometry may also suggest ongoing viral infection in high-risk patients. Systemic infections with CMV, EBV, VZV, human T-lymphotropic virus (HTLV-3), and hepatitis B virus have all resulted in inverted ratios of helper to suppressor T lymphocytes, with relatively decreased helpers and increased suppressors. The specificity of these alterations remains to be determined, but such assays may give the clinician important information on the presence of viral infections involving circulating lymphocytes. Morphologic alterations of cells by viruses in urine, sputum, and tissue specimens have already been used to identify certain viruses (herpes group, papovavirus group) and will likely be used more in the future. Nonspecific bacterial adherence to virus-infected cells has also been suggested as a possible diagnostic test of the future.

In conclusion, let us peer into our crystal ball to predict how the two cases discussed earlier may be diagnosed in the not-too-distant future:

CASE 1: A 63-year-old obstetrician is seen at the hospital outpatient department with the same symptom complex (fever, malaise, nausea, and vomiting) following coronary artery surgery as in the case at the beginning of the chapter. Measurement of peripheral blood lymphocyte ratios shows increased suppressor cells and decreased helper cells. Molecular hybridization and enzyme-linked immunosorbent assays demonstrate CMV DNA and antigens in blood and urine. At the same time, an IgM assay shows CMV-specific antibodies in his serum. The patient is reassured and sent home to convalesce or, alternatively, treated with a CMV-specific antiviral drug.

CASE 2: A 10-year-old boy is admitted with a headache, fever, and inappropriate behavior. Cerebrospinal fluid is tested for HSV1 early protein antigens using a monoclonal antibody in a sensitive and specific enzyme-linked immunosorbent assay procedure. The assay is strongly positive. Antiviral drugs are started within hours of admission and no brain biopsy is required. Recovery is complete.

Although some of the assays mentioned in these two cases are not available as this chapter is written, by the end of this decade they may be in routine use. O brave new world, that has such assays in it!

Recommended Readings

Almeida J: Practical aspects of diagnostic electron microscopy. *Yale J Biol Med* 53:5–18, 1980.

Baumgarten A: Viral immunodiagnosis. *Yale J Biol Med* 53:71–83, 1980.

Chonmaitree T, Menegus M, Powell K: The clinical relevance of "CSF viral culture," a two-year experience with aseptic meningitis in Rochester, N.Y. *JAMA* 247:1843–1847, 1982.

Deinhardt F: Predictive value of markers of hepatitis virus infection. *J Infect Dis* 141:299–305, 1980.

Doane F: Virus morphology as an aid for rapid diagnosis. *Yale J Biol Med* 53:19–25, 1980.

Gardner P, McQuillin J: *Rapid Virus Diagnosis: Application of Immunofluorescence*, 2nd ed. London, Butterworth, 1980.

Henle W, Henle G: Seroepidemiology of the virus. In Epstein M, Achong B (eds): *The Epstein-Barr Virus*. New York, Springer-Verlag, 1979.

Hsiung GD: *Diagnostic Virology*, 3rd ed. New Haven, CT, Yale University Press, 1982.

Hsiung GD: Laboratory diagnosis of viral infections: General principles and recent developments. *Mt Sinai J Med* 44:1–26, 1977.

Hsiung GD, Fong C, August M: The use of electron microscopy for diagnosis of virus infections: An overview. *Prog Med Virol* 25:133–159, 1979.

Lennette E, Lennette D: Immune adherence hemagglutination: Alternative to complement-fixation serology. *J Clin Microbiol* 7:282–285, 1978.

Lennette E, Schmidt N: *Diagnostic Procedures for Viral, Rickettsial and Chlamydial Infections*, 5th ed. Washington, DC, American Public Health Association, 1979.

Ray, CG, Minnich L: Regional diagnostic virology services: Are satellite laboratories necessary? *JAMA* 247:1309–1310, 1982.

Richman DD, Cleveland PH, Redfield DC, et al: Rapid viral diagnosis. *J Infect Dis* 149:298–310, 1984.

Schattner A, Merlin G, Levin S, Wallach D, Hahn T, Revel M: Assay of an interferon-induced enzyme in white blood cells as a diagnostic aid in viral diseases. *Lancet* 2:497–500, 1981.

Schmidt NJ: Laboratory diagnosis of viral infections, Chapter 5 in Galasso GJ, Merigan TC, Buchanan RA (eds): *Antiviral Agents and Viral Diseases of Man*. New York, Raven Press, 1984.

WHO Scientific Group Report: *Rapid Laboratory Techniques for the Diagnosis of Viral Infections*. World Health Organization Technical Report Series 661. Geneva, World Health Organization, 1981.

Yolken R: Enzyme immunoassays for the detection of infectious antigens in body fluids: Current limitations and future prospects. *Rev Infect Dis* 4:35–68, 1982.

14

The Status of Antiviral Therapy

FRED ALLISON, JR.

Despite the ubiquity of viruses and their remarkable capacity to produce dramatic epidemic illness in humans, specific treatment to either moderate or cure viral infections with pharmacologic agents has lagged far behind progress achieved in prevention of viral diseases using immunologic measures. Fortunately, active and, to some extent, passive immunization procedures have provided valuable means for control of viral diseases. As an example, smallpox, a disease that caused major epidemics and loss of life throughout the world for many centuries, was eradicated when active immunization was combined with a well-conceived case-finding program. The World Health Organization certified in 1980 that the smallpox virus was no longer a threat to humans on any continent, although the variola virus is preserved solely for research purposes in a handful of restricted laboratories. This accomplishment represents a truly successful application of basic public health and immunologic principles to the problem of disease control on a global scale.

Although many empiric attempts have been made to develop non-immunologic therapeutic agents to prevent, limit, or cure viral infections in humans, meaningful progress has been limited to recent years—a time characterized by rapid increases in the understanding of the biochemical and metabolic properties of viruses and infected human cells. Even as this information was being assembled, medical scientists began a series of systematic clinical trials with antiviral substances that interfered with viral metabolism long enough to allow the intrinsic immune mechanisms of humans to eradicate or contain an infection. From these early studies, however, it became apparent that some drugs—cytosine arabinoside, for example—damaged the immune system and could make the viral infection worse.

The first specific antiviral agent that was clearly applicable to humans was developed by Kaufman and colleagues when they instilled the nucleic acid analog, idoxuridine, directly onto the ocular conjunctiva of human patients and thereby dramatically cured herpetic keratitis. The Kaufman group made this startling observation after earlier work in animals had shown unequivocally that certain nucleic acid derivatives had specific inhibitory activity against herpes virus multiplication in vivo. A favorable outcome in humans, therefore, represented a major therapeutic breakthrough and provided the first steppingstone to guide development of additional antiviral substances.

CURRENT STATUS OF ANTIVIRAL AGENTS

Antiviral compounds have been developed for use in chemoprophylaxis and chemotherapy of a variety of infections in humans, including those caused by influenza viruses, respiratory syncytial virus, and herpesviruses. The efficacy of several of these compounds has been demonstrated in rigorously controlled trials. A listing of drugs used for treatment of virus infections in human beings is provided in Table 14.1. The use of these drugs for treatment of specific viral infections in humans is summarized in Table 14.2 and discussed in the sections that follow.

Variola Virus (Smallpox)

For some years, smallpox, caused by variola virus, has been known to respond to the antagonistic action of certain of the thiosemicarbazones. Limited use of these agents, primarily methisazone (Marboran), modified and lessened the severity of clinical smallpox when given early in the disease. This agent was never widely used or extensively evaluated because eradication of the disease was achieved through active immunization using attenuated strains of vaccinia (cowpox) virus.

Influenza A Virus

AMANTADINE

For the last 15 years, amantadine hydrochloride (Symmetrel) has been intensively studied as an agent to counteract influenza A virus infection. Neither influenza B nor C viruses are inhibited by this substance. Amantadine and closely related compounds are believed to function by preventing the uncoating of the viral particle after penetration into the cytoplasm of susceptible cells. Unable to escape from its capsid or outer protective coat, the virus cannot replicate intracellularly as it ordinarily does by subverting the ribonucleic acid (RNA) machinery of the cell cytoplasm.

TABLE 14.1 Antiviral Drugs

Chemical Name	Generic Name	Trade Name(s)	Clinical Results
N-Methylisatin-β-thiosemicarbazone	Methisazone	Marboran	Of limited use in modifying smallpox
1-Aminoadamantane hydrochloride	Amantadine	Symmetrel	Effective for prevention and treatment of infection by influenza virus but not effective against influenza viruses B or C
α-Methyl-1-adamantanemethylamine	Rimantadin		Similar to amantadine but has less CNS toxicity in influenza A; may be useful against influenza B
Leukocyte interferon (HUIFN-α1)			Effective against varicella-zoster virus in immunocompromised, hepatitis B virus with vidarabine; not effective for rhinovirus, influenza A, B viruses
Human interferon alpha-2 from E. coli (HUIFN-α2)			May be effective against rhinovirus types 4, 13, 19
5-Iodo-2'-deoxyuridine (IDU)	Idoxuridine	Stoxil	Modifies keratitis by reducing virus shedding and promotes healing; does not prevent recurrence
Trifluorothymidine (TFT)	Trifluridine	Viroptic	Similar to idoxuridine but may be superior as a therapeutic agent
Adenine arabinoside (araA)	Vidarabine	Vira-A	Useful in varicella-zoster virus infection in immunocompromised patients. Effective in herpetic infections, particularly in encephalitis
Adenine arabinoside monophosphate			When compared to vidarabine, may have less gastrointestinal and central nervous system side effects
9-(2-Hydroxyethoxymethyl)-guanine (ACV)	Acyclovir	Zovirax	Clinical trials in varicella-zoster patients. Accepted for use in genital herpes simplex virus infection. Under clinical trial for other herpetic infections, cytomegalovirus, and hepatitis B infection
E-5 (2-bromovinyl)-2'-deoxyuridine (BVDU)			Clinical trials in herpetic infections underway
5-Iodo-5'-amino-2', 5'-dioxyuridine (AIU)			Clinical testing in progress against herpes simplex virus infections

TABLE 14.1 Antiviral Drugs (*continued*)

Chemical Name	Generic Name	Trade Name(s)	Clinical Results
1-(2-Deoxy-2-fluoro-B-D-arabinofuranosyl)-5-iodocytosine (FIAC)			A research drug, being tested against herpes simplex virus
Trisodium phosphoxoformate		Foscarnet	Clinical trials in herpes labialis enhanced healing but caused significant irritation; available only in Europe
Phosphonoacetate (PAA)			Trials for labial herpes infections underway; was less irritating to skin
	Tromantadine	Virimenz	Used only in Europe topically for ocular herpes infections
Iododeoxycytidine		Cebe-Viran	For topical treatment of ocular herpes; available in Europe only
5-Ethyl-2'-deoxyuridine		Aedurid	Available in Europe only for ocular herpes infection; a useful topical preparation
2-Deoxy-D-glucose			Has not proved effective against herpetic labial or genital infections
2-Amino-1-(isopropylsulfonyl)-6-benzimidazole phenyl betone oxime	Enviroxime		Clinical trials in rhinovirus type 9 infection suggest modification of the disease
1-β-D-ribofuranosyl-1, 2, 4-triazole-3-carboxamide	Ribavarine		Administered in small-particle aerosol modifies influenza A, may inhibit bone marrow synthesis of red blood cells
Inosine dimethylaminoisopropanol-*p*-acetamidodibenzoate		Inosiplex Isoprinosins	Large, poorly controlled clinical trials suggest a significant reduction in mortality in SSPE, a disease with an unpredictable course
L-tetramisole	Arildone		A broad-spectrum antiviral currently under study
Transfer factor	Levamisole		No effect on SSPE
			No evidence to substantiate a beneficial therapeutic result in viral diseases
Ascorbic acid			Although advocated as a preventive for the common cold, rigorous clinical studies do not substantiate claims

TABLE 14.2 Diseases Modified by Antiviral Drugs

Disease	Virus	Drug	Status[a]	Toxicity	Result
Influenza	A	Amantadine	Approved (Symmetrel)	Mild CNS symptoms	Effective—does not block antibody response to vaccine
					Modifies symptoms if given within 48 hours of onset
	A and B	Rimantadine	Experimental	Mild CNS symptoms	
	A	Ribavarin	Experimental	Less than amantadine	Same as amantadine, but may also prove useful for influenza B
	A and B	Interferon	Experimental	Fever	Modifies severe infection
					Not effective
Other Respiratory Diseases					
Acute respiratory infections	Rhinovirus type 9	Enviroxine	Experimental	Nasal stinging, nausea, and vomiting	Lessened severity of symptoms
	Rhinovirus types 4, 19, 13	Interferon α2	Experimental	Fever	Prevented colds and reduced virus shedding
Common cold		Vitamin C	Experimental	None	Inclusive for prevention and treatment
Smallpox	Variola	Marboran	Restricted		Lessened severity of disease when given early
Chickenpox/ shingles	Varicella-zoster	Adenine arabinoside (araA)	Approved (Vidarabin Vira-A)	Leukopenia, thrombocytopenia	Best results in patients under 38 years of age with reticuloendothelial neoplasms
		Interferon	Experimental	Fever	Reduced appearance of new lesions in immunocompromised cancer patients in all groups
		Acyclovir	Experimental	Renal function must be monitored	Reduced pain and shedding of virus in zoster, severity of neuralgia at 3 months after treatment not influenced

TABLE 14.2 Diseases Modified by Antiviral Drugs (continued)

Disease	Virus	Drug	Status[a]	Toxicity	Result
Herpetic infections					
Congenital/ neonatal herpes	Herpes simplex type 2	Adenine arabinoside (araA)	Experimental	Leukopenia, thrombocytopenia	A significant reduction in mortality of most severely ill babies
Fever blisters	Herpes, type 1 or 2	Trisodinar- phosphornate	Experimental	Mild local irritation	Effective in recurrent fever blisters
		Phosphono- formic acid	Experimental	Less irritation	Results not known
		Acyclovir			Effective when applied within 9 hours of lesions
Keratitis	Herpes simplex type 1	Idoxuridine	Approved (Stoxil)	Local irritation	Reduced viral shedding, enhanced corneal healing
		Adeninine arabinoside	Approved (Vidarabin; vira-P)	Local irritation	Enhanced corneal healing
		Trifluorothy- midine	Approved (Viroptic)	Local irritation	Enhanced healing
		Acyclovir	Experimental	Local irritation	Enhanced healing. None of the above used locally prevents recurrence
Genital herpes	Herpes simplex Types 1 or 2	Idoxuridine	Experimental	Local burning and contact dermatitis	Hastened healing but disease recurred
		Acyclovir	Approved (Zovirax)	Minimal	In primary disease, hastened healing. In recurrent disease, no effect in women, some benefit to men.

Encephalitis	Herpes simplex	Adenine arabinoside	Approved (Vidarabin; Vira-A)	Leukopenia, thrombocytopenia	A substantial reduction in morbidity and mortality when given to young patients early in the disease
Chronic active (aggressive)	Hepatitis B	Interferon with adenine arabinoside	Experimental	Drugs given sequentially to lessen severity of complicating symptoms	A significant drop in evidence of virus growth
		Acyclovir	Experimental	Transient renal impairment	Evidence for interference of virus assembly
Chorio-retinitis	Cytomegalovirus	Adenine arabinoside	Experimental		Promising
CMV infection in the immunocompromised	Cytomegalovirus	Acyclovir	Experimental		Promising
Subacute sclerosing panencephalitis (SSPE)	Paramyxovirus, (? measles virus)	Inosiplex	Experimental	None	Promising

[a] Approved by FDA with trade name in parentheses.

Prophylaxis. It is now clear that amantadine can be used clinically in several ways. First, it may be administered to protect nonimmune persons during community outbreaks of influenza A virus infection. Amantadine is advocated particularly for patients known to be immunocompromised, the elderly, persons with chronic obstructive pulmonary disease, individuals with degenerative diseases including congestive heart failure, and patients allergic to egg products contained in influenza vaccine. Given twice daily in 100-mg doses for 10 days, amantadine has been found to be remarkably effective in preventing evidence of infection. Because amantadine does not interfere with active immunization, it may be given along with influenza virus vaccine. When administered in this fashion to patients with intact immune systems, amantadine may not be needed on subsequent exposure to influenza A virus during the same epidemic season. Persistence of immunity in the face of subsequent outbreaks in succeeding years, it must be remembered, is unpredictable because of the problem of antigenic "drift and/or shift" that may appear with each new epidemic strain of the influenza A virus. Furthermore, the killed virus antigen used in most influenza vaccines fails to induce long-lasting, protective levels of circulating and secretory antibody. At this time, it is not clear whether the newer viral component vaccines will overcome this deficiency and thereby result in longer-lasting levels of protective immunoglobulin.

Treatment. Amantadine was shown to modify influenza A virus infections unequivocally when used within the first 24 to 48 hours after symptoms emerge. The drug should be given for about 5 days in this situation and at the same dosage—200 mg daily—or until fever and symptoms have subsided. Recent evidence indicates that this maneuver will not only modify the febrile course of influenza but also reduce, for many patients, the hazard of developing serious lower respiratory system infection due to secondary bacterial invasion.

RIMANTADINE

Rimantadine, a compound closely related to amantadine, is undergoing clinical trials but has yet to be released by the Food and Drug Administration for general use. This substance is not only effective against influenza A virus infections but has promise of efficacy against influenza B virus infection as well. A recent comparative study of amantadine and rimantadine was a placebo-controlled, double-blind, randomized trial in a young adult population. Both drugs given 100 mg twice daily for 6 weeks were clearly effective in preventing not only influenzalike illnesses but also laboratory evidence of influenza A virus infection. Side effects (chiefly insomnia, increased nervousness, and inability to concentrate) were encountered in 13% of the amantadine recipients compared with 6% of the rimantadine recipients. The side effects lasted no more than 48 hours. The CNS side effects of rimantadine recipients were no greater than those of placebo recipients. Thus, rimantadine may eventually prove to be the drug of choice for prophylaxis of influenza

A infections in young adults. It is not known at this time, however, if a similar low rate of side effects will be encountered in the elderly or other individuals at high risk for influenza A virus infections. Further trials on a broader population base must be carried out with both drugs as prophylaxis and for treatment of the earliest stages of influenzalike illnesses in both young and older patients.

RIBAVIRIN

A similar compound, ribavirin, when administered by small-particle aerosol, has been found to be effective in modifying severe influenza infections in humans, especially influenzal pneumonia, and warrants further clinical evaluation. Ribavirin remains in the investigational category and has not been released for clinical use.

Fortunately, all strains of influenza A virus have been found to be susceptible to these drugs and their effectiveness has not changed despite antigenic drift or shift of viral strains recovered from year to year.

Rhinovirus

Many forms of acute respiratory infection, including common colds, are caused by rhinoviruses. Rhinovirus infections are usually mild and self-limiting but may be associated with relapses of chronic bronchitis in adults and wheezy bronchitis in children. Vaccination offers little prospect for the control of rhinovirus infection because there are about 120 serotypes. Several compounds active against rhinoviruses in vitro including interferon inducers have been tested in volunteers, but only two have proved of clinical value.

ENVIROXIME

The compound 2-amino-1-(isopropyl sulphonyl)-6 benzimidazole phenyl ketone oxime (enviroxime) has a high therapeutic ratio in tissue culture, appears to be virus-specific, and is highly active against rhinovirus replication in organ cultures of human embryonic nasal epithelium. In a double-blind, placebo-controlled study of volunteers inoculated with human rhinovirus type 9, treatment with enviroxime was given by nasal spray and orally both before and for 5 days after virus challenge. The clinical score of symptoms and the volume of nasal secretions were significantly less in volunteers treated with the drug. Unfortunately, the spray produced significant nasal discomfort in the form of stinging and the oral form caused nausea and vomiting presumed to be of central origin. Usage in the general population has not been approved.

INTERFERON

Interferon has been used in several studies in attempts to prevent rhinoviral infection. Partially purified human leukocyte interferon (HUIFN-α) was used to protect human volunteers against infection with human rhinoviruses (HRV) types 4, 9, or 13. Unfortunately, the limited supply of HUIFN-α prohibits widespread use for persons at special risk of infection during outbreaks of disease. Recently, human interferon 2 (HUINF-α2), derived by recombinant DNA techniques, has been evaluated in volunteers challenged with HRV-9. The results from a placebo-controlled but small study indicated that HUINF-α2 prevented development of clinical colds and also sharply limited virus production as measured in nasal secretions. Side effects from HUINF-α2 were "trivial" when it was sprayed intranasally three times daily for 4 days, even though the dose of interferon used was about six times higher that that used in previous successful trials. As a consequence of these promising results, the authors suggested that HUINF-α2 may prove a safe and cheap way to prevent colds in humans and should therefore be studied further to determine the minimum dose needed, the best schedule of administration, and the full spectrum of side effects, if any.

Varicella-Zoster Virus

Chickenpox (varicella) and shingles (herpes zoster) are now known to be caused by the same agent, namely, the varicella-zoster virus (VZV). This virus usually causes a relatively benign, primary eruptive infection in childhood that is of limited duration although it may reemerge in adulthood as shingles, frequently at a time of reduced or compromised immunity. Unfortunately, there is no easy way to test for previous infection, that is, for the existence of circulating antibody or intact cell-mediated immunity in a child or an adult with a history of recent exposure to an infectious case of VZV.

ADENINE ARABINOSIDE

The first agent to show convincing evidence for control of VZV infections was adenine arabinoside (araA) after disappointing earlier clinical trials with other purine analogs, namely, cytosine arabinoside, and idoxuridine. It has been shown that the lesions of herpes zoster in immunosuppressed patients were significantly ameliorated when araA was given within 6 days of onset of symptoms. The drug was administered intravenously over a 12-hour interval at a dose of 10 mg/kg/day in concentrations not in excess of 0.7 mg/ml saline. Because most adults require as much as 500 ml solution per day, hospitalization is essential. Well-tolerated with low levels of toxicity, the drug was most effective in patients under 38 years of age with reticuloendothelial neoplasms, persons at greatest risk for serious complications of VZV infection. The drug seemed to hasten clearance of localized lesions and, to a lesser extent, speed the healing of disseminated lesions. Unfortunately, no clear evidence for a beneficial effect could be found for the prevention of disseminated infection or for the elimination of the late complication, postherpetic neuralgia.

LEUKOCYTE INTERFERON

Interferon has been reported to be of value for treatment of VZV infections. The rationale for its use is supported by evidence that treatment of malignant disease with either radiation or chemotherapy depresses both cell-mediated immunity and interferon production. This suppression of specific antiviral responses seems to be the most important predisposing factor for recurrence of VZV infections specific in lymphoma patients. High doses of purified interferon have prevented the occurrence and produced a decrease in cutaneous spread of herpes zoster lesions. High-dose interferon treatment reduced new vesicle formation in the primary dermatome, acute pain with vesicular lesions, the severity of postherpetic neuralgia, and visceral complications. Treatment with interferon, particularly the higher dose (3.5×10^5 units/kg/day for 72 hours), has also been found to significantly reduce the appearance of new lesions and morbidity from visceral dissemination in children with cancer. It appears that antiviral therapy should be continued for 5 days more in patients with persistent or recurrent lesions or viremia. Although these results seem to be comparable to the effect of araA given intravenously to hospitalized immunocompromised cancer patients with varicella, patients can be treated with interferon on an outpatient basis, thereby appreciably reducing the risk of nosocomial infection. Side effects of leukocyte interferon consisted chiefly of febrile reactions after administration, with little or no evidence of disturbance in liver function.

ACYCLOVIR

The nucleoside, acyclovir (ACV), a guanine derivative [9-(2 hydroxyethoxymethyl)-guanine], is selectively phosphorylated in HSV-infected cells by the virus-specific thymidine kinase (TK). Subsequent phosphorylation to the triphosphate derivative specifically inhibits the DNA polymerase of herpes simplex virus. Fortunately, because the host-specific TK cannot phosphorylate ACV, uninfected mammalian cells in culture tolerate concentrations of ACV several hundredfold higher before DNA synthesis is significantly inhibited. ACV seems to shorten the duration of fever, reduce acute pain, shorten the period of viral shedding, and cause a resolution of skin lesions, especially in patients treated early in the disease. However, no demonstrable benefit in preventing the occurrence of postherpetic pain has been noted with this drug. No serious side effects have been encountered in liver, hematopoietic, and renal function by administration of the drug as long as patients have remained well hydrated.

Herpes Simplex Virus

Interest in the natural history and management of infections in humans caused by herpes simplex virus (HSV) types 1 and 2 has grown

**TABLE 14.3   Compounds with Antiviral
Activity Against HSV1[a]**

Compound

9-(2-Hydroxyethoxymethyl) adenine
Vidarabine
9-(2-Hydroxyethoxymethyl)-2, 6-diaminopurine
Trifluorothymidine
Idoxuridine
Acyclovir

[a]Drugs in decreasing order of effectiveness against
HSV1 in vitro.

greatly in the 1980s, particularly in view of the suspected oncogenic potential of these organisms in human cancer. As a result of an increased
awareness of the varied manifestations that can be seen in either primary or recurrent infections in immunocompromised as well as normal
patients, substantial improvements in specific treatment have been made.
Listed in Table 14.3 are sensitivities of HSV1 to drugs now being used.
Other drugs, thymidine kinase inhibitors, are being evaluated, as noted
in Table 14.1.

CONGENITAL AND NEONATAL INFECTIONS
   Congenital HSV infections acquired in utero are rare and are associated with significant birth defects. By contrast, neonatal infection, acquired during vaginal delivery of women with genital herpes, has an
attack rate of about 50%. If it is untreated, the overall mortality is about
85%, with better prognosis in the localized form. About 80% of infections are in the localized form, but of these, as many as 70% may go on
to dissemination. Treatment with araA of neonates with localized skin
infections resulted in a decrease in overall mortality from 85% in the
placebo group to 57% in the treated recipients. A decrease in mortality
was seen most clearly in babies receiving the drug who had localized
involvement of the central nervous system, skin, eye, or mouth. Acyclovir is being evaluated, but results remain to be determined.

HERPES LABIALIS
   Long recognized as a common, relatively benign complication of other
illnesses, exposure to sunlight, stress, or menstruation, "fever blisters"
caused by primary or reactivated HSV1 or HSV2 have been treated with
a variety of empiric forms of therapy. Enthusiastic and usually uncontrolled anecdotal modes of treatment advocated have included such
things as large doses of vitamin C, a variety of topical agents including
vital dyes, some combined with ultraviolet irradiation, herpes virus vac-

cines, and even repeated administration of cowpox (vaccinia) vaccine. None has proved successful.

Trisodium Phosphornate. A study using trisodium phosphornate in a 3% cream in a double-blind, placebo-controlled schedule suggested an antiviral effect in patients with recurrent herpes labialis. Animal studies indicated substantial accumulation in bone, a finding of unknown significance. Mild local irritation was encountered and no data on systemic side effects were available.

Acyclovir. Topical 5% acyclovir (ACV) in polyethylene glycol has been shown to have a significant antiviral effect when applied to early lesions and when applied frequently for 5 days. If treatment of lesions is delayed 9 hours or more, there is no difference between ACV and placebo. As with other topical forms of therapy, failure to show improvement may reflect poor penetration of the striatum corneum, a need for more frequent application as well as earlier initiation of therapy, or requirement of higher concentrations of the drug with a change in the vehicle to increase flux through the skin.

OCULAR INFECTIONS

Herpetic keratoconjunctivitis was the first virus infection in humans to show convincing evidence of response to a pharmacologic agent, 5-iodo-2'-deoxyuridine (IDU). Since then, a number of nucleoside analogs have been evaluated using IDU as the "gold standard." Both 2% trifluorothymidine (TFT) and 3% acyclovir (ACV) have been found to reduce viral shedding and promote corneal healing in 95% of patients; this is better than the 76% rate with IDU. Use of 3% araA ointment produced healing rates falling between ACV and TFT. All three produce toxic side effects of conjuctival hyperemia, edema, punctal occlusion, and superficial punctate keratopathy. In addition, collagen content and stromal wound healing may be reduced. None of the three drugs administered locally prevents superficial or deep stromal keratitis and, perhaps of even greater importance, they do not prevent HSV from reaching the nerve ganglia of the eye. As a direct consequence, the ganglia become reservoirs from which latent HSV reinfects the cornea and conjuctiva and causes chronic as well as destructive changes. Recovery of latent HSV from ganglia has been reduced in animals by systemic administration of ACV. In view of minimal toxicity encountered in humans with parenteral administration of ACV, intravenous administration must be examined as a means of treatment not only to control active infection but also to block or eliminate the latency state.

GENITAL INFECTIONS

HSV2 causes 80 to 90% of male and female genital viral infections in the United States and HSV1 causes the remainder. Primary infections in men tend to be milder, especially if antibody is present, whereas recurrent infections seem to be shorter, less severe, and less likely to be associated with systemic involvement in either sex. However, symptoms are likely to be more severe in women. Primary genital infection with HSV1 is less likely to be followed by recurrence than primary infection with HSV2. Although women may have less recurrent HSV disease than men, the severity of the primary infection with antibody production may in part determine the risk of developing recurrent disease. It is not clear what role, if any, antibody, viral latency with reactivation, or reinfection plays in recurrence of the process.

Idoxuridine. There is conflicting evidence regarding the effectiveness of IDU in the treatment of either primary or recurrent HSV genital infection in men or women. Concentrations of IDU (0.1%) found to promote healing in patients with keratitis failed to control mucocutaneous infections and led to studies using substantially greater amounts (5 to 30%) suspended in vehicles that enhanced tissue penetration. Although earlier authors reported favorable responses to treatment with 5 to 20% concentrations of IDU applied locally, subsequent studies utilizing dimethyl sulfoxide (DMSO) to promote penetration have yielded disappointing results. When topically applied, 30% IDU in DMSO shortened the duration of viral shedding in both primary and recurrent HSV infection compared with 30% IDU in saline. Nonetheless, 30% IDU in DMSO had no effect on duration of symptoms, new lesion formation, healing time, or primary or recurrent infection. Side effects have included local burning and generalized contact dermatitis. One patient with a total of seven recurrences of HSV developed vulvar carcinoma in situ 3 months after the sixth recurrence had been treated with a 7-day course of 30% IDU in DMSO. Until a more effective route or means of administration can be identified, IDU should not be recommended for management of either primary or recurrent genital HSV infection.

Acyclovir. Intravenous ACV decreases the median time required for healing, duration of vesicles, new lesion formulation, viral shedding, and all other symptoms associated with infection. In primary infections, topical 5% ACV shortened viral shedding and time required for crusting over of lesions. The 5% ointment is marketed for use under the trade name Zovirax. An oral preparation of Zovirax is now available for treatment of all forms of cutaneous and mucocutaneous herpes virus infections.

CENTRAL NERVOUS SYSTEM INFECTIONS

Convincing evidence is now in hand to show that herpetic encephalitis, a life-threatening infection of the central nervous system, will respond to antiviral therapy. Herpetic encephalitis is the most common cause of fatal, sporadic encephalitis in the United States. Fortunately, it causes no more than 2 to 4% of all cases of viral encephalitis. Even so, a 70% mortality may be expected if the infection is untreated. Thus, herpetic encephalitis represents not only an extremely lethal but a damaging form of viral infection and, because of the serious neurologic residua found after nonfatal infection, an appropriate form of therapy is highly desirable.

Because success in treatment depends heavily on early intervention, laboratory results are vital for establishing the correct diagnosis. Unfortunately, spinal fluid obtained by lumbar puncture is usually not diagnostic because it may have the same cellular and chemical characteristics as are found in other aseptic meningitides. Flow and computerized tomography (CT) scans may show evidence of a mass effect in a segregated portion of brain, which, although highly suggestive of HSV infection, may not be different from brain abscess. Thus, obtaining a biopsy specimen from the brain has been advocated to expedite reaching an early and specific diagnosis.

Adenine arabinoside. Treatment with araA has substantially reduced morbidity and mortality of patients proved to have herpetic encephalitis. Patients under 30 years of age generally have a better prognosis with less residual damage. Mortality after treatment with a placebo, cytosine arabinoside, or idoxuridine has been reported to be 70%, whereas mortality after treatment with araA was only 40%, a highly significant difference. These significant figures serve as a sound basis for justification of both brain biopsy and treatment. AraA must be administered intravenously. When biopsy results are negative, treatment with araA should be discontinued after the fifth day. In this way, complications of araA therapy that affect from 10 to 15% of the patients treated may be minimized. To reduce problems arising from araA treatment, brain biopsy can be justified on the basis of the low rate of complications of the procedure.

Acyclovir. Clinical trials with ACV for the treatment of herpetic encephalitis are underway and may prove it to be a less-toxic drug. Whether brain biopsy will continue to be a mandatory step for establishing the correct diagnosis remains to be seen as new, nontoxic, antiviral agents become available and better diagnostic measures are devised.

INFECTIONS IN IMMUNOCOMPROMISED PATIENTS

Reports of immunocompromised recipients of bone marrow transplants indicate that ACV can be used not only prophylactically to prevent reactivation of latent herpes but also for treatment of patients with chronic mucocutaneous HSV infections. After intravenous administration, herpetic lesions become negative for virus isolation more rapidly on ACV therapy, and the lesions heal more quickly and there is less pain. Once treatment with ACV was completed, recurrence of viral shedding was commonly found and exacerbation of infection occurred until immunocompetency was restored. HSV resistant to ACV has been reported, but only after completion of therapy, and without any change in the total spectrum of disease. The significance of this event is not yet clear.

RESISTANCE OF HSV

In view of the preceding, it comes as no surprise to learn that drug-resistant strains of HSV have been isolated from bone marrow transplant patients treated with ACV. The most common mechanism for resistance appears to be a deficiency in thymidine kinase, as shown by isolation of thymidine-kinase deficient mutants within a week after termination of therapy. Even so, lesions continued to heal despite continued shedding of resistant virus. It remains to be seen whether resistant HSV associated with ACV and similar antiviral drugs will cause either more or less virulent disease and thereby dictate modification of treatment.

Hepatitis B Virus

Human infection with hepatitis B virus (HBV) usually causes a relatively benign, acute, self-limited disease. On occasion, however, fatal fulminant hepatic necrosis may occur, or persistent chronic infection may cause prolonged morbidity, progressive, and irreversible damage to the liver that may eventually prove fatal in the form of postnecrotic cirrhosis or increase the risk of prolonged infectivity with persistent viremia or the development of hepatocellular carcinoma. Recently it was shown that most patients with benign, self-limited acute viral hepatitis produce increased amounts of interferon, whereas, patients with fulminant hepatitis fail to produce circulating interferon. Thus, particular interest has been directed to the treatment of persistent hepatitis and chronic active (aggressive) hepatitis with leukocyte interferon alone or in combination with araA, because such patients are most likely to progress to fatal postnecrotic cirrhosis and a few may develop hepatocellular carcinoma.

INTERFERON AND araA

Results of continued interferon and araA treatment of patients with chronic HBV hepatitis have been mixed. Some patients respond with permanent loss of DNA polymerase activity (a sensitive index of viral replication), HB_eAg, and HB_sAg from serum. Other patients respond with permanent loss of DNA polymerase activity but with lower levels of persistent HB_sAg in serum, with HB_eAg becoming undetectable over a varying interval. Finally, a third group of patients respond with transient decreases in DNA polymerase activity, persistent HB_eAg, and unchanged levels of HB_sAg. Generally, patients treated with either agent alone do not seem to respond as well.

Patients with chronic active hepatitis and lower initial DNA polymerase levels responded better than other patients. There was a strong association between a successful response to therapy and a rise in serum aspartate aminotransferase values during treatment. This rise in enzyme level may indicate host immune responses working in conjunction with the antiviral agents. In general, patients with chronic active hepatitis, women patients, and patients with a history of recent steroid therapy seemed to respond to the antiviral agents better.

Widespread use is not warranted in view of the side effects, including fever, general malaise, decrease in neutrophil platelet counts, and gastrointestinal and neurologic side effects.

ACYCLOVIR

Preliminary clinical trials with ACV in patients with chronic liver disease due to HBV infection suggest that the drug interferes with complete viral particle assembly in some way as yet unknown. At present, nothing can be said about its effect on the disease itself. However, side effects noted include a transient impairment in renal function when the drug is used in high doses.

Other Viral Diseases

CYTOMEGALOVIRUS VIRAL INFECTIONS

Chorioretinitis–araA. Chorioretinitis due to cytomegalovirus (CMV) infections has sufficiently distinct characteristics to suggest the diagnosis to a skilled observer. Studies performed in the 1970s indicated at least a partial response of CMV ocular syndromes to araA, but later evaluation failed to support further trials of the substance.

Immunocompromised patients. Infections in neonates and immunocompromised adults by CMV results in serious and often fatal disease that has not responded to treatment with either leukocyte interferon, araA, or combinations of both drugs. Although no resistance to ACV in vitro has been identified in CMV isolates (CMV does not encode for a viral thymidine kinase but the CMV-DNA polymerase is sensitive to ACV), excretion of virus persisted and viremia recurred in about 20% of immunocompromised patients after treatment stopped. This demonstration of relative resistance of CMV to ACV is probably the most significant fact to come out of these studies.

EPSTEIN-BARR INFECTIONS

Treatment of EBV infection has been reported in only a few patients. From this limited experience, it was thought that ACV might be effective only on those B lymphocytes with a productive EB virus infection, a situation in which virus titers are high and also, perhaps, immunologic integrity required for recovery is intact.

SUBACUTE SCLEROSING PANENCEPHALITIS

Subacute sclerosing panencephalitis (SSPE) is an uncommon, progressive CNS disease of childhood through young adulthood. It is thought to be caused by a paramyxovirus closely related to rubeola, although the pathogenesis is not clear, despite the isolation of virus from

CNS cells and persistence of high levels of serum and cerebrospinal fluid antibody to measles virus. Since immunologic mechanisms have also been considered central to the pathogenesis of the disease, either antiviral substances or drugs to moderate the immune mechanism have been tested. Of these, trials of levamisole and transfer factor to alter the immune response have failed to provide convincing evidence of benefit. Likewise, several studies of the antiviral drug amantadine have not clearly moderated the unpredictable course of SSPE. On the other hand, inosiplex (Table 13.1), a drug with both direct antiviral and immunomodulatory properties, has been used with some evidence of success.

Inosiplex. A multicenter study of this drug in patients with SSPE suggested that inosiplex prolonged life significantly. Unfortunately, the patient groups from the United States and Canada were not randomized, and the control data were derived from patients with SSPE who were either untreated or received other forms of treatment. Although the survival rate of inosiplex-treated patients was better than that of untreated patients, the methods of analysis may be open to challenge because of the unavoidable introduction of bias in the control groups. Other than difficulties encountered in administration, no serious complications from treatment were recorded.

Recommended Readings

Bauer DJ: Clinical experience with the antiviral drug Marboran (1-methysatin 3-thiosem-icarbazone). *Ann NY Acad Sci* 130:110–117, 1965.

Becker Y (ed): *Antiviral Drugs and Interferon.* Hingham, MA, Martinus Nijhoff Publishing, 1984.

Dolin R, Reichman RC, Madore P, Maynard R, Linton PM, Weber-Jones J: A controlled trial of amantadine and rimantadine in the prophylaxis of influenza A infection. *N Engl J Med* 307:580–584, 1982.

Dyken PR, Swift A, DuRant RH: Long-term follow-up of patients with subacute sclerosing panencephalitis treated with inosiplex. *Ann Neurol* 11:359–364, 1982.

Kaufman HE: Clinical cure of herpes simplex keratitis by 5-iodo-2'-deoxyuridine. *Proc Soc Exp Biol Med* 109:251–252, 1962.

Stiehm ER, Kroneuberg LH, Rosenblatt HM, Bryson Y, Merigan TC: Interferon: Immunobiology and clinical significance. *Ann Intern Med* 96:80–93, 1982.

Whitley RJ, Soong SJ, Hirsch MS et al: Herpes simplex encephalitis: Vidarabine therapy and diagnostic problems. *N Engl J Med* 305:313–318, 1981.

15

Immunization Against Viral Agents

JAMES D. CHERRY

HISTORY

Aside from social progress, such as the development of sewer systems and clean water supplies, the most cost-effective gains in the prevention of morbidity from infectious diseases have been through the development and use of vaccines. In particular, vaccines against viral agents have had a profound effect on the prevalence and epidemiology of selected diseases.

Although one commonly associates immunization against viral diseases historically with the latter half of the twentieth century, its history actually spans about 15 centuries. The earliest recorded forms of preventive medicine involved attempts to prevent death and disfigurement from smallpox. Inoculation (variolation) with variola virus is recorded in Chinese history during the Sung Dynasty (A.D. 590). Inoculation was also practiced in India and Africa in ancient times.

Inoculation to prevent smallpox was not used in Europe until early in the eighteenth century, when, in 1721, a leading British intellectual, Lady Mary Wortley Montague introduced it to the British Royal family after observing its practice in Turkey. Vaccination as used today has a history of almost 200 years. In 1796, Edward Jenner took material from a cowpox sore on the hand of a milkmaid and inoculated it into the arm of an 8-year-old boy. Despite the considerable controversy that surrounded smallpox vaccination for several years, the procedure eventually gained worldwide use, leading ultimately to the eradication of the disease from the planet in October 1979.

Although worldwide eradication of other viral diseases for which effective vaccines are available is not now in sight, the reduction of morbidity and mortality in countries where vaccine use has been widespread has been truly remarkable. In 1981, in the United States, only

3032 cases of measles and 6 cases of paralytic poliomyelitis occurred; in contrast, before widespread immunization about 425,000 cases of measles and 40,000 cases of paralytic poliomyelitis occurred yearly in the United States.

PROPERTIES OF VIRAL VACCINES

The ideal viral vaccine is one that would produce lifelong protection from disease on exposure and be free of both initial and long-term side effects. Such a vaccine does not exist. Desirable characteristics of vaccines include the following:

- The ability to induce protective humoral antibodies, local secretory IgA antibody, and cell-mediated immunity.
- Freedom from troublesome side reactions; no back-mutation and reversion to virulence, with a live-virus preparation.
- In most instances, no spread of live-virus vaccine from vaccinees to others.
- No production of a latent infection with the possibility of later oncogenicity.
- No content of adventitious infectious agents.
- Simple mode of administration acceptable to both the public and the health care provider.
- Justifiability on the basis of a cost-benefit analysis relating to the cost of the vaccine and its administration compared to the cost of disease in terms of morbidity, mortality, and emotional impact.

Viral vaccines can be divided into two major classifications: (1) live-virus (replicating, attenuated) and (2) killed-virus (nonreplicating, inactivated). Killed-virus vaccines can be subdivided into whole-virus vaccines or subunit and split-virus preparations. Table 15.1 lists the viral vaccines available in the United States by type.

The route of vaccine administration depends on its type and, to some degree, the immune response desired. In general, the ideal vaccine would be a live-virus product administered by the route of natural infection; its multiplication within the body would be similar to that occurring during clinical infection with wild-type virus. Oral poliovirus vaccine and yellow fever virus vaccine are examples of vaccines given by routes of natural infection. The greatest advantage of a vaccine given by the natural route is that it allows an immune response to occur at the initial site (i.e., IgA secretory antibody in the gastrointestinal tract in oral poliovirus vaccine recipients).

Unfortunately, the process of attenuation has diminished the ability of some live-virus vaccines to cause infection by the normal route. With measles, mumps, and rubella vaccines, subcutaneous administration is necessary. With these three vaccines, parenteral administration results

TABLE 15.1   Viral Vaccines Available in the United States, 1982

Live Virus (Replicating; Attenuated)	Killed Virus (Nonreplicating; Inactivated)	
	Whole	Subunit or Split
Adenovirus types 4 and 7[a]	Influenza	Hepatitis B
Measles	Poliovirus	Influenza
Mumps	Rabies	
Poliovirus		
Rubella		
Smallpox		
Yellow fever		

[a]For use in the military.

in viremia, multiplication of virus at distant sites in the body, and, in some instances, the production of local secretory antibody as well as generalized humoral antibody.

All presently available killed-virus vaccines are administered parenterally.

PRINCIPLES OF IMMUNIZATION

Target Groups

Selection of the target group to be vaccinated is contingent on the size and composition of the susceptible population at risk, the prevalence of the infectious agent, and the biologic behavior of the vaccine in relation to certain host factors, such as age. For example, yellow fever vaccine is not used routinely in the United States because the disease is not an endemic threat. The target group for this vaccine comprises mainly travelers to endemic areas. Similarly, rabies vaccine is used in the preexposure setting only among selected persons such as veterinarians and certain laboratory workers. Its chief application is in postexposure prophylaxis.

The target group of measles, mumps, rubella, and polio vaccines is young children, because the attack rates for the diseases are greatest in children. Influenza is similar to the above-mentioned viruses in that the attack rate is also greatest in children; logically, therefore, children should be the major target group for immunization. However, because immunity from influenza vaccines is short-lived and repeated immunizations are necessary, the present target for influenza vaccination comprises the older age groups who are at the greatest risk of having complications from, or dying of, influenza.

Herd Immunity

The concept of "herd immunity" is a useful and valid indicator of population resistance to epidemic disease. Unfortunately, in recent years, considerable misunderstanding relating to herd immunity has led many persons to question the efficacy of some vaccines and the strategies of specific vaccine programs. Herd immunity occurs in a population (herd) when the number of persons immune (resistant) to a given infectious agent is so great that epidemic spread of the disease will not occur if an animal or human with the contagious disease is introduced to the herd.

With many animal diseases, under experimental conditions, when 80% or more of the herd is immune, spread of the disease is clearly curtailed. However, factors other than the percentage of immunes in a population affect herd immunity. Obviously the specific degree of contagiousness of the viral agent is important, as well as the mobility of persons in the group. In a hypothetical situation, a particular agent might not spread at all in a group in which only 70% of the individuals were immune if all the animals were immobilized. On the other hand, even if 95% of the animals were immune, the disease might still spread among the few remaining susceptibles if the animals had extensive contact with each other.

The chief misconception with regard to herd immunity in human diseases comes from disregarding the fact that persons are not restricted to one herd. The human population comprises many subpopulation groups that gather in many different clusters throughout a given period. Therefore, although a survey of immune status might show that 90% of a given population is immune to measles, when that immunity level is analyzed by age, all of those less than 15 years old and those more than 30 years old may be immune but only 80% of the population 16 to 29 years of age may be immune.

In the United States today, immunity in one herd does not generally prevent epidemic disease in another herd, because the interaction within and between herds is so great.

Risks and Benefits

Because no vaccine is absolutely safe, the risk-benefit ratios for all viral vaccines must be reassessed at intervals. Early in the twentieth century in the United States, for example, it was obvious that the benefit of immunization against smallpox (by vaccination with vaccinia virus) far outweighed its risks. By 1949, however, smallpox had been eliminated from this country. For the next two decades, although no mortality or morbidity from the disease occurred, serious reactions (generalized vaccinia, postvaccinal encephalitis) and death associated with vaccination were still occasionally reported. By 1970, the risk of immunization was clearly greater than the risk of disease, and routine immunization against smallpox in the United States was therefore discontinued.

Added to the success of the smallpox vaccine has been the remarkable success in controlling both paralytic poliomyelitis and measles in the United States. In contrast to smallpox, however, if we were now to discontinue use of either polio or measles vaccines, the epidemic diseases would soon reappear. Unfortunately, because of vaccine successes in eradicating epidemic disease, excessive attention is being focused on risks of vaccine use rather than on risk-benefit ratios. Many current and recent lawsuits have been filed because of damage from polio vaccine; frequently these suits are settled not on the basis of public health good and highly favorable benefit-risk ratios but of the failure to warn the public about the risks. This trend, if it continues, will most certainly result in a decreased use of effective vaccines, and may even lead to the discontinuance of vaccine manufacture.

Informed Consent

Physician use of most viral vaccines up until recent years has been controlled to a great degree by strong public health principles and a considerable degree of paternalism. Public health laws have in many instances dictated who should be immunized, and potential vaccinees or their parents have expected the health care provider to make judgments regarding risks and benefits and carry out the necessary immunizations. At the present time the overwhelming majority of consumers still expect the system to function in the same way. However, in the current era of consumer awareness, there has been considerable effort by a minority of persons to insist on participation in making decisions related to individual immunizations. As a result of this pressure, the official position of the American Academy of Pediatrics and the United States Public Health Service since about 1977 has been that informed consent be obtained prior to administration of all vaccines. Unfortunately, to truly elicit informed consent takes considerable time and therefore most health providers are still doing what is sound public health practice and what is good for the patient and only providing vaccine information to patients if it is specifically asked for. However, the present litigious climate is such that a change is necessary. Informed consent must clearly not give undue emphasis to vaccine reactions otherwise vaccine utilization will decrease and epidemic disease will reoccur. At the present time information sheets (consent forms) are available and necessary for all vaccines supplied by the United States government.

PREVENTION OF SPECIFIC VIRAL DISEASES BY IMMUNIZATION

Poliomyelitis

BACKGROUND

Epidemic paralytic poliomyelitis is caused by poliovirus types 1, 2, or 3 and occasionally other enteroviruses. With the advent of tissue culture techniques in the late 1940s and the demonstration by Enders, Weller, and Robbins that poliovirus would grow in non-nervous tissue, the way was paved for the development of both killed and live poliovirus vaccines. All three types of poliovirus grew readily and to high titer in monkey kidney tissue culture. Inactivated poliovirus vaccine was prepared by growing poliovirus in tissue culture, harvesting the virus-containing tissue culture fluids, inactivating the virus with formalin, and, after clarification filtration, combining the three types of virus in a single preparation. After extensive testing, inactivated polio vaccine was licensed for use in the United States in 1957.

Oral, live poliovirus vaccines were developed from virus grown in monkey kidney tissue culture. Attenuated strains were selected on the basis of their lack of neurovirulence when injected intraspinally or intracerebrally into monkeys. Initial vaccines were monovalent, but investigators soon found that by careful attention to specific concentrations of each virus type, effective trivalent vaccines could be prepared and administered together in one dose. Oral poliovirus vaccines were first licensed for use in the United States in 1961 and 1962, and mass vaccination campaigns were carried out in 1963. For the last 20 years, live-virus vaccine has been used for almost all poliovirus immunization in the United States.

In Figure 4.3 shown in Chapter 4, the incidence of paralytic poliomyelitis in the United States is plotted. Before the availability of poliovirus vaccine, the incidence of paralytic poliomyelitis was about 10 per 100,000 population. After the introduction of inactivated poliovirus vaccine, the incidence fell about 20-fold. Since the introduction and widespread use of oral poliovirus vaccine, the incidence of paralytic poliomyelitis had a further decrease from 0.5 per 100,000 to about 0.005 cases per 100,000 population.

INACTIVATED POLIOVIRUS VACCINE (IPV)

General. When the recommended schedule is followed (Table 15.2), IPV stimulates high levels of humoral antibody without the development of secretory IgA antibody. Protection with IPV is due to circulating antibodies that prevent viremia and consequent spread to the central nervous system. Infection of the gastrointestinal tract is not prevented.

TABLE 15.2   A Schedule for Active Immunization of Children and Adults Living in the United States Not Initially Immunized at Recommended Times in Early Infancy Against Selected Viral Diseases

Category	Timing	Vaccine	Comments
Children	1st visit	OPV[a]	If there is a threat of measles, mumps, or rubella, then MMR should be given on 1st visit and schedule adjusted so that OPV doses are 2 months apart
	1 month after 1st visit	MMR[b]	MMR should usually not be given before age 15 months
	2 months after 1st visit	OPV	
	4 months after 1st visit	(OPV)	This dose suggested in endemic areas and in areas where polio might be imported
	18 months after 1st visit	OPV	
	2–4 years after 1st visit	OPV	
Adults ≥ 18 years of age	1st visit	IPV[c]	If threat of measles, mumps, or rubella, then MMR should be given on 1st visit
	1 month after 1st visit	MMR	
	2 months after 1st visit	IPV	
	3–4 months after 1st visit	IPV	
	10–14 months after 1st visit	IPV	
	Booster doses every 5 years	IPV	

[a] Oral, attenuated poliovirus vaccine (contains poliovirus types 1, 2, and 3). See Table 14.3 for contraindications.

[b] Live measles, mumps, and rubella viruses in a combined vaccine. See manufacturer's literature for contraindications.

[c] Inactivated polio vaccine (contains poliovirus types 1, 2, and 3).

In countries in which immunization is virtually universal, IPV vaccine efficacy is close to 100%. In Sweden, for example, where only IPV has been used, the control of poliomyelitis has been similar to that in the United States in which oral poliovirus vaccine has been the immunizing agent of choice. Although the use of IPV theoretically allows the

TABLE 15.3   Limitation on the Use of Oral Poliovirus Vaccine

CONTRAINDICATIONS—OPV IS CONTRAINDICATED

In persons with immune deficiency states such as combined immunodeficiency, hypogammaglobulinemia, and agammaglobulinemia; thymic abnormalities; leukemia,
lymphoma, or other generalized malignancy; and lowered resistance due to therapy
with corticosteroids, alkylating drugs, antimetabolites, or radiation

In families with immunodeficient members

PRECAUTIONS—OPV SHOULD USUALLY NOT BE GIVEN

In pregnancy
To persons 18 years of age or older
Shortly after the administration of immune serum globulin
During a febrile illness

---

spread of wild polioviruses in the population by gastrointestinal infection, in fact, this does not seem to occur. In Sweden and the Nether-
lands, where IPV is also routinely used, the circulation of wild poliovirus strains has decreased considerably.

Adverse Reactions.   Over the last two decades, IPV has been an extremely safe vaccine. Possible side effects include allergic reactions in
persons sensitive to antibiotics used in cell culture or perhaps also to
residual monkey tissue antigens in the vaccine. Guillain-Barré syndrome and other neurologic vaccine-related illnesses are also remote
possible complications.

Use.   IPV is not recommended for routine use in the United States
at present. However, it is clearly indicated for use in persons in whom
oral poliovirus vaccine is contraindicated and is usually indicated when
oral poliovirus vaccine precautions exist. Contraindications and precautions of oral poliovirus vaccine are listed in Table 15.3. IPV should also
be offered to persons who have refused oral poliovirus vaccine for personal reasons.

LIVE, ORAL POLIOVIRUS VACCINE (OPV)

General.   OPV is the vaccine currently used for routine immunization of children in the United States and in most other countries
throughout the world. Oral administration of vaccine stimulates both a
local (gastrointestinal tract) IgA immune response and a serum-
circulating-antibody response. Two doses of OPV produce an antibody
response to all three viral types in more than 90% of individuals to whom
it is administered.

Adverse Reactions.   Paralytic disease following OPV ingestion occurs
very rarely either in vaccinees or in persons who have had close contact
with vaccinees. The following rates of vaccine-associated paralytic dis-

TABLE 15.4 Schedule for Active Immunization against Viral Diseases in Normal Infants and Children in the United States

Age	Vaccine	Comments
2 months	OPV[a]	In areas of poliomyelitis endemicity, vaccine can be given at an earlier age
4 months	OPV	A 2-month interval between doses is desirable to avoid interference
6 months	(OPV)	This dose suggested in endemic areas and in areas where poliovirus might be imported
15 months	MMR[b]	
18 months	OPV	
4–6 years	OPV	

[a] Oral, attenuated poliovirus vaccine (contains poliovirus types 1, 2, and 3). See Table 15.3 for contraindications.

[b] Live measles, mumps, and rubella viruses in a combined vaccine. See manufacturer's literature for contraindications.

ease were compiled from 1969 to 1976 data: 1 case in a vaccine recipient (mostly children less than 10 years of age) per 11.5 million vaccinations; 1 case in a household contact (mostly parents 30 to 40 years of age) per 3.9 million vaccinations; and 1 case in a community contact (mostly playmates of vaccinees less than 10 years of age) per 22.9 million immunizations.

Use. Unless contraindicated (Table 15.3), OPV is the vaccine currently recommended in this country for the routine immunization of children. The schedule of immunization is presented in Tables 15.2 and 15.4.

PRESENT PROBLEMS RELATED TO POLIO IMMUNIZATION; OPV VERSUS IPV

In recent years, more cases of poliomyelitis in the United States have been associated with vaccine administration than natural wild-type viral infection. Because of this, some authorities have pushed to change the vaccine policy in the United States. The possibility of a change in policy has been carefully evaluated by health authorities, and the position that OPV is generally recommended for routine immunization in the United States has been reaffirmed. This decision is based on the extremely small risk associated with OPV, the increased expense associated with a vaccination program requiring booster doses throughout life, and, most important, the lack of a method to ensure continued booster immunization once school age is passed.

Susceptible Adults (Particularly Parents of Young Children). Because of the herd immunity due to the general success of our immunization program, a substantial number of young adults have grown up having had neither poliovirus immunization nor the development of immunity due to natural infection. These young adults are at definite risk to naturally acquired poliomyelitis if exposed and also at some risk on exposure to vaccine virus. Therefore, adults who anticipate travel to poliovirus-endemic areas and those working in laboratories with poliovirus should be immunized. Immunization should be carried out with IPV, as indicated in Table 15.2.

The problem of the unimmunized parent is more complex. Ideally, it would be best to immunize the adults in a household with IPV before starting an OPV regimen in a child. However, if such a procedure will decrease the likelihood of the child to receive OPV, the child should be immunized irrespective of the immune status of the adult family members.

Problems with OPV in the Tropics. In tropical and semitropical areas, seroconversion rates among vaccinated susceptible children are much lower than similar rates in the United States. Interference due to gastrointestinal tract infection with other enteroviruses is one cause of the decreased seroconversion rate, but apparently it is not the major cause. In recent field studies, an inhibitory substance that prevents viral multiplication was found in the throats of children. To increase seroconversion rates in children in tropical areas, the following may be useful: start immunization shortly after birth, increase the vaccine virus concentration by a factor of 10, and use monovalent rather than trivalent vaccines.

Measles

BACKGROUND

This disease is caused by the singular measles (rubeola) virus, the isolation of which in cell culture in 1954 set the stage for vaccine development. The virus was first adapted to growth in chicken embryos and then cultivated in chicken embryo tissue culture cells. After attenuation by chicken embryo tissue culture passage, extensive vaccine trials were done from 1958 through 1962.

In 1963, attenuated live measles virus (Edmonston B) vaccine was licensed for use in the United States. At the same time a killed measles virus vaccine was also licensed. The initial Edmonston B live measles vaccine was associated with a high rate of febrile reactions. Because of these reactions, this vaccine was usually given in association with a small dose of immune serum globulin. Because measles was a threat to all children older than 6 months, immunization was usually given at about 9 months of age and frequently as early as age 6 months. By 1964, the vaccine failure rate was unacceptably high among children immunized during the first year of life, and the recommendation was therefore changed to have vaccination done at about the first birthday.

Killed measles vaccine enjoyed only limited popularity, although it was preferred by some because it was not associated with febrile reactions. However, to elicit an antibody response with killed vaccine, multiple doses of vaccine were necessary. The protection resulting from killed-vaccine administration was found to be only short-lived; one year after immunization, exposed children would get a mild modified case of measles. Because of the short duration of immunity after killed-vaccine use, children were often given a dose of live vaccine after a killed-vaccine series. From 1963 through 1967, about 600,000 to 1 million children received one or more doses of killed vaccine.

In 1965 and 1967, two further attenuated strains of live measles vaccine became available, eliminating the considered need to administer immune serum globulin with measles vaccination.

Initial licensure of a vaccine in 1963 had little effect on the epidemiologic pattern of measles in the United States. However, in 1965 the Vaccine Assistance Act provided money for public immunization campaigns, and a decided reduction in measles occurred by 1967. Since 1967, modest increases in reported measles cases in the United States occurred in 1971 (75,290 cases) and 1977 (57,345 cases). The single greatest contributor to the 1971 epidemic was vaccine failure due to initial immunization practices that were later found to be suboptimal. Almost one half of all vaccine failures occurred in children immunized during the first year of life. Other vaccine failures were due to excessive amounts of immune serum globulin administered with measles vaccine and poor vaccine storage practices.

The 1977 epidemic was unique in that about 65% of the cases occurred in persons over 10 years of age who were for the most part unimmunized children and young adults protected from natural measles for a number of years by the high level of vaccine-induced immunity (herd immunity) in young children.

One form of vaccine failure that was particularly alarming occurred in some previous recipients of killed measles vaccine. This vaccine failure, which is called "atypical measles," was frequently noted in the 1971 outbreak; it has continued to be a problem. Atypical measles is a severe disease. It is characterized by an incubation period of 1 to 2 weeks followed by the onset of high fever and diffuse body pains and headache. On about the third day, a rash occurs that is peripheral in distribution (suggesting Rocky Mountain spotted fever). Most patients with atypical measles also have a pneumonia characterized by nodular pulmonary infiltrates. The acute illness lasts 2 weeks or more, but residual pulmonary lesions have persisted for a year or more after clinical recovery. This illness seems to be due to persistent measles-specific lymphocyte reactivity in the absence of serum measles antibody.

MEASLES VIRUS VACCINE

General. The presently available vaccine contains a strain of virus that is more attenuated than the attenuated Edmonston strain from which it is derived. The licensed vaccine is prepared in chick embryo fibro-blast cultures; each dose contains 1000 tissue culture infectious doses ($TCID_{50}$) of virus and about 20 μg of neomycin per dose. Vaccination results in a mild or inapparent, noncommunicable infection that induces measles antibody in more than 95% of susceptible persons. Antibodies resulting from immunization have a pattern suggestive of life-long immunity. Vaccine failures have occasionally been noted in apparently successfully immunized persons, but this waning immunity occurs so infrequently that it is not thought to be of epidemiologic importance.

Precautions, Contraindications, and Complications. Measles immunization should not be given at times of febrile illness or when interference due to another viral infection might reduce the efficacy of the measles vaccine. Vaccination should be deferred for 3 months after the administration of immune serum globulin or other antibody containing human blood products. Because measles vaccine is prepared in chick embryo cell cultures, allergic reaction in egg-sensitive persons is a remote possibility. Until recently, no adverse reactions had been noted in egg-sensitive children, and, therefore, immunization was routinely carried out. However, recently, three children with a history of anaphylaxis due to egg protein allergy have had allergic reactions to measles vaccine. Because of these recent observations, it is probably wise not to immunize any child with a history of egg-associated anaphylaxis and to be extremely careful with vaccine administration in other egg-sensitive persons.

Measles vaccine should not be administered to pregnant women because there is an undocumented but theoretical risk to the fetus. Vaccine should also not be given to immunocompromised persons because progressive infections resulting in death have occurred. In normal children, measles vaccine is exceedingly safe. Neurologic disease such as encephalitis, Reye's syndrome, cranial nerve palsy, cerebellar ataxia, and Guillain-Barré syndrome after immunization occur at a rate of about 1 case per 1 million immunizations in the United States.

Use of Measles Vaccine. Measles vaccination is recommended for all susceptible persons who are 15 months of age or older in whom immunization is not contraindicated.

PRESENT PROBLEMS RELATED TO MEASLES IMMUNIZATION

Optimum Age for Immunization in the United States. In 1976, the age of immunization was changed from 12 months to 15 months or more because of poorer seroconversion rates in vaccinated 12-month-olds. Other studies have shown no advantage in delaying immunization to more than 15 months of age. Subsequently, this recommendation has remained because changing again would cause considerable confusion. One must realize, however, that compliance with routine child health care is not as great in the second year as in the first year. Therefore, unless extra effort through follow-up systems is made, it is possible that fewer children will be immune to measles with the present regimen than with the previous schedule. Indeed, in a recent study, the measles vaccine compliance in one middle-class population was worse in 1979 than in 1976.

Immunization During the First Year of Life. In many areas of the world, measles vaccine is given early in life (age 6 to 9 months) because epidemic measles is still a reality and the morbidity and mortality of measles in infancy are great. In addition, in the United States it has been recommended that when outbreaks of measles occur in a community, vaccine should be given to all unimmunized children more than 6 months of age. By this policy, it was assumed that a substantial number of infants would be afforded protection and those vaccine failures due to residual maternal antibody who did not get natural measles could be reimmunized at age 15 months or more.

Recent reimmunization studies of children initially vaccinated during the first year of life and found later to be seronegative have had disturbing results. Almost 50% of these previously immunized children have had apparent vaccine failure on revaccination. Further studies with more sensitive neutralizing antibody techniques have determined that many of the children who were apparent vaccine failures after early immunization did, in fact, have low levels of antibody. What percentage of these children with low antibody levels are protected against natural challenge is unknown. It is possible that many of these children will be susceptible to measles in the future as their low levels of antibody wane. If this is true, there is cause for concern, because reimmunization is apparently ineffective or results in only transitory boosts in antibody levels. At present in the United States, immunization before the first birthday should be avoided except under unusual circumstances.

Measles Immunization of Adults. About 20 years ago, serologic studies in military recruits showed that 99% of young adults had measles antibody. The recent epidemiologic pattern of measles in the United States, with a far greater percentage of cases in young adults, suggests that the number of susceptible young adults has increased. Recent studies in Los Angeles indicate that about 10% of young adults do not have measurable hemagglutination-inhibiting measles antibody. Therefore, young adults without a history of measles or measles immunization by a presently acceptable vaccination method should be immunized. It is generally recommended that the susceptible young adult age range includes all those born after 1957. However, experience in Los Angeles in 1981 and calculations based on age-related measles before the present vaccine era clearly indicate that the cutoff date for measles immunization should be a birthdate after 1950 rather than after 1957.

Because measles is a serious but preventable disease in adults as it is in children, extensive efforts should be made to immunize young adults. Immunization programs should be vigorous among populations of young adults such as the military, college students, and industrial groups. Unfortunately, those who care for adults rarely report measles when it occurs, so the public health importance of measles in this population group goes unrecognized.

Booster Immunizations. Because of vaccine failure in some persons in whom immunization was carried out by acceptable methods, some have suggested a booster measles immunization. However, no evidence exists today either that waning immunity vaccine failure is of epidemiologic importance or, more important, that booster immunizations will do anything but cause transitory rises in antibody titers. Furthermore, some case report data suggest sensitization results in persons receiving multiple doses of live measles vaccine.

Rubella

BACKGROUND

The disease rubella (German measles) is caused by rubella virus, an agent of one serologic type. The virus was first recovered in tissue culture in 1962 and subsequently was found to grow in many different tissue cultures, such that by 1967 several candidate attenuated vaccines had been developed. After extensive testing in the United States and in other countries, three vaccines were licensed in the United States in 1969: HPV 77 DE 6, HPV 77 DK 12, and the Cendehill strain. The first two vaccines were attenuated in African green monkey kidney tissue culture and then passed five times in duck embryo tissue culture (HPV 77 DE 5) or 12 times in dog kidney tissue culture (HPV 77 DK 12), and the Cendehill strain was grown and attenuated in primary rabbit kidney cell culture. Because of the occurrence of peripheral nerve syndromes, use of the dog kidney cell culture was discontinued.

Both Cendehill strain and HPV 77 DE 5 vaccines induced serum antibody in more than 95% of vaccinees, and vaccine virus infection was associated with nasal viral vaccine shedding in most instances. In 1978, the RA 27/3 attenuated strain rubella virus vaccine was licensed for use in the United States, and today this is the only available vaccine in this country. The RA 27/3 vaccine was attenuated and is prepared in W138 human embryonic lung (diploid) cell culture. Antibody titers resulting from immunization with this vaccine tend to be higher than with the other rubella vaccines; seroconversion rates are about 98%, and immunization leads to a secretory antibody response as well as a serum antibody response that is more akin to that following natural infection with wild-type virus. Pandemic rubella has not occurred since 1964–1965, and in 1981 a record low number of cases was reported.

At the time of licensure of rubella vaccine in the United States, considerable controversy arose as to how best to use the vaccine. Because rubella in children is usually a mild disease, the sole rationale for immunization is to prevent congenital rubella by preventing infection in pregnant women. To approach this goal in the United Kingdom, it was decided to allow infection to occur in children and to concentrate the vaccination effort on adolescent girls and nonpregnant women. In contrast, the approach in the United States was to immunize all prepubertal children. After the original immunization push in the United States, secondary attention was directed to the immunization of adolescent and young adult women. However, this latter aspect of the immunization program has had less than optimal compliance. Since 1969, well over 100,000 million doses of rubella vaccine have been administered.

RUBELLA VACCINE

General. The RA 27/3 rubella vaccine contains 1000 $TCID_{50}$ of rubella virus and about 25 μg neomycin. Antibodies resulting from immunization persist in a pattern suggestive of lifelong immunity, and vaccine failure has been exceedingly rare.

Precautions, Contraindications, and Complications. Rubella immunization should not be given at times of febrile illness or when interference due to another viral infection might cause vaccine failure. Because of the possible, but as yet undemonstrated, risk to the fetus, rubella vaccine should not be given to pregnant women; pregnancy should be deferred for 3 months after immunization. Vaccine should not be given to immunocompromised persons.

In normal children, rubella vaccine is exceedingly safe; symptoms similar to natural rubella (such as lymphadenopathy, rash, and low-grade fever) rarely occur. Arthralgia and arthritis of varying degree and duration occur occasionally; joint reactions occur in 10 to 20% of adult women, but in fewer than 3% of children after immunization.

Use of Rubella Vaccine. Rubella vaccination should be carried out in all susceptible persons more than 12 months of age, in whom immunization is not contraindicated.

PRESENT PROBLEMS RELATED TO RUBELLA IMMUNIZATION

Since the vaccine was licensed in 1969, the number of reported rubella cases has declined steadily, but, at least through 1977, the number of cases in persons of childbearing age (older than 15 years) remained constant (Fig. 15.1). Experience in Los Angeles during 1982 suggests that the record low number of recent cases is not an expression of true vaccine effect, but is due to lack of disease reporting. Because a high percentage of physicians who care for adults generally fail to recognize clinical rubella and seem to be less aware of public health measures of disease control, outbreaks of rubella in adults involving two or three generations of illness are occurring in places of employment without being recognized.

Recent studies suggest that about one third of young adults who have not been immunized are susceptible to rubella. This figure is particularly alarming because few adults over age 20 have been immunized,

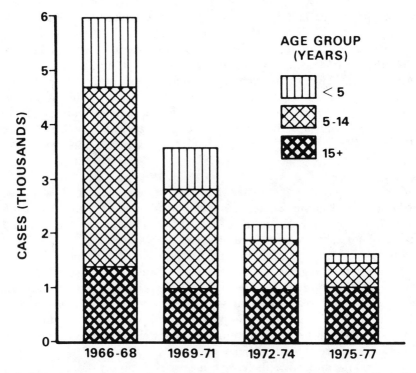

FIG. 15.1. Average number of reported rubella cases in Massachusetts, New York City, and Illinois, by age group and selected period (3-year averages) from 1966 to 1977. (*Source:* Center for Disease Control: *Rubella Surveillance,* Jan. 1976–Dec. 1978.)

and, therefore, it can be expected (and serologic survey has shown) that overall about 25% of young adult women are susceptible to rubella. In the prevaccine era, only 10 to 15% of young adults were susceptible to rubella; now, because of the success of our program in children and its effects on herd immunity, we actually have a greater number of susceptible adults.

It is imperative that vaccination programs be extended to reach all susceptible young adults. Both men and women should be immunized, because pregnant women often acquire disease from their husbands or their male associates in the workplace. Rubella immunization or proof of immunity should be requirements for entrance into college, employment in industry, and military service.

Mumps

BACKGROUND

Mumps is a disease caused by a paramyxovirus (mumps virus) that is antigenically stable, that is, there is but a single serotype. The virus grows in chick embryos and chick embryo fibroblasts and in a number of other cell culture systems. The mumps vaccine is prepared from the Jeryl Lynn strain of mumps virus, grown in chick embryo fibroblast culture. In contrast with measles and rubella vaccines, mumps vaccine was licensed for use in the United States in 1967 without much fanfare. It is usually incorporated into the triple vaccine along with attenuated measles and rubella vaccine viruses (MMR). By 1980, 92% of children entering school for the first time had had mumps vaccine, and today the disease is a distinct rarity in the pediatric age group.

MUMPS VACCINE

Mumps virus vaccine contains 5000 $TCID_{50}$ of attenuated virus and 25 μg neomycin in each dose. About 95% of susceptible persons develop an antibody response after immunization, and studies to date indicate antibody persistence for at least 12 years.

Precautions and contraindications of mumps vaccine are similar to those of measles vaccine, and the literature supplied by the manufacturer should be examined for details. All susceptible persons who are more than 12 months of age should receive mumps vaccination.

PRESENT PROBLEMS RELATED TO MUMPS IMMUNIZATION

Although not yet as clearly identified, a paradox similar to that with measles and rubella exists, in that the number of susceptible young adults is increased due to the effectiveness of the vaccine in children. Because of the virtual elimination of epidemic disease in children and the almost exclusive use of vaccine in young children, a large number of adolescents and young adults have grown up without having had either natural disease or immunization. Because mumps is generally a more severe disease in adults than in children, it is important now to institute vaccination programs in adolescent and young adult populations.

Rabies

BACKGROUND

The most feared disease of animals that can be transmitted to humans, rabies is a unique disease because of its exceedingly long incubation period and short, rapidly fatal, clinical course. Because of the long incubation period, active postexposure immunization can be carried out. Almost 100 years ago, Pasteur introduced postexposure immunization for rabies with a vaccine prepared from dried spinal cords from deliberately infected animals.

During most of this century, most rabies vaccine has been prepared in animal nervous tissue. Because of this preparation method and the fact that multiple doses were necessary, the serious neurologic reaction rate to the immunization procedure was high. In 1957, vaccine prepared in duck embryo cells became available, and with this the neurologic reaction rate was reduced from about 1 per 2000 vaccinees to 1 per 24,000. From 1958 to 1975, there were 21 serious neurologic reactions associated with duck embryo vaccine (DEV) in the United States.

A major advance in rabies immunization was realized in June 1980, when, after extensive testing, the human diploid-cell rabies vaccine (HDCV) was licensed for use in the United States. The antibody response with this vaccine was 10 to 20 times higher than that produced by DEV, even though the number of doses required was reduced by 75%. At the time of licensure, no serious reactions had been associated with HDCV administration. In 1981, the sale of DEV ended in the United States.

In addition to postexposure use, HDCV may be used for preexposure immunization in persons at high risk, such as veterinarians, animal handlers, certain laboratory workers, and persons—especially children—living in, or visiting, countries where rabies is a constant threat.

RABIES VACCINE

Human diploid cell rabies vaccine (HDCV) is an inactivated virus vaccine prepared from fixed rabies virus grown in WI38 or MRC-5 human diploid cell tissue culture. It is inactivated with either tri-n-butyl or beta-propiolactone.

IMMUNIZATION

Postexposure Prophylaxis   Five 1-ml doses should be given on days 3, 7, 14, and 28 after the first dose. In addition to vaccine, human rabies immune globulin should be administered once at the beginning of antirabies prophylaxis. Up to half of the dose of rabies immune globin should be infiltrated in the area of the animal bite or wound and the rest administered intramuscularly. The wound should be immediately and thoroughly washed with soap and water.

Preexposure Immunization.   Three 1-ml injections of HDCV should be given intramuscularly, the first two a week apart, the third 2 to 3 weeks later. Persons with continuing risk of exposure should receive a booster dose (1 ml) every 2 years or have their serum tested for antibody to rabies virus every 2 years and, if the titer is inadequate, receive a booster dose.

COMMENT

The present availability of HDCV vaccine is a major advance; however, rabies and its prevention should not be taken lightly. At this time, about 30,000 persons are immunized each year in the United States, and of these only about 20% are actually at risk of acquiring rabies. Because the rabies risk differs greatly from one geographical area to another, local or state public health officials and published comprehensive immunization guides should generally be consulted before beginning rabies prophylaxis.

Influenza

BACKGROUND

Epidemic influenza has been recognized for at least 1500 years, and it remains a great plague of humankind. From 1968 to 1981, more than 150,000 influenza-related deaths occurred in the United States. During the last 20 years, efforts to prevent or control influenza in the United States have been aimed at protecting those at greatest risk of serious illness or death.

Since 1977, the epidemiologic pattern of influenza has apparently changed in that, for the last 9 years, we have had two different influenza A types (H_3N_2 and H_1N_1) circulating concurrently. Present vaccines are trivalent and contain recent isolates of both H_1N_1 and H_3N_2 influenza A strains and influenza B viruses. (See Chapter 1 for explanations of terminology.)

INFLUENZA VACCINE

Available influenza vaccines are inactivated products prepared from the allantoic fluids of infected chick embryos. Both whole-virion and subvirion (split-virus) preparations are available. The most recent vaccines contain 15 μg each of A/Chile/83 (H_1N_1), A/Philippines/82 (H_3N_2), and B/USSR/100/83 hemagglutinin antigens in each 0.5-ml dose. The split virus vaccine is recommended for those less than 13 years of age and the dose is halved (0.25 ml) in children less than 3 years of age.

Annual vaccination is recommended for all persons who are at increased risk of adverse consequences from infection of the lower respiratory tract: persons with acquired or congenital heart disease, chronic disorders with compromised pulmonary function, chronic renal disease, diabetes mellitus and other metabolic diseases with increased susceptibility to infection, chronic, severe anemia, and conditions that compromise the immune mechanism. Vaccination is also generally recommended for older persons, particularly those over age 65, the age group in which excess mortality occurs during outbreaks of influenza.

In 1976, Guillain-Barré syndrome was noted to occur with excess frequency among persons who had received the A/New Jersey/76 (swine) influenza vaccine. However, since 1978, immunization surveillance data have shown no clear association between influenza vaccination and the Guillain-Barré syndrome.

COMMENT

Although the immunoprophylactic approach to influenza obviously is exceedingly more complex than, for example, that to measles or poliomyelitis, one would think that in 1985 we could be more successful. As suggested in 1977 by Glezen, the strategy of stressing immunization of "high-risk" patients should be reassessed. Because the highest attack rates of influenza occur in young children, vaccination programs aimed at children (such as those with poliovirus and measles virus) might protect high-risk patients more than the present approach. Because repeated immunization with killed-virus vaccines would be logistically difficult, present research should be directed toward live-virus vaccines. During the past 10 years, several promising candidate live-virus vaccines have been developed; however, widespread trial in children has not been done because of the present climate, which appears to be more concerned with potential risks rather than clearly achievable benefits. We need a bolder approach to influenza immunization.

Hepatitis B

BACKGROUND
The following events led to the development of hepatitis B virus (HBV) vaccine:

• The demonstration in 1968 by electron microscopy of hepatitis B surface antigen (HB$_s$Ag) particles.
• The observation of the complete hepatitis B virus (Dane particle) in 1970.
• The large-scale purification of the HB$_s$Ag in 1969 to 1971.
• The development of sensitive hepatitis B virus markers (1969–73).
• The demonstration that boiled HB$_s$Ag-positive plasma was noninfectious, immunogenic, and induced protection against subsequent HBV challenge in 1971.
• And, last, the development of a chimpanzee model for the study of virus infectivity.

The first experimental hepatitis B subunit vaccines were studied in chimpanzees from 1975 to 1978. Trials in humans were begun in 1977.

HEPATITIS-B VACCINE
Hepatitis B virus vaccine is a suspension of inactivated, alum-adsorbed, 22-nm surface antigen particles that have been extracted from human plasma. The vaccine contains 20 μg/ml HB$_s$Ag protein. In field trials, vaccine efficacy has been 80 to 90%. The duration of protection and the need for booster doses is as yet unknown.

Primary adult vaccination consists of three intramuscular doses of 1.0 ml each, with the second and third doses 1 and 6 months after the first dose. Hemodialysis patients and other immunosuppressed patients should receive three 2-ml doses of vaccine. For children less than 10 years of age, three similarly spaced doses of 0.5 ml (10 μg) are sufficient.

Side effects among 12,000 recipients of HBV vaccine observed to date have been limited to soreness and redness at the injection site.

Persons at substantial risk of HBV infection should be vaccinated. The risk of hepatitis B infection, as demonstrated by serologic markers in selected categories, is presented in Table 15.5. Health care workers in high-risk categories should be vaccinated as soon as possible after be-

TABLE 15.5   Hepatitis B Virus Infection (Past or Chronic) in Various U.S. Population Groups

Category	Prevalence of Serologic Markers of Infection (%)
HIGH RISK	
Immigrants/refugees from areas of high hepatitis B endemicity	70–85
Clients in institutions for the mentally retarded	35–80
Users of illicit parenteral drugs	60–80
Homosexually active males	35–80
Household contacts of HBV carriers	30–60
Patients of hemodialysis units	20–80
INTERMEDIATE RISK	
Prisoners (male)	10–80
Staff of institutions for the mentally retarded	10–25
Health care workers who risk contact with human blood (e.g., blood bank workers)	15–30
LOW RISK	
Health care workers with no or infrequent contact with human blood	3–10
Healthy adults (first-time volunteer blood donors)[a]	3–5

[a] Viral markers present at time blood is taken.

ginning work in the risk area. Vaccination is also recommended for clients and staff of institutions for the mentally retarded, hemodialysis patients, homosexually active males, illicit users of injectable drugs, repeated recipients of certain blood products, and household and sexual contacts of HBV carriers.

Hepatitis B virus carriers and those having antibody from previous infection need not be vaccinated, but serologic screening to detect such persons before vaccination may or may not be cost-effective. In general, the higher the risk group to which a potential vaccinee belongs, the greater the cost-effectiveness of serologic screening.

COMMENT

The availability of an HBV vaccine is truly a significant advance. The expense of the present vaccine (about $100 for a three-dose program), however, makes its widespread use impossible in some settings and of a lower priority in others. Certainly, every effort should be made to deliver vaccine to high-risk groups because even this limited use will reduce the overall carrier pool and the risk to others outside the high-risk groups.

Some caution should be observed with this vaccine because of the precedent set by other inactivated viral vaccines (measles and respiratory syncytial viruses). It is possible that the present hepatitis-B vaccine could sensitize in a manner similar to killed measles vaccine and lead in future years to an atypical hepatitis illness on natural exposure to wild-type virus. No data are available that rule out this possibility.

Yellow Fever

BACKGROUND

The most recent indigenous case of yellow fever in the United States was in 1911, and the last reported imported case was in 1924. However, cases of yellow fever continue to occur in Africa and South America, and vaccination is often a requirement for international travel.

VACCINE

The yellow fever vaccine available in the United States is a live, attenuated virus product prepared in chick embryo with a fixed passage level of the 17D strain of virus. Vaccination is recommended for persons 6 months of age or older traveling or living in areas where yellow fever virus infection exists and for laboratory personnel who might be exposed to virulent yellow fever virus.

Vaccinated international travelers should have an international certificate of immunization filled in, signed, and validated with the stamp of the center where the vaccine was given.

Primary immunization for both adults and children is a single subcutaneous injection of 0.5 ml of reconstituted vaccine. Yellow fever immunity after vaccination with 17D strain virus persists for more than 10 years; the International Health Regulations do not require revaccination more than every 10 years. The manufacturer's package insert should be consulted for information regarding precautions and contraindications.

Adenoviral Acute Respiratory Disease

Acute respiratory disease (ARD) is the leading cause of hospitalization of military recruits undergoing basic combat training in the United States. This acute febrile respiratory disease in recruit populations is caused primarily by two viruses, adenovirus type 4 and adenovirus type 7. Because of the morbidity and economic cost associated with ARD, adenovirus types 4 and 7 vaccines were developed. The vaccine viruses are cultivated in human diploid tissue cultures and the vaccines are administered in enteric-coated capsules. Immunization results in asymptomatic gastrointestinal infection and the development of specific neutralizing antibody.

Routine simultaneous use of types 4 and 7 beginning in 1970 significantly decreased rates of ARD. In recent years, in spite of the immunization program, ARD has again become more prevalent in recruit camps. Recent outbreaks have been due to adenovirus 21, and, therefore, studies with a type 21 vaccine are being carried out.

Smallpox

Because smallpox has been eliminated from the world, smallpox vaccine should not be used or its use should be limited. Vaccine usage, however, is still widespread, although in most instances it is inappropriate. In the United States, smallpox vaccination of civilians is recommended only if they are laboratory workers directly in contact with smallpox or other closely related orthopox viruses. Smallpox vaccine is ineffective in the treatment of any disease; it should not be used to treat herpes simplex virus infections.

The 34th World Health Assembly of the World Health Organization by resolution amended the International Health Regulations to remove smallpox from the diseases subject to the Regulations, effective 1982. All of the countries of the world except Chad, in Africa, have indicated that smallpox vaccination is no longer required for any traveler.

In the United States, vaccination is still carried out in the military; more than 1 million military personnel are vaccinated each year.

EXPERIMENTAL VACCINES AND VIRAL ILLNESSES FOR WHICH VACCINES SHOULD BE DEVELOPED

Respiratory Viruses

Respiratory viral infections are responsible for staggering morbidity and considerable mortality. The serologic diversity of viral agents and the fact that similar recurrent symptomatic disease is caused by any of a number of distinct serotypes has led to a general disillusionment concerning practical vaccine development. However, certain viruses are responsible for more serious disease than others, and therefore selected vaccine development and use would seem to be a worthwhile goal.

RESPIRATORY SYNCYTIAL VIRUS

This singular virus is the cause of epidemic bronchiolitis in infants and is also the leading cause of pneumonia during the first year of life.

Both bronchiolitis and pneumonia are the result of primary infections; recurrent infections in older children and adults are usually manifest as the common cold. Initial trials with an alum-precipitated, formalin-inactivated vaccine in the 1960s resulted in high seroconversion rates in vaccines but lack of protection from disease. More important, on subsequent natural challenge exaggerated disease occurred, suggesting an aberrant immunologic response similar to that in atypical measles.

Recent trials with a parenterally administered live-virus vaccine indicated lack of efficacy compared with a control group. Several trials with intranasally administered cold-adapted and temperature-sensitive mutant live-virus, attenuated vaccines have been carried out and have shown promise. However, in the target population (infants less than 6 months of age) some respiratory symptoms occurred, and one vaccine strain was genetically unstable and revertant virus was shed by some vaccinees.

PARAINFLUENZA VIRUSES

Parainfluenza virus types 1 and 2 are the leading cause of laryngotracheitis in children, and parainfluenza type 3 is an important cause of pneumonia in young children. A trivalent, killed-virus vaccine was found to be immunogenic but not protective. Trial with an attenuated type 3 virus vaccine in children resulted in serum antibody in the seronegative children, but none developed specific nasal (secretory IgA) antibody. A live attenuated intranasal vaccine has shown promise in bovine trials.

RHINOVIRUSES

Because there are more than 89 antigenically distinct types of rhinoviruses, the possibility of prevention of infections by immunization seems bleak. A recent trial with vaccine containing 10 formalin-killed viruses elicited antibody to only 30% of the antigens, and a live-virus vaccine administered by the enteric route proved disappointing. However, as recent evidence indicates that many rhinoviruses are antigenically interrelated, there is some hope for the possible development of multivalent vaccines.

Herpesviruses

The herpesviruses (cytomegalovirus [CMV], herpes simplex virus [HSV] Types 1 and 2, varicella-zoster virus, and Epstein-Barr virus [EBV]) are unique in that primary human infections are usually followed by latent infections that exist throughout the life of the host. Because current evidence links latent infection with two herpesviruses, HSV type 2 and EBV, and human cancer, the consideration of immunization with herpesviruses is cause for considerable deliberation.

VARICELLA-ZOSTER (VZV)

Although chickenpox is almost always a benign disease of children, its occurrence in adults is associated with significant morbidity and mortality. In addition, in the immunocompromised patient, the risk of death due to disseminated natural infection is extremely high. Since the development of the OKA strain of live VZV vaccine in 1975, numerous trials have been, and are being, carried out in Japan and elsewhere. In trials in the United States in healthy children, the seroconversion rate has been excellent, clinical reactivity minimal, and viral shedding and transmission absent. Trials in children with leukemia are now in progress. Preliminary results indicate good antibody responses in most vaccinees, but mild rashes develop in a significant percentage of vaccine recipients. In follow-up studies, vesicular zosterlike rashes have occurred in two vaccinated children.

CYTOMEGALOVIRUS

Congenital infection with CMV results in about 3700 neurologically damaged children per year in the United States. Because about 35% of the women of childbearing age are seronegative to CMV, the selected use of an effective vaccine could considerably reduce the incidence of congenital CMV disease. Recently, the candidate Towne 125 vaccine strain was administered to ten seronegative women, and both cellular and humoral immune responses resulted.

Acquired CMV infections are associated with considerable morbidity and mortality in seronegative renal transplant and other immunocompromised patients. Studies with attenuated vaccine are presently being conducted in immunocompromised patients, and results to date are encouraging.

HERPES SIMPLEX AND EPSTEIN-BARR VIRUSES

Recently developed glycoprotein candidate vaccines have shown some promise in animal studies. However, whether either vaccine is of any practical value in the prevention of human primary or recurrent disease remains a question.

Enteric Viruses

SELECTED ENTEROVIRUSES

Since the advent and success of polioviruses immunization, the enteroviruses are all but forgotten. However, they are still responsible for staggering morbidity and considerable mortality. Because only 15 serotypes have been responsible for most morbidity and mortality during the last 20 years, it would seem that attenuated vaccines could be developed that could reduce this disease burden. Because mortality is greatest with Coxsackie virus B types 1 to 5 and central nervous system morbidity is greatest with echo virus types 4, 6, and 9, these would appear to be primary targets for vaccine development.

ROTAVIRUSES

Rotaviruses are the most important cause of gastroenteritis in infants in developed countries, and they are also of major importance in developing countries, where diarrhea is the leading cause of death. Obviously, the development of a rotavirus vaccine is of the highest priority.

The type 2 human rotavirus has been adapted to growth in tissue culture. Because resistance to rotavirus disease depends strongly on local intestinal immunity, a live-virus vaccine that can be administered orally will probably be necessary.

HEPATITIS A VIRUS

Infection and disease due to hepatitis A virus is worldwide. Because hepatitis A virus has recently been adapted to several cell culture systems and because only one major serologic type seems to exist, the development of either live- or killed-virus vaccines is promising. One candidate vaccine is now available in the United States.

CONCLUSION

When immunization is to be performed with an unfamiliar vaccine, the physician should consult the manufacturer's package insert for more complete information. Useful vaccine information is also available in the 1982 Report of the Committee of Infectious Diseases (19th edition of the *Red Book*) of the American Academy of Pediatrics and in various reports of the Advisory Committee on Immunization Practices of the U.S. Public Health Service, which are published by the Centers for Disease Control in their *Morbidity and Mortality Weekly Report*.

Recommended Readings

Alter HJ: The evolution, implications, and applications of the hepatitis B vaccine. *JAMA* 247:2272–2275, 1982.

American Academy of Pediatrics: *Report of the Committee of Infectious Diseases* (1986 Red Book), edition 20. Evanston, IL, in press.

Arbeter AM, Starr SE, Weibel RE, Plotkin SA: Live attenuated varicella vaccine: Immunization of healthy children with the OKA strain. *J Pediatr* 100:886–893, 1982.

Boyer KM, Cherry JD: Influenza viruses. In Feigin RD, Cherry JD (eds): *Textbook of Pediatric Infectious Diseases*, pp 1298–1316. Philadelphia, Saunders, 1981.

Cherry JD: The "new" epidemiology of measles and rubella. *Hosp Pract* 15:49–57, 1980.

Glezen WP: Influenza prophylaxis for children. *Am J Dis Child* 131:628–630, 1977.

Institute of Medicine, National Academy of Sciences, Division of Health Promotion and Disease Prevention: *New Vaccine Development: Establishing Priorities*, vol. 1: Diseases of importance in the United States. Washington, DC, National Academy Press, 1985.

Kahn C: History of smallpox and its prevention. *Am J Dis Child* 106:597–609, 1963.

Kapikian AZ, Wyatt RG, Greenberg HB et al: Approaches to immunization of infants and young children against gastroenteritis due to rotaviruses. *Rev Infect Dis* 2:459–469, 1980.

Krause PJ, Cherry JD, Deseda-tous J et al: Epidemic measles in young adults. Clinical, epidemiologic, and serologic studies. *Ann Intern Med* 90:873–876, 1979.

Nightingale EO: Recommendation for a national policy on poliomyelitis vaccination. *N Engl J Med* 297:249–253, 1977.

Plotkin SA: Rabies vaccination in the 1980s. *Hosp Pract* 15:65–72, 1980.

Report of the Committee on Infectious Diseases of the American Academy of Pediatrics, 19th edition of the Red Book, 1982.

Robinson RG, Dudenhoeffer FE, Holroyd HJ, Baker LR, Bernstein DI, Cherry JD: Rubella immunity in older children, teenagers and young adults: A comparison of immunity in those previously immunized with those unimmunized. *J Pediatr,* 101:188–191, 1982.

Glossary

attenuated: reduced in virulence (e.g., live poliovirus vaccine).

B cells: bone-marrow-dependent lymphocytes that bear surface immunoglobulin(s) and that, on specific interaction with antigen, undergo blastogenesis and differention into plasma cells; the latter produce antibody reactive with the same antigen.

capsid: the outer protein shell of a viral particle in close association with nucleic acid. In viruses having cubic symmetry, the capsid is made up of *capsomers;* in viruses displaying helical symmetry, protein is bound to viral nucleic acid inside the viral envelope.

capsomer: the morphological protein subunit of a viral *capsid;* see *hexon* and *penton.*

cellular oncogene: see *proto-oncogene.*

clone: the progeny of a single cell or microorganism.

dalton: the unit used to express molecular weight. It is equal to the mass of the hydrogen atom (1.67×10^{-24} gm).

defective particle: an incomplete (i.e., noninfectious) viral particle.

denaturation: the loss of the natural structure and/or function of a macromolecule. For example, heat treatment may cause *double-stranded* DNA to dissociate into two separate strands of DNA.

derepress: to allow a gene to undergo *transcription* by release from a nonfunctioning (repressed) state.

dimer: a complex comprising two identical subunits or monomers.

double- or single-stranded: the numerical designation of the configuration of covalently linked nucleotides (DNA or RNA), either as one thread of bases or two intertwined strands of base pairs. If double-stranded, the strands are complementary in adjacent purine and pyrimidine base sequence, guanine pairing with cytosine, and adenine with thymidine or uracil.

"early" or "late" (genes): viral genes coding for viral mRNA transcribed before (early) or after (late) viral DNA replication begins. "Early" and "late" may also be applied to the proteins of these genes.

envelope: the layer surrounding the *capsid* derived from the host's cytoplasmic membrane; it contains virus-coded proteins (peplomers).

epizootic: pertaining to a disease affecting many animals of a species simultaneously, analogous to an epidemic.

eukaryote: an organism in which each cell has a membrane-enclosed nucleus.

genetic code: the arrangement of nucleotides in the chromosomal polynucleotide chain governing the transmission of genetic information by *transcription* to mRNA and *translation* into proteins. Genetic information is encoded in the nucleic acid by the specific arrangement of four bases (two purines: adenine and guanine, and two pyrimidines: thymine and cytosine). In this language, each sequence of three adjacent bases (a codon) determines the insertion of a specific amino acid in the polypeptide. The genetic code is universal in that it appears to be the same for all forms of life.

genome: the complete chromosome set of an organism bearing the entire complement of genes.

genotype: the entire genetic makeup of an organism; the genes present at specific loci.

"helper" virus: a virus that aids the replication of a defective virus by restoring or supplying the activity of a viral gene required for *lytic infection*.

heterologous: pertaining to different species or to lack of correspondence in structure, as between an antigen and an unrelated antibody. In the latter example, either may be said to be heterologous to the other.

hexon: configuration of proteins in the adenoviral capsid, in which each *capsomer* has six contiguous identical neighbors.

homologus: pertaining to the same species or correspondence in structure, as two chromosomes, or an antigen and its specific antibody.

homology: the fundamental similarities among structures, e.g., macromolecules in different organisms derived from a common ancestor.

hybrid: the progeny resulting from mating of genetically dissimilar parents.

hybridization of nucleic acid: formation of a double helix by annealing one chain of DNA to another chain of DNA or RNA by means of hydrogen-bonded complementary base pairs.

icosahedron: the geometric form composed of 20 equilateral triangular faces (the shape of cubic viruses).

immunosuppression: the prevention or diminution of the immune response, as by irradiation or administration of antimetabolites, antilymphocyte serum, or specific antibody, or as in the course of certain viral infections, e.g., measles, AIDS.

inclusion body: the specific acidophilic or basophilic heterogeneous virus-associated mass resulting from viral infection and visible by light microscopy in cytoplasm and/or nucleus. Usually composed of massed virions or viral subunits.

lytic infection: viral infection leading to synthesis of new virus, death of the infected cell, and release of progeny viral particles.

messenger RNA (mRNA): an RNA molecule formed by *transcription* that is complementary to a DNA cistron and that functions as a *translation* template.

minus or plus strand: pertaining to the ability or inability of a nucleic acid sequence to function directly as messenger in translation. "Plus" strand refers to viral RNA that can itself function as messenger (e.g., picornaviral RNA); "minus" or "negative" strand viruses require synthesis of a complementary (plus) strand, from which mRNA is then made, for translation to occur (e.g., paramyxoviruses, sometimes called "negative strand" viruses).

mutant: an organism bearing an altered gene that may or may not express itself in the *phenotype* of the organism.

mutation: a permanent transmissible change in the genetic material, usually in a single gene. The change may be in the form of a loss, gain, or exchange of one or more nucleotides in the nucleic acid comprising the gene.

naked: lacking an envelope, as certain viruses.

nonreplicating subunit vaccine: a viral vaccine composed of viral particles that have been disrupted and are therefore no longer infectious (e.g., influenza virus killed by treatment with ether).

nucleocapsid: a unit of viral structure, consisting of a nucleic acid with associated protein that, in some virus groups, may be the *capsid*. A virus my be a *naked* nucleocapsid of cubic symmetry, or, in either cubic or helical form, it may be enclosed in an essential *envelope*.

nucleoprotein: the combination of nucleic acid and attached basic proteins.

oncogene: a gene that codes for a protein that potentially can transform a normal cell into a malignant cell; may be transmitted in a virus (viral oncogene).

oncogenic: capable of inducing neoplasia, benign or malignant.

penton: *capsomer* of icosahedral (cubic) virions including five protein subunits *(protomers);* pentons comprise the 12 vertices of icosahedral structures.

peplomer: virus-coded proteins in the envelope of certain viruses responsible for cell attachment and certain enzymatic functions.

permissive: relating to a cell type or conditions (e.g., temperature) that allow full expression of the viral genome, i.e., support *lytic infection.*

phage (bacteriophage): a virus that infects bacteria.

phenotype: the observable physical, biochemical, and/or physiological properties of an organism produced by the *genotype* as it interacts with the environment.

plaque: a clear area in a monolayer of cells produced by destruction (lysis) of contiguous cells or fusion of indicator cells by several cycles of viral proliferation originating in a single infected cell.

plasmid: an extrachromosomal genetic element capable of autonomous replication in the bacterial cell.

polymerase: any of several enzymes that catalyze the formation of DNA or RNA from preexisting purines and pyrimidines, in the presence of DNA or RNA, which act as a template.

proto-oncogene: a host gene that is homologous to that found in a virus but can induce *transformation* only after mutation.

provirus: the viral gene(s) integrated into the host-cell chromosome and transmitted from one cell generation to another without causing lysis of the host cell. By the application of any of several physicochemical techniques, provirus may be rescued from its integrated state and produce infectious virus. Formerly referred to as prophage, provirus generally is also now used to describe oncogenic virus that has been integrated into the *genome* of a mammalian cell concomitantly with *transformation.*

pseudovirion: a *virion* that contains fragments of cellular DNA in addition to viral DNA encapsidated by viral proteins.

recombination: formation of progeny having combinations of genes different from those of either parent. Applied recombinant technology is used in the production of hepatitis vaccine (HBs antigen in yeast cells) and of insulin in *E. coli.*

restriction endonucleases: components of the restriction-modification cellular defense system against foreign nucleic acids. These enzymes cleave uncod-ified (e.g., methylated) DNA from other organisms.

reverse transcriptase: RNA-dependent DNA polymerase, characteristic of RNA tumor viruses (retroviruses). RNA acts as a *template* for the enzyme in syn-thesis of DNA.

serotype: a subgroup within a species, e.g., microbial or viral, determined by the kinds and combinations of constituent antigens present in the microor-ganism and detectable with microorganism antibodies of known specificity.

"strandedness": refers to whether nucleic acid exists as a single or double strand of purine and pyrimidine bases, e.g., single-stranded DNA.

T antigen (tumor antigen): an early virus-specific antigen found in the nucleus associated with cellular transformations.

T cell: thymus-dependent lymphocyte that mediates cellular (as opposed to hu-moral) immune responses and in mice possesses the surface phi (ϕ) anti-gen, and in humans is identified with monoclonal antibodies.

temperature-sensitive mutant: a conditional *mutant* that expresses one *phenotype* at certain temperatures but not at other temperatures because of inactiva-tion of protein(s).

template: the macromolecular model that guides the synthesis of a similar or related macromolecule.

transcription: the formation of *mRNA* through the attachment (to DNA) of in-dividual nucleotides by base pairing (guanine with cytosine and adenine with uracil), and synthesis catalyzed by RNA polymerase.

transformation: the changes in the biologic function and antigenic specificity of a cell that result from integration of, and subsequent regulation by, viral genetic sequences in the cellular genome and that confer on the infected cell certain properties characteristic of neoplasia.

translation: the formation of polypeptide chains with a specific amino acid se-quence that corresponds to genetic information transcribed to mRNA from DNA.

virion: an individual, complete viral particle, comprising all components re-quired to initiate infection.

viroid: a highly resistant, nonimmunogenic pathogenic agent thought to con-sist of a short RNA molecule in close association with cellular membrane.

virus: infectious disease-causing agent that is smaller than a bacterium, re-quires host-cell biosynthetic machinery for replication, and contains either DNA or RNA as its genetic component.

wild-type: the reference genotype or typical form of an organism as ordinarily encountered in nature.

Index

Acetominophen, 129
Acquired immune deficiency syndrome (AIDS), 63, 76, 87, 226, 233
Acute yellow atrophy, 153
Acyclovir, 147, 185, 218, 283, 285–87, 291–97
Adenine arabinoside. *See* Arabinoside, adenine
Adeno-associated virus, 9
Adenoviridae, 9, 261
Adenoviruses, 8, 9, 11, 69, 78, 126–28, 170, 243, 245, 266, 267, 320
Adsorption, 21, 24
Aedurid, 284
African swine fever virus, 10
AIDS. *See* Acquired immune deficiency syndrome
Alanine aminotransferase, 164
Aleutian mink disease, 9
Alphaherpesvirinae, 10
Alphavirus, 13
Amantadine, 130, 131, 282, 283, 285, 288
Animal exposures, 105
Animal models, 14, 221, 231, 232
Antibiotics, 130
Antibody, 50, 52, 57, 83, 85, 235, 237, 257, 275, 279
Antigen, 47–49, 83, 213, 272, 277
  binding site (Fab), 53
  hepatitis B surface, 154
  herpes simplex viral, 84
Antiviral therapy, 5, 44, 281
Aphthovirus, 12
"Apollo disease," 138
araA. *See* Arabinoside, adenine
Arabinoside, adenine, 147, 185, 218, 285–87, 290, 295–97
Arbovirus, 93, 141, 145
Arcus senilis, 20
Arenaviridae, 15, 262
Arenaviruses, 8, 15, 267
Arildone, 284
Assays, virus, 17–24, 66, 67, 70, 77, 80, 85, 168, 264, 265, 269–76
Astrovirus, 170, 266
Attachment, 54
Atypical pneumonia, 120
Australia antigen, 154, 191
Aviadenovirus, 9
Avipoxvirus, 11

B lymphocytes, 51
Betaherpesvirinae, 10
Biostatistics, 99
Biosynthesis, viruses, 31–36
BK virus, 243, 245, 261

Blood-brain barrier, 135, 136
Bronchiolitis, 112, 115, 116
Bronchitis, 14
Bunyaviridae, 14, 261
Bunyavirus, 8, 14, 267
Bunyawera, 14
Burkitt, Denis, 252
Burkitt's lymphoma, 206, 211, 212, 215, 252, 253

Cadang-Cadang disease, 225
Caliciviridae, 12
Calicivirus, 12, 170, 266
Cancer, 4, 74, 103, 252–54
  human, 4, 252
  viruses and, 240
Canine distemper, 14
  virus, 197
Capripoxvirus, 11
Capsid, 6, 8, 12
Capsomers, 6, 9
Carcinoma. *See* Cancer
Cardiovirus, 12
Catarrh, 112
Cebe-Viran, 284
Cell culture, 77, 260
Cell fusion, 73
Cell lysis, 71
Central nervous system (CNS), diseases of the, 82, 133, 135, 149, 226, 267, 294
Chandipura, 14
Chemotherapy, 46, 105, 281 ff
Chickenpox. *See* Varicella virus
Childhood disease, 139
Chloridovirus, 10
Chordopoxvirinae, 11
Choriomeningitis, 15
Chorioretinitis, 287, 297
Chronic diseases, 229, 237
Chrysanthemum stunt disease, 225
C-J disease. *See* Creutzfeldt-Jacob disease
Classification, 5, 16, 17, 264
CMV. *See* Cytomegalovirus
CNS. *See* Central nervous system, diseases of the
Cohort studies, 102
Cold, common, 112–15, 285
Cold sores, 185
Colorado tick fever, 12, 97, 105
  virus, 97
Common cold. *See* Cold, common
Complement-fixation, 67, 269
Composition, 5
C-onc. *See* Oncogene
Condyloma accuminata, 244
Congenital infections, 173, 267, 292

Congenital rubella syndrome, 178
Conjunctivitis, 137
Coronaviridae, 14, 262
Coronavirus, 14, 114, 115, 170
Cowpox virus. *See* Vaccinia
Coxsackie virus, 12, 133, 137
   A, 79, 188, 261
   B, 79, 186–88, 261, 323
Creutzfeldt-Jacob (C-J) disease, 221, 222,
   229
Crick, Francis, 8
Crimean-Congo, 14
Croup, 112, 115
Cuffing, perivascular, 82
Culture methods, 262, 263
Cytolysis, 56, 57
Cytomegalovirus (CMV), 10, 73, 78, 172,
   174–77, 205, 206, 210–18, 243, 261,
   267, 287, 297, 323
Cytotoxicity, 57, 58

Dane particle, 155, 158, 192
Defective-interfering virus particles, 39
Definition, 3–5
Delayed hypersensitivity, 52
Delta agent, 159, 233, 234
Demyelination, 232
Dengue, 13
Densonucleosis virus, 9
Densovirus, 9
Dependovirus, 9
Detection. *See* Assays, virus
Diabetes, 230, 231
Diarrhea, 3, 152, 165, 168
Disease vs. infection, 67
Diseases modified by antiviral drugs, 285
Drugs, antiviral. *See* Chemotherapy

Eastern encephalitis virus, 144
EBV. *See* Epstein-Barr virus
Echovirus, 12, 79, 133, 137, 186, 187
Edmonston strain, 309
Electron microscopy, 80, 275
ELISA. *See* Enzyme-linked immunosorbent
   assays
EMC virus of mice, 12
Encephalitis, 13, 14, 93, 133–35, 139–46,
   149, 198, 210, 218, 286, 294
Encephalopathy, transmissible mink. *See*
   Transmissible mink encephalopathy
Endocytosis, 27
Enterovirus, 12, 70, 72, 79, 93, 96, 137,
   186–89, 261, 262, 266, 267, 323
Entomopoxvirinae, 11
Envelope, 6, 16, 43, 66
Enviroxime, 284, 285, 289
Enzyme-linked immunosorbent assays
   (ELISA), 23, 168, 274, 276
Enzymes, virus-specific, 23, 81
Epidemic, 93, 166
Epididymo-orchitis, 201
Epiglottitis, 115
Epstein-Barr virus (EBV), 10, 205, 206, 209,
   211–15, 218, 243, 253, 261, 267, 297,
   323
Evans's postulates, 242
Exanthem, 196, 267

Fc binding, 56
Fetus, 172
Fever blisters, 286, 292
"Filterable virus," 4, 241
Flavivirus, 13
Flow cytometry, 279
Fluorescent antibody, 270
Focus assay, 18
Foot and mouth disease, 12
Foscarnet, 284
Fowl pox, 11
Frog virus, 3, 10
Fusion, 27

Gajdusek, D. Carleton, 221, 223
Gammaherpesvirinae, 10
Gastrointestinal disease, 165, 170, 266
Genital infections, 181, 209, 214, 286, 292,
   293
Genitourinary disease, 267
German measles. *See* Rubella
Guillain-Barré syndrome, 317

HA. *See* Assays
HAV. *See* Hepatitis A
HBV. *See* Hepatitis B, virus vaccine
Hemagglutination (HA). *See* Assays
Hemorrhagic fever, 14, 234, 267
Hepatitis
   A (HAV), 87, 154, 162–64, 262, 266, 324
   B (HBV), 154–61, 191–93, 234, 243–47,
     254, 255, 261, 266, 296, 318, 319
   B, antibody, 155, 157, 158
   B, antigen, 155, 158, 233
   B, epidemiology and transmission, 97, 159
   B, virus vaccine, 193, 318
   cytomegalovirus, 165
   delta agent. *See* Delta agent
   Epstein-Barr virus, 165
   mouse, 14, 232
   non-A, non-B, 154, 155, 164
   posttransfusion, 164
Herd immunity, 302
Herpes
   extragenital, 185
   genital, 209, 286
   labialis, 209, 214, 292
   simplex, 7, 10, 78, 172, 182–85, 243, 252,
     261, 278, 291, 296, 323
   simplex virus (HSV), types 1 and 2, 205,
     206, 216, 267, 292
Herpes zoster, 210, 285, 290
Herpesviridae, 10, 261
Herpesvirus, 205, 207–18, 243, 246, 322
   saimiri, 10, 246
Herpetic infections, 214, 216, 286
Hexon, 11
Hog cholera virus, 13
Host range, 25
Host response, 46, 66, 81, 261
HPV. *See* Papillomavirus
HSV. *See* Herpes simplex virus, types 1 and
   2
Human foamy virus, 15
Human T-cell leukemia viruses, 243
Hybridization, 235–37
Hyperplasia, 4, 74
Hypogammaglobulinemia, 49

Icosahedron, 9
Idoxuridine, 282, 283, 286, 292–94
IgA antibody, 52
IgG antibody, 50
IgM antibody, 50, 275
Immune aggregation, 54
Immune clearance, 59
Immune response, 46–49, 51–56, 63, 214
Immunization against viral agents, 299–307
Immunocompromised patients, 295, 297
Immunoenzyme analysis, 273
Immunofluorescence, 83, 84, 264–66, 273, 276, 279
Immunologic techniques, 83, 235
Immunopathology, 83
Immunoperoxidase, 83, 85
Immunoprophylaxis, 46
Immunosorbent assay, 168, 271, 274
Immunosuppression, virus-induced, 60, 63
Inclusions, 71
Infected cells, elimination of, 55
Infection
    congenital, 174
    vs. disease, 67
    latent, 106, 107
    persistent, 107
    recurrent, 106
Infectious virus, elimination of, 52
Inflammatory response, 85
Influenza, 3, 120, 124, 125, 285, 317
    A, 78, 261, 282
    B, 78, 261
    C, 261
    vaccine, 131, 317
    virus, 14, 68, 93, 120–25, 266
Inoculation, 299
    animal, 259
    egg, 260
Inosiplex, 284, 287, 298
Insect viruses, 10
Interferon, 61, 62, 279, 285, 287, 289, 291, 296
Intrauterine infections, 172–74
Iridescence viruses of arthropods, 10
Iridoviridae, 10
Iridovirus, 10
Isolation, 259
Isoprinosins, 284

JC virus, 9, 226–28, 243, 245, 261

K cells. See Killer cells
Kaposi's sarcoma, 233
Kaufman, Herbert E., 282
Keratitis, 282, 286
Keratoconjunctivitis, herpetic, 293
Killer cells (K cells), 51, 57
Koch's postulates, 241
Kuru, 221, 223

Laboratory, clinical virology, 257
LaCrosse virus, 141, 142
Laryngotracheobronchitis, 115
Lassa fever, 15, 262
Lentivirinae, 15
Leporipoxvirus, 11
Leukemia, 15, 233, 243, 255
    T-cell, 233, 255
    virus, 243

Leukoencephalopathy. See Progressive multi-focal leukoencephalopathy
Levamisole, 284, 298
Lymphadenitis, 201
Lymphocryptovirus, 10
Lymphocytic choriomeningitis virus, 10, 148, 262, 267
Lymphoma. See Burkitt's lymphoma
Lyssavirus, 14

Macrophages, 58, 60
Macular/papular disease, 267
Maedi viruses, 15
Malignancy. See Cancer
Mammary tumor virus, 15, 243
Marboran, 282, 283, 285
Marek's disease, 10, 246
Mason-Pfizer virus, 15
Mastadenovirus, 9
MC29 (retrovirus), 243
Measles, 14, 93, 96, 98, 107, 196, 197, 308
    in altered host, 199
    atypical, 309
    complications and mortality, 198
    with encephalitis, 139
    immunization, 310, 311
    vaccine, 303, 308–12
    virus, 79, 197, 228, 261, 267
Membrane, viral, 16
Meningitis, aseptic, 133, 137, 201
Methisazone. See Marboran
Milker's nodules, 11
Miopapovirus, 9
Molluscum contagiosum, 11
Monkey virus, 15
Monoclonal antibodies, 81, 85, 235, 237, 257, 279
Mononucleosis, 211, 213, 215, 267
Montague, Lady Mary Wortley, 299
Morbillivirus, 14
Multiple sclerosis, 231–33
Mumps, 14, 199–201, 314, 315
    meningoencephalitis, 139
    vaccine, 315
    virus, 79, 261, 267
Muromegalovirus, 10
Myxoma, 11

Nairovirus, 14
"Naked" viruses, 68
Nasopharyngeal carcinoma, 206, 212
Natural killer (NK) cells, 56–58
Negative-strand, 5
Neonatal infections, 292
Neoplasia. See Cancer
Neoplastic transformation, 39, 74, 240–42, 248
Nephritis, 201
Neuraminidase, 25
Neutralization, 52–55, 67, 276
Neva, 202
Newborn virus, 172
NK cells. See Natural killer cells
Nonenveloped virus, 68
Norwalk agent, 12, 266
Norwalk virus, 166–67
Norwalklike virus, 166–68
Nosocomial infections, 130
Nucleic acid hybridization, 278

Nucleocapsid, 12, 39, 43
Null cells, 57

Ocular infections, 293
Oncogene, 241, 248–50
Oncogenic viruses, 243
Oncornavirus, 15
Oncovirinae, 15
OPV. *See* Poliovirus, vaccine
Orbivirus, 12, 262
Orthomyxoviridae, 14, 121, 261
Orthopoxvirus, 11
Orthoreovirus, 12

Panencephalitis. *See* Subacute sclerosing pan-
    encephalitis
Papillomavirus (HPV), 9, 75, 76, 243, 244
Papovaviridae, 9, 261
Papovavirus, 243–45, 267
Parainfluenza virus, 79, 115–18, 261, 266,
    322
Parainfluenzae, 14
Paramyxoviridae, 14, 117, 261
Paramyxovirus, 14, 67, 314
Parapoxvirus, 11
Parotitis, 201
Parvoviridae, 9
Parvovirus, 7, 9
Pasteur, Louis, 315
Pathology, 65, 71, 72, 77
Penetration, 26–29, 54
Penton, 11
Perinatal infections, 172, 181, 267
Persistent infection, 55
Pestivirus, 13
PFU. *See* Plaque-forming units
Pharyngitis, 112
Phlebovirus, 14
Phosphonoformic acid, 218, 286
Picornaviridae, 12, 261
Picornavirus, 69
Plant viroids, 220
Plaque assay, 17, 18, 80
Plaque-forming units (PFU), 18
PML. *See* Progressive multifocal leukoence-
    phalopathy
Pneumonia, 74, 112, 115, 116, 121
Pneumovirus, 14
Pock assay, 19
Poliomyelitis, 12, 95, 133, 138, 139, 304, 307
Poliovirus, 8, 79, 89, 138, 261, 304, 306
    vaccine (OPV), 303–8
Polkilovirus, 10
Polykaryocytes, 73
Polymorphonuclear leukocytes, 59
Polyomavirus, 9, 243, 244
Positive-strand, 8, 33
Potato spindle tuber disease, 225
Powassan encephalitis virus, 145
Poxviridae, 11, 261
Poxviruses, 77, 11
Prion, 225
Progressive multifocal leukoencephalopathy
    (PML), 226–29
Proteins
    nonvirion, 47
    virus-coded, 47
Protomer 9, 11
Protooncogenes, 249

Pseudocowpox, 11
Pseudorabies, 10

Rabies, 14, 315
    encephalitis, 148
    vaccine, 316
    virus, 135, 147, 262
Radinovirus, 10
Radioimmunoassay (RIA), 23, 274, 276
    enzyme-linked, 271, 274
Ranavirus, 10
Receptors, 24, 25
Recombination, 7, 235
Reoviridae, 12, 262
Reovirus, 12, 262, 266
Replication, virus, 24, 34, 39, 260
Respiratory disease, 128–30, 285, 320
Respiratory syncytial virus (RSV), 74, 79,
    118, 119, 261, 266, 321
Respiratory viruses, 93, 112, 321
Retroviridae, 15, 262
Retrovirus, 12–15, 41, 42, 75, 243, 247–49,
    255
    type C, 255
Reverse transcriptase, 23, 32
Reye's syndrome, 104, 126, 129
Rhabdoviridae, 14, 262
Rhinitis, 112
Rhinovirus, 12, 113, 114, 261, 266, 289, 322
RIA. *See* Radioimmunoassay
Ribavarin, 284, 285, 289
Rigoni-Stern, 252
Rimantadine, 283, 285, 288
Rinderpest virus, 197
Rivers, 4, 5
RNA viruses, 12–15, 41, 42, 243. *See also* Re-
    trovirus
Rocio encephalitis, 145
Rotavirus, 12, 168, 169, 262, 266, 324
RSV. *See* Respiratory syncytial virus
Rubella, 13, 92, 179, 180, 312, 314
    congenital, 172
    encephalitis virus, 140
    epidemics, 91
    immunization, 313, 314
    infections in pregnancy, 178, 180
    vaccine, 313
    virus, 67, 68, 174, 178, 179, 202, 261, 267
Rubeola. *See* Measles
Rubivirus, 13

St. Louis encephalitis virus, 93, 143, 145
Sandfly fever, 14
Sarcoma, 15
    virus, 243
Scrapie, 221, 223
Serologic tests, 150, 268
Sheep pox, 11
Shingles. *See* Herpes zoster
Single-stranded RNA, 8
Skin, 47
Smallpox, 11, 281, 285, 299, 302, 321
    vaccine, 303, 321
Smokers, 103
Specimen collection and handling, 266–68
Spongiform encephalopathies, 220–25
Spumavirinae, 15
Status spongiosus, 220
Stomatitis, 14

Stoxil, 283
Structure, 5, 6
Subacute sclerosing panencephalitis, 107,
    108, 109, 139, 198, 228, 229, 287, 297,
    298
Subviral particle, 36
Suipoxvirus, 11
Surveillance, 87
SV40, 9, 243–45
Swine pox, 11
Symmetrel, 130, 282, 283
Symmetry, icosahedral and helical, 9
Syncytial giant cells, 73, 74
Syncytial virus pneumonia, 74

T antigens, 244
T cells, 50
T lymphocytes, 56
Tacaribe complex viruses, 262
T$_c$ lymphocytes, 58, 59
Theiler's viral RNA, 236
Thetalymphocryptovirus, 10
Tissue culture host systems, 262
TMV. See Tobacco mosaic virus
Tobacco mosaic virus (TMV), 4, 5, 13
Togaviridae, 13, 261
Togavirus, 267
Tonsillitis, 112
TORCH, 172
Toxoplasma gondii, 174
Toxoplasmosis, congenital, 172
Tracheobronchitis, 112
Transcriptase, reverse. See Reverse transcrip-
    tase
Transcription, 37
Transformation. See Neoplastic transforma-
    tion
Transmissible mink encephalopathy, 221
Transmission, modes, 89, 90
Transposon, 247
Trifluorothymidine, 286, 292, 293
Trifluridine, 283
Trisodinar-phosphornate, 286
Trisodium phosphornate, 293

Tromantadine, 284
Tropism, 25

Uncoating, 30
Uukuvirus, 14

Vaccine, 256, 300–304, 309, 321
Vaccinia, 11, 261, 267, 282
Varicella virus, 140, 189, 190, 206, 210, 267,
    285, 290, 322
Varicella-zoster virus (VZV), 78, 89, 181,
    189, 190, 205, 206, 212, 214, 218, 236,
    243, 261, 290, 322
Variola virus, 282, 299
Venezuelan encephalitis virus, 145
Verruca vulgaris, 244
Vesicular disease, 267
Vesicular exanthem of swine, 12
Vesiculovirus, 14
Vidarabine, 185, 283, 292
Vira-A, 283
Viremia, 136
Virimenz, 284
Virion, 6, 16, 39, 40, 43, 47
Viroids, 5, 225
Virolysis, 54
Viropexis, 27
Viroptic, 283
Visna, 15
    canine distemper, 232
Vitamin C, 131, 285, 292
V-onc. See Oncogene
VZV. See Varicella-zoster virus

Wart virus, 9, 244
Watson, James B., 8
Western encephalitis virus, 144
Western equine encephalitis, 93
Willowbrook State School, 154, 159, 161, 164
Winter-vomiting disease, 166

Yellow fever vaccine, 320
Yellow virus, 4, 13, 165, 320

Zovirax, 283, 294